AF686497

Small Satellite Missions for Earth Observation

Rainer Sandau · Hans-Peter Röser ·
Arnoldo Valenzuela
Editors

Small Satellite Missions for Earth Observation

New Developments and Trends

Springer

Editors
Dr. Rainer Sandau
German Aerospace Center (DLR)
Rutherfordstr. 2
12489 Berlin
Germany
rainer.sandau@dlr.de

Prof. Hans-Peter Röser
Universität Stuttgart
Institute of Space Systems
Pfaffenwaldring 31
70569 Stuttgart
Germany
roeser@irs.uni-stuttgart.de

Dr. Arnoldo Valenzuela
Media Lario International S.A.,
23842 Bosisio Parini LC
Località Pascolo
Italy
valenzuela@media-lario.it
arnoldo_valenzuela@hotmail.com

ISBN 978-3-642-03500-5 e-ISBN 978-3-642-03501-2
DOI 10.1007/978-3-642-03501-2
Springer Heidelberg Dordrecht London New York

Library of Congress Control Number: 2009940671

© Springer-Verlag Berlin Heidelberg 2010
This work is subject to copyright. All rights are reserved, whether the whole or part of the material is concerned, specifically the rights of translation, reprinting, reuse of illustrations, recitation, broadcasting, reproduction on microfilm or in any other way, and storage in data banks. Duplication of this publication or parts thereof is permitted only under the provisions of the German Copyright Law of September 9, 1965, in its current version, and permission for use must always be obtained from Springer. Violations are liable to prosecution under the German Copyright Law.
The use of general descriptive names, registered names, trademarks, etc. in this publication does not imply, even in the absence of a specific statement, that such names are exempt from the relevant protective laws and regulations and therefore free for general use.

Cover illustration: Image of the model of the TET-satellite, taken in the framework of an R&D project, Ref. No 50 RV 0801, implemented on behalf of the Federal Ministry of Economics and Technology. Image of the Model of the Rapid-Eye satellite: RapidEye AG.

Cover design: WMXDesign GmbH, Heidelberg

Printed on acid-free paper

Springer is part of Springer Science+Business Media (www.springer.com)

Preface

This book was compiled from contributions given at the 7th IAA Symposium on Small Satellites for Earth Observation, May 4–8, 2009, Berlin (IAA – International Academy of Astronautics). From the 15 sessions for oral presentations and two poster sessions, 52 contributions were selected which are representative for the new developments and trends in the area of small satellites for Earth observation. They reflect the potentials of a diversity of missions and related technologies. This may be based on national projects or international co-operations, single satellites of constellations, pico-, nano-, micro- or mini-satellites, developed by companies, research institutions or agencies. The main focus is on new missions to monitor our Earth's resources (Part I), and the environment in which our Earth is embedded (Part II). Part III deals with distributed space systems, a unique feature of small satellites and in most cases impractical to do with large satellites. Here we concentrate on constellations of satellites with focus on future missions relying on co-operating satellites. For all the new developments and projects we need well educated specialists coming from the universities. Many universities included already the development and implementation of small satellites in their curriculum. The university satellites chapter (Part IV) shows the high quality which is already reached by some of the universities worldwide.

To achieve high performance Earth observation mission on the basis of small satellites, instruments and technologies are essential for both high performance and miniaturization. Part V gives insight into new developments on these sectors. The last two parts (Parts VI and VII) deal with subjects, necessary to make use of the data coming from the satellite systems: attitude and position. But high quality attitude control and navigation systems are essential not only for geocoding of high resolution spatial and spectral data. They are also of very high importance when performing formation flying missions as addressed in Part III.

We would like to thank the Symposium and Program Coordinator, Bernd Kirchner, assisted by Ute Dombrowski and Karl-Heinz Degen, without whose effort this book would not have been possible.

Berlin, Germany — Rainer Sandau
Stuttgart, Germany — Hans-Peter Röser
Bosisio Parini, Italy — Arnoldo Valenzuela

Contents

Contributors

Miguel Aguirre ESA/ESTEC, European Space Agency, Keplerlaan 1, 2200 AG Noordwijk, The Netherlands, Miguel.Aguirre@esa.int

Naveed Ahmed Surrey Space Centre (SSC), University of Surrey, GU2 7XH Guildford. UK, N.Ahmed@surrey.ac.uk

R. Amini Delft University of Technology, Delft, The Netherlands

Manuel Angulo INTA, Dto. de Programas Espaciales y Ciencias del Espacio, Torrejon de Ardoz 28850, Madrid, Spain, angulom@inta.es

Yoshihide Aoyanagi Hokkaido Institute of Technology, Maeda 7-15, Teine-ku, Sapporo, Hokkaido, Japan, q07601@hit.ac.jp

Franziska Arlt Institute of Aeronautics and Astronautics, Technical University of Berlin, 13507 Berlin, Germany

Gianni Babini Rheinmetall Italia, Via Affile 102, 00131 Roma, Italy

G. Bakker Department of Aerospace Design, Integration & Operations, Delft University of Technology, Kluyverweg 1, 2629 HS, Delft, The Netherlands

Thérèse Barroso CNES – DCT/ OP/M2, 31401 Toulouse, France

Maria Libera Battagliere Scuola di Ingegneria Aerospaziale, "Sapienza" University of Roma, Via Eudossiana 16, 00184, Rome, Italy, battagliere@gmail.com

D. Bindel Center of Applied Space Technology and Microgravity, University of Bremen, Bremen, Germany, bindel@zarm.uni-bremen.de

Laszlo Bodnar BL Electronics, Solymár, Hungary

Maurice Borgeaud Space Center EPFL, Station 11, Ecole Polytechnique Fédérale de Lausanne, CH-1015 Lausanne, Switzerland, maurice.borgeaud@epfl.ch

Klaus Brieß Institute of Aeronautics and Astronautics, Technical University of Berlin, 13507 Berlin, Germany, Klaus.Briess@ilr.tu-berlin.de

G. Brouwer Delft University of Technology, Delft, The Netherlands

M. Buhl Institute of Aeronautics and Astronautics, Technical University of Berlin, Marchstraße 12, D-10587 Berlin, Germany, matthias.buhl@ilr.tu-berlin.de

Enrico Canuto Politecnico di Torino, Corso Duca degli Abruzzi 24, 10129 Turin, Italy

Xinghui Cao Research Center of Satellite Technology, Harbin Institute of Technology, Harbin 150001, China, caoxinghui1982@163.com

Chantal Cappelletti Scuola di Ingegneria Aerospaziale, "Sapienza" University of Rome, Rome, Italy, chantal.cappelletti@gmail.com

Warren F. Case SGT Inc., Greenbelt, MD 20770, USA

A. Cawthorne Surrey Satellite Technology Ltd (SSTL), Tycho House, Surrey Space Centre, Surrey Research Park, Guildford, Surrey GU2 7YE, UK

Stefano Cesare Thales Alenia Space Italia, Strada Antica di Collegno 253, 10146 Turin, Italy, stefano.cesare@thalesaleniaspace.com

François Châteauneuf INO, 2740, rue Einstein, QC, G1P 4S4, Canada, francois.chateauneuf@ino.ca

Ted Choueiri Space Center EPFL, Station 11, Ecole Polytechnique Fédérale de Lausanne, CH-1015 Lausanne, Switzerland

Q.P. Chu Department of Aerospace Design, Integration & Operations, Delft University of Technology, Kluyverweg 1, 2629 HS, Delft, The Netherlands

Andrea Cisbani Galileo Avionica, Via A. Einstein 35, 50013 Florence, Italy

Marline Claessens Verhaert Space, Hogenakkerhoekstraat 99150, Kruibeke, Belgium

M. Collon cosine Research B.V., Niels Bohrweg 11, 2333 CA Leiden, The Netherlands

Stefania Cornara DEIMOS Space S.L., Madrid, Spain, stefania.cornara@deimos-space.com

Philippe Crebassol Centre National d'Etudes Spatiales, 31401 Toulouse, France, philippe.crebassol@cnes.fr

Manuel Czech Institute of Astronautics at the Technische Universität München, Boltzmannstraße 15, 85748 Garching, Germany, m.czech@lrt.mw.tum.de

Simone D'Amico DLR, German Space Operations Center, D-82234 Munich, Germany, simone.damico@dlr.de

Marco D'Errico Department of Aerospace and Mechanical Engineering, Second University of Naples, Via Roma 29, 81131 Aversa (CE), Italy, marco.derrico@unina2.it

A. da Silva Curiel Surrey Satellite Technology Ltd (SSTL), Tycho House, Surrey Space Centre, Surrey Research Park, Guildford, Surrey GU2 7YE, UK, A.da-Silva-Curiel@sstl.co.uk

Zhao Dan Research Centre of Satellite Technology, Harbin Institute of Technology, Harbin, China, zhaodan_hit@yahoo.com.cn

M. Ann Darrin The Johns Hopkins University Applied Physics Laboratory, Laurel, MD 20723-6099, USA

C.W. de Boom TNO Science and Industry, Stieltjesweg 1, 2628 CK Delft, The Netherlands, info@tno.nl

Cristina de Negueruela Alemán GMV, Isaac Newton 11, P.T.M. Tres Cantos, 28760 Madrid, Spain

L. de Vos OIP, Westerring 21, 9700 Oudenaarde, Belgium, ldv@oip.be

Gérard Dedieu Centre National d'Etudes Spatiales, 31401 Toulouse, France, gérard.dedieu@cnes.fr

Lars Dornburg Institute of Aeronautics and Astronautics, Technical University of Berlin, 13507 Berlin, Germany

Jens Eickhoff EADS Astrium GmbH, 88039 Friedrichshafen, Germany

Werner Enderle European Commission, Galileo Unit G3, Rue De Mot 28, Brussels, Belgium, werner.enderle@ec.europa.eu

Segundo Esteban Universidad Complutense de Madrid, Madrid, Spain, segundo.esteban@ucm.es

Giancarmine Fasano Department of Aerospace Engineering, University of Naples "Federico II", Piazzale Tecchio 80, 80125 Napoli, Italy, g.fasano@unina.it

Csaba Ferencz Eötvös University, Budapest, Hungary

Pierric Ferrier Centre National d'Etudes Spatiales, 31401 Toulouse, France, pierric.ferrier@cnes.fr

Andreas Fleischner Institute of Astronautics at the Technische Universität München, Boltzmannstraße 15, 85748 Garching, Germany

Bertrand Fougnie Centre National d'Etudes Spatiales, 31401 Toulouse, France, bertrand.fougnie@cnes.fr

Michael Fritz Institute of Space Systems, Universität Stuttgart, Pfaffenwaldring 31, 70569 Stuttgart, Germany

Claudio Galeazzi Agenzia Spaziale Italiana (ASI), Via le Liegi 26, 00198 Roma, Italy

Kristof Gantois ESTEC, Postbus 299, NL 2200 AG Noordwijk, The Netherlands

P. Garner Surrey Satellite Technology Ltd (SSTL), Tycho House, Surrey Space Centre, Surrey Research Park, Guildford, Surrey GU2 7YE, UK

E. Gill Department of Earth Observation and Space Systems, Delft University of Technology, Kluyverweg 1, 2629 HS, Delft, The Netherlands, E.K.A.Gill@tudelft.nl

Bernhard Grafmueller Astrium GmbH, 88039 Friedrichshafen, Germany

Michele Grassi DIAS – Department of Aerospace Engineering, University of Naples "Federico II", P.le Tecchio 80, 80125 Naples, Italy, michele.grassi@unina.it

Filippo Graziani Scuola di Ingegneria Aerospaziale, "Sapienza" University of Roma, Via Eudossiana 16, 00184, Rome, Italy, filippo.graziani@uniroma1.it

J. Roselló Guasch ESA/ESTEC, European Space Agency, Keplerlaan 1, 2200 AG Noordwijk, The Netherlands, Josep.Rosello@esa.int

B. Gunter Department of Earth Observation and Space Systems, Delft University of Technology, Kluyverweg 1, 2629 HS, Delft, The Netherlands

J. Guo Department of Earth Observation and Space Systems, Delft University of Technology, Kluyverweg 1, 2629 HS, Delft, The Netherlands

Olivier Hagolle Centre National d'Etudes Spatiales, 31401 Toulouse, France, olivier.hagolle@cnes.fr

David Hall Astrium Ltd., Anchorage Road, Portsmouth, PO3 5PU, England

S. Hannemann cosine Research B.V., Niels Bohrweg 11, 2333 CA Leiden, The Netherlands

J. Harris Swiss Space Technology, Route de Chavalet 2, 18349 Champery, Switzerland

A. Haslehurst Surrey Satellite Technology Ltd (SSTL), Tycho House, Surrey Space Centre, Surrey Research Park, Guildford, Surrey GU2 7YE, UK

Jacob Herscovitz Rafael Advanced Defense Systems Ltd., Haifa 31021, Israel, jacobh@rafael.co.il

D. Ivanov Moscow Institute of Physics and Technology, State University, Moscow, Russia, danilivanov@mail.ru

J.F. Jamoye Nanoshape, Rue des Chasseurs Ardennais, 4031 Angleur, Belgium, jf@nanoshape.be

Zhang Jinxiu Research Centre of Satellite Technology, Harbin Institute of Technology, Harbin, China, jinxiu@hit.edu.cn

Fabien Jordan Space Center EPFL, Station 11, Ecole Polytechnique Fédérale de Lausanne, CH-1015 Lausanne, Switzerland

D. Kajon Università degli Studi di Roma "La Sapienza", Rome, Italy

Angelita C. Kelly NASA Goddard Space Flight Center, Code 428 Earth Science Mission Operations (ESMO) Project, Mission Validation and Operations Branch, Greenbelt, MD 20771, USA

O. Khromov Keldysh Institute of Applied Mathematics, Russian Academy of Sciences, Moscow, Russia

Stanislav Klimov Space Research Institute (IKI) of RAS, 117997 Moscow, Russia, sklimov@iki.rssi.ru

Kay Köhler Institute of Aeronautics and Astronautics, Technical University of Berlin, 13507 Berlin, Germany

M. Kohling ESA, Keplerlaan 1, 2201 AG Noordwijk, The Netherlands

Valeriy Korepanov Lviv Centre of Institute of Space Research, Lviv, Ukraine

Götz Kornemann Aerospace Institute (ILR), Technische Universität Berlin (TU Berlin), Marchstrasse 12, 10587 Berlin, Germany, goetz.kornemann@ilr.tu-berlin.de

E. Kroesbergen Mecon Engineering B.V., Koopmanslaan 25, 7005 BK Doetinchem, The Netherlands

M. Kruijff Delta-Utec, Middelstegracht 89 g, 2312 TT, Leiden, The Netherlands

Toshinori Kuwahara Institute of Space Systems, Universität Stuttgart, Pfaffenwaldring 31, 70569 Stuttgart, Germany, kuwahara@irs.uni-stuttgart.de

E. Laan TNO Science & Industry, Stieltjesweg 1, 2600 AD, Delft, The Netherlands

Philippe Landiech CNES, Toulouse, France, philippe.landiech@cnes.fr

Giovanni Laneve CRPSM (Centro di Ricerca Progetto S. Marco) – Sapienza University of Rome, Via Salaria 851, 00138, Rome, Italy, laneve@psm.uniroma1.it

L. Perez Lebbink Delft University of Technology, Delft, The Netherlands

J. Leijtens TNO Science and Industry, Stieltjesweg 1, 2628 CK Delft, The Netherlands, info@tno.nl

Bruno Leone ESA-ESTEC, Keplerlaan 1, 2201 AZ, Noordwijk ZH, The Netherlands

Janos Lichtenberger Eötvös University, Budapest, Hungary

S. Lievens VITO, Flemish Institute for Technological Research, Boeretang 200, BE-2400, Mol, Belgium

Adam Loverro Jet Propulsion Laboratory, California Institute of Technology, Pasadena, CA 91109-8099, USA

D. Maessen Department of Earth Observation and Space Systems, Delft University of Technology, Kluyverweg 1, 2629 HS, Delft, The Netherlands

Olaf Maibaum German Aerospace Center, Simulation and Software Technology (SISTEC), D-38108 Braunschweig, Germany, olaf.maibaum@dlr.de

Chistophe Maréchal CNES – DCT/ OP/M2, 31401 Toulouse, France

L. Maresi ESA/ESTEC, Keplerlaan 1, 2201 AG Noordwijk, The Netherlands, Luca.Maresi@esa.int

Andriy Marussenkov Lviv Centre of Institute of Space Research, Lviv, Ukraine

Gabriele Mascetti Agenzia Spaziale Italiana, Viale Liegi 26, 00198, Rome, Italy, gabriele.mascetti@asi.it

Luca Massotti ESA/ESTEC, European Space Agency, Keplerlaan 1, 2200 AG Noordwijk, The Netherlands, Luca.Massotti@esa.int

S. Matyugov FIRE RAS, Russian Federation, Moscow, Russia

Z. Ming Center of Space Technology and Microgravity, Bremen University, Bremen, Germany

W. Moelans OIP, Westerring 21, 9700 Oudenaarde, Belgium

Elisa Molina INTA, Dto. de Programas Espaciales y Ciencias del Espacio, Torrejon de Ardoz 28850, Madrid, Spain, molinaae@inta.es

Oliver Montenbruck DLR, German Space Operations Center, D-82234 Munich, Germany, oliver.montenbruck@dlr.de

S.G. Moon cosine Research B.V., Niels Bohrweg 11, 2333 CA Leiden, The Netherlands, s.moon@cosine.nl

V. Moreau AMOS, Rue des Chasseurs Ardennais, 4031 Angleur, Belgium, Vincent.Moreau@amos.be

Sergio Mottini Thales Alenia Space Italia, Strada Antica di Collegno 253, 10146 Turin, Italy

C. Müller Technical University of Berlin, Germany, christiane.i.mueller@gmail.com

Fabio Musso Thales Alenia Space Italia, Strada Antica di Collegno 253, 10146 Turin, Italy

Thomas Nagler ENVEO, ICT Technologiepark, Technikerstrasse 21a, A-6020 Innsbruck, Austria

Akihiro Nakamura AIDMA Inc., Sapporo, Hokkaido, Japan

J. Rodriguez Navarro Center of Applied Space Technology and Microgravity, University of Bremen, Bremen, Germany

Alfred Ng Canadian Space Agency, 6767, Route de l'Aéroport, St-Hubert, QC, J3Y 8Y9, Canada, marleen.vanmierlo@asc-csa.gc.ca

Muriel Noca Space Center EPFL, Station 11, Ecole Polytechnique Fédérale de Lausanne, CH-1015 Lausanne, Switzerland

Denis Novikov Space Research Institute (IKI) of RAS, 117997 Moscow, Russia, dnovikov@iki.rssi.ru

Claas Olthoff Institute of Astronautics at the Technische Universität München (TUM), Boltzmannstraße 15, 85748 Garching, Germany, office@lrt.mw.tum.de

Emiliano Ortore School of Aerospace Engineering – Sapienza University of Rome, Via Eudossiana 18, 00184, Rome, Italy

M. Ovchinnikov Russian Institute of Space Device Engineering, Moscow, Russia, ovchinni@keldysh.ru

Fabian Pacholke Aerospace Institute (ILR), Technical University of Berlin (TU Berlin), Marchstrasse 12, 10587 Berlin, Germany, fabian.pacholke@ilr.tu-berlin.de

Jordi Palau AD Telecom, Barcelona, Spain, palaucj@inta.es

Fabrizio Paolillo Scuola di Ingegneria Aerospaziale, "Sapienza" University of Rome, Rome, Italy, pfabrizio@fastwebnet.it

Manlio Parisch Thales Alenia Space Italia, Strada Antica di Collegno 253, 10146 Turin, Italy

M. Pastena Satellites Project, Carlo Gavazzi Space SpA, Via Tiengo snc, 82100 Benevento, Italy, mpastena@cgspace.it

A. Pavelyev FIRE RAS, Russian Federation, Moscow, Russia

Larry J. Paxton The Johns Hopkins University Applied Physics Laboratory, Laurel, MD 20723-6099, USA, larry.paxton@jhuapl.edu

Linh Ngo Phong Canadian Space Agency, 6767, Route de l'Aéroport, St-Hubert, QC, J3Y 8Y9, Canada, marleen.vanmierlo@asc-csa.gc.ca

Fabrizio Piergentili DIEM, University of Bologna Alma Mater Studiorum, Bologna, Italy, fabrizio.piergentili@uniroma1.it

M. Pointer Surrey Satellite Technology Ltd (SSTL), Tycho House, Surrey Space Centre, Surrey Research Park, Guildford, Surrey GU2 7YE, UK

Manuel Prieto Universidad de Alcalá, Dpto. de Automática, Escuela Politécnica, Alcalá de Henares, Madrid, 28871, Spain, mpm@aut.uah.es

Giuseppina Pulcrano Agenzia Spaziale Italiana, Viale Liegi 26, 00198, Rome, Italy, giuseppina.pulcrano@asi.it

Ralf Purschke Institute of Astronautics at the Technische Universität München (TUM), Boltzmannstraße 15, 85748 Garching, Germany, office@lrt.mw.tum.de

Nadège Quéruel CNES – DCT/ OP/M2, 31401 Toulouse, France

Christian Raschke Astro- und Feinwerktechnik Adlershof GmbH, D-12489 Berlin, Germany, c.raschke@astrofein.com

E. Razzano Satellites Project, Carlo Gavazzi Space SpA, Via Tiengo snc, 82100 Benevento, Italy, erazzano@cgspace.it

Alfredo Renga DIAS – Department of Aerospace Engineering, University of Naples "Federico II", P.le Tecchio 80, 80125 Naples, Italy

U. Renner Institute of Aeronautics and Astronautics, Technical University of Berlin, Marchstraße 12, D-10587 Berlin, Germany

Jens Rießelmann Institute of Aeronautics and Astronautics, Technical University of Berlin, 13507 Berlin, Germany, Jens.Riesselmann@ilr.tu-berlin.de

B. Rievers Center of Applied Space Technology and Microgravity, University of Bremen, Bremen, Germany, rievers@zarm.uni-bremen.de

Alice Robert Astrium SAS, 31 rue des Cosmonautes, Z.I. du Palays, 31402 Toulouse, France

Paul Rodrigues CNES, Toulouse, France, paul.rodrigues@cnes.fr

Oscar Rodriguez Universidad de Alcalá, Dpto. de Automática, Escuela Politécnica, Alcalá de Henares, Madrid, 28871, Spain, opolo@aut.uah.es

Hans-Peter Röser Institute of Space Systems, Universität Stuttgart, Pfaffenwaldring 31, 70569 Stuttgart, Germany

Guillaume Roethlisberger Space Center EPFL, Station 11, Ecole Polytechnique Fédérale de Lausanne, CH-1015 Lausanne, Switzerland

Aaron Q. Rogers The Johns Hopkins University Applied Physics Laboratory, Laurel, MD 20723-6099, USA, Aaron.Rogers@jhuapl.edu

A. Romanov FSUE "RISDE", Russian Federation, Moscow, Russia

Helmut Rott ENVEO, ICT Technologiepark, Technikerstrasse 21a, A-6020 Innsbruck, Austria

Andrea Sacchetti Carlo Gavazzi Space, Via Gallarate 150, 20151 Milano, Italy, asacchetti@cgspace.it

Takeshi Sakamoi Planetary Plasma and Atmospheric Research Center, Tohoku University, Aoba 6-6-01, Sendai 980-8579, Japan

Yuji Sakamoto Department of Aerospace Engineering, Graduate School of Engineering, Tohoku University, Aoba 6-6-01, Sendai 980-8579, Japan

B. Sanders TNO Defense, Security and Safety, Delft, The Netherlands

Fabio Santoni Scuola di Ingegneria Aerospaziale, "Sapienza" University of Roma, Via Eudossiana 16, 00184, Rome, Italy, fabio.santoni@uniroma1.it

Shin Satori Hokkaido Institute of Technology, Maeda 7-15, Teine-ku, Sapporo, Hokkaido, Japan, satori@hit.ac.jp

Noémy Scheidegger Space Center EPFL, Station 11, Ecole Polytechnique Fédérale de Lausanne, CH-1015 Lausanne, Switzerland

Gianfranco Sechi Thales Alenia Space Italia, Strada Antica di Collegno 253, 10146 Turin, Italy

V. Selin Roscosmos, Russian Federation, Moscow, Russia

A. Selivanov FSUE "RISDE", Russian Federation; Keldysh Institute of Applied Mathematics, Russian Academy of Sciences, Moscow, Russia

Laura Seoane INTA, Dto. de Programas Espaciales y Ciencias del Espacio, Torrejon de Ardoz 28850, Madrid, Spain, seoanepl@inta.es

Tony Sephton Astrium Ltd., Gunnels Wood Road, Stevenage, SG1 2AS, England, tony.sephton@astrium.eads.net

Lan Shengchang Research Centre of Satellite Technology, Harbin Institute of Technology, Harbin, China

Pierluigi Silvestrin ESA/ESTEC, European Space Agency, Keplerlaan 1, 2200 AG Noordwijk, The Netherlands, Pierluigi.Silvestrin@esa.int

Nicolas Steiner Space Center EPFL, Station 11, Ecole Polytechnique Fédérale de Lausanne, CH-1015 Lausanne, Switzerland

Karsten Strauch ESTEC, Postbus 299, NL 2200 AG Noordwijk, The Netherlands

Fulin Su School of Electronic and Information Technology, Harbin Institute of Technology, Harbin 150001, China

M. Taccola ESA/ESTEC, Keplerlaan 1, 2201 AG Noordwijk, The Netherlands, Matteo.Taccola@esa.int

Yukihiro Takahashi Department of Geophysics, Graduate School of Science, Tohoku University, Aoba 6-6-01, Sendai 980-8579, Japan

Yusuke Takeuchi Hokkaido Satellite Inc., Sapporo, Hokkaido, Japan, hsc-take@hit.ac.jp

Urbano Tancredi Department for Technologies, University of Naples "Parthenope" Centro Direzionale, Isola C4, 80133 Naples, Italy

Thomas Terzibaschian German Aerospace Center, Institute of Robotics and Mechatronics, Optical Information Systems, D-12489 Berlin, Germany, thomas.terzibaschian@dlr.de

Francesc Tinto Centre National d'Etudes Spatiales, 31401 Toulouse, France, francesc.tinto@cnes.fr

Tsuyoshi Totani Hokkaido University, Sapporo, Hokkaido, Japan

Carlo Ulivieri CRPSM (Centro di Ricerca Progetto S. Marco) – Sapienza University of Rome, Via Salaria 851, 00138, Rome, Italy

Craig I. Underwood Surrey Space Centre (SSC), University of Surrey, GU2 7XH Guildford. UK, C.Underwood@surrey.ac.uk

Marleen van Mierlo Canadian Space Agency, 6767, Route de l'Aéroport, St-Hubert, QC, J3Y 8Y9, Canada, marleen.vanmierlo@asc-csa.gc.ca

Jan Vermeiren XenICS, Ambachtenlaan 44, 3001 Leuven, Belgium, Jan.Vermeiren@xenics.com

J. Versluys OIP, Westerring 21, 9700 Oudenaarde, Belgium

A. Vinogradov FSUE "RISDE", Russian Federation, Moscow, Russia

V. Vishnyakov FSUE "RISDE", Russian Federation, Moscow, Russia, Vishnyakov@rniikp.ru

Huu Quan Vu Aerospace Institute (ILR), Technische Universität Berlin (TU Berlin), Marchstrasse 12, 10587 Berlin, Germany, huu.quan.vu@ilr.tu-berlin.de

Ulrich Walter Institute of Astronautics at the Technische Universität München, Boltzmannstraße 15, 85748 Garching, Germany

T. Walter Center of Space Technology and Microgravity, Bremen University, Bremen, Germany

Tippawan Wanwiwake Geo-Informatics and Space Technology Development Agency, Bangkok, Thailand; Surrey Space Centre, University of Surrey, Guildford, UK, T.Wanwiwake@surrey.ac.uk

Jana Weise Institute of Aeronautics and Astronautics, Technical University of Berlin, 13507 Berlin, Germany

K. Wielinga Mecon Engineering B.V., Koopmanslaan 25, 7005 BK Doetinchem, The Netherlands

Roland Winklmeier Institute of Astronautics at the Technische Universität München (TUM), Boltzmannstraße 15, 85748 Garching, Germany, office@lrt.mw.tum.de

Alex Wishart Astrium Ltd., Gunnels Wood Road, Stevenage, SG1 2AS, England

Cao Xibin Research Centre of Satellite Technology, Harbin Institute of Technology, Harbin, China

Guodong Xu Research Center of Satellite Technology, Harbin Institute of Technology, Harbin 150001, China

O. Yakovlev FIRE RAS, Russian Federation, Moscow, Russia

Yoram Yaniv ISRAEL Aerospace Industries Ltd., MBT Space Division, Industrial Zone, Yehud 56000, Israel, yyaniv@iai.co.il

Toshihiko Yasunaka Uematsu Electric Co. Ltd, Akabira, Hokkaido, Japan

Zizung Yoon Department of Aeronautics and Astronautics, Berlin Institute of Technology, D-10587 Berlin, Germany, zizung.yoon@tu-berlin.de

Kazuya Yoshida Department of Aerospace Engineering, Graduate School of Engineering, Tohoku University, Aoba 6-6-01, Sendai 980-8579, Japan, yoshida@astro.mech.tohoku.ac.jp

Keisuke Yoshihara Japan Aerospace Exploration Agency, 2-1-1, Sengen, Tsukuba, Ibaraki, 305-8505, Japan, keisuke.yoshihara@jaxa.jp

B. Zandbergen Delft University of Technology, Delft, The Netherlands, b.t.c.zandbergen@tudelft.nl

I. Zaramenskikh The Keldysh Institute of Applied Mathematics of RAS, Moscow, Russia

G.T. Zheng School of Aerospace, Tsinghua University, 100084, Beijing, China

Claas Ziemke Institute of Space Systems, Universität Stuttgart, Pfaffenwaldring 31, 70569 Stuttgart, Germany, ziemke@irs.uni-stuttgart.de

Part I
New Earth Observation Missions

Overview on CNES Micro Satellites Missions: In Flight, Under Development and Next

Philippe Landiech and Paul Rodrigues

Abstract This chapter gives a programmatic and technical overview of the MYRIADE micro satellites line of product MYRIADE developed by CNES. The characteristics and mission topics of satellites under CNES responsibility are presented, for in flight, and under development systems. The main drivers of the roadmap for next years are addressed as a Conclusion.

1 Introduction

CNES initiatives. in micro and mini satellites since 1996 have allowed to build, through MYRIADE and PROTEUS, operational two lines of product allowing scientific and operational missions in low earth orbit, for payloads ranging from 50 to 350 kg. From the beginning, these lines of products have been devised so as to offer efficient access to space in terms of schedule and costs. Although significant differences in terms of satellites requirements have appeared along the missions, the benefits of the line of products approach has been constantly highlighted in the 2 mini and micro cases, thanks to a well mastered technical definition and validated associated means. Platform high maturity allows in both cases to start satellite activities at the latest, while schedule is driven by the payload development itself, relying on well known interfaces. In parallel to platforms development, a common ground segment has been settled, which minimizes adaptation effort from a mission to the next one and allows to standardize operations. This chapter focuses on MYRIADE product line.

An overview on missions already in flight will be given, with associated lessons learnt: DEMETER, PARASOL.

Benefit of the approach for missions currently under development will also be highlighted: PICARD, TARANIS, MICROSCOPE.

P. Landiech (✉)
CNES, Toulouse, France
e-mail: philippe.landiech@cnes.fr

This chapter is also written by all CNES mini and micro satellites teams

R. Sandau et al. (eds.), *Small Satellite Missions for Earth Observation*,
DOI 10.1007/978-3-642-03501-2_1, © Springer-Verlag Berlin Heidelberg 2010

2 MYRIADE Genesis

MYRIADE initiative started with the Arcachon scientific seminar held in 1998. It allowed to define a consistent panel of mission objectives, and permitted to offer to scientists the basis of a versatile tool for testing small payload instruments in the range 60 kg–60 W, for low duration missions (typically 2 years), with short development schedule and reduced costs.

The Arcachon scientific seminar held in 1998 allowed to define a panel of missions in this range, which allowed to refine major specifications. DEMETER and PARASOL, the first 2 MYRIADE launched were among them.

Development started in 1999. In an alternate way to PROTEUS, an internal CNES development was followed, with limited engineering support from industry.

Another difference is the search for very low cost equipment units, which lead to select a majority of commercial components which offer lower prices but also allow to take advantage of more integrated functionalities. These components however sustained a ground qualification to space environment radiations.

The OBC (On Board Computer) developed in house, has limited redundancies restricted to critical functions such as TMTC. As it is not fully immune to SEU nor SEL, a specific FDIR function, which combines hardware watchdogs and software functions has been developed to overcome SEL, SEU, corrupted data, interrupted data transfers or dead processes. This FDIR function has been extensively tested and verified on dedicated ground test bench.

The design of the bus structure (Fig. 1) is very simple (quasi cubic – 60 × 60 × 55 cm, aluminium and honey-comb aluminium, ...) to limit development and recurring costs. The –X panel includes the launcher adapter and the propulsion subsystem with its hydrazine tank.

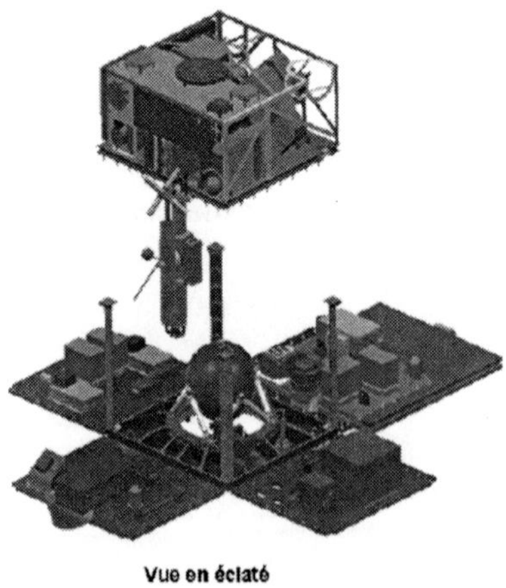

Fig. 1 MYRIADE platform layout

The payload is located on the X+ side of the bus. The power subsystem is based on a single wing solar generator with 2 rigid panels with AsGa cells, rotated by a Solar Array Drive Mechanism controlled by the AOCS through the OBC.

A Power Conditioning and Distribution Unit (PCDU) is in charge of:

- launcher separation detection to connect the main non regulated bar to the battery
- battery regulation
- power distribution to equipment and payload (regulated voltages or voltages 22–37 V)
- thrusters and magnetotorquers commands
- pyro lines distribution

The battery is Li-Ion type.

AOCS design is rather classic. It uses solar sensor, 3 axis magnetometer, high accuracy star sensor, gyros and GPS (option) as sensors, and magnetotorquers, reaction wheels as actuators. Hydrazine propulsion (4 × 1 N thrusters) (option) is only used for orbit control.

Four AOCS modes are used: acquisition/safe mode, transition mode, normal mode and orbit control mode.

On board data management and control/command perform the following main functions:

- satellite configuration management
- mission plan management
- storage of house keeping and payload data, and transmission to S band station
- implementation of AOCS

The architecture is centralized: processing is achieved by one single OBC with direct links with PCDU, Solar generator rotating system, AOCS equipment, RX and TX, payload.

Thermal control is based on use of passive systems (paints, MLI, SSM coatings, ...) and SW controlled heaters.

Table 1 gathers MYRIADE main characteristics.

Table 1 MYRIADE characteristics synthesis

MYRIADE characteristics/Basic performances		
Structure	Alu structure, honeycomb panels	600 × 600 × 800 mm, 130 kg (with payload) payload 600 × 600 × 350 mm, 60 kg max.

Table 1 (continued)

MYRIADE characteristics/Basic performances		
Power	Solar panels (*ThalesAlenia Space*) AsGa cells (*Spectrolab*) Battery Li-ion 14 Ah (*AEA*) PCDU (*ETCA*) Solar Array Drive (*OERLIKON*)	2 panels, 0.9 m^2 total, rotating (200 W peak) more than 90 W total permanent in SSO (60 W permanent – even during eclipse – for payload)
AOCS	Sun sensors (*Astrium*) Magnetometer (*IAI/Tamam –Israël*) Star sensor (*TUD Denmark*) Gyros < 6°/h (*Litef – D*) Magnetoactuators (IAI/Tamam – Israël) Reaction wheels 0.12 Nms (Teldix – D) Propulsion: 4 × 1 N thrusters, hydrazine system Isp 210 s (EADS Gmbh)	*Demonstrated Performances nominal mode*: 1 axis, 3 axis, ... pointing capability A priori pointing: <0.02° (1σ) each axis Pointing stability: <0.02°/s DeltaV available: 80 m/s for 120 kg satellite
Localization/Orbit determination	Performed by Control Center Option: GPS TOPSTAR 3000 (*Alcatel A. S.*)	By Doppler measurements: Position restitution/prevision at 3 σ: ±350/±575 m along the track and less than ±10 m ⊥ to the track or in altitude (idem for prevision) Localisation by GPS: <±1 m
On board data management and Command/Control	OBC with μprocessor T 805 CNES design (STEEL manufacturing) Flight software: (CSSI)	5 Mips, 1 Gb memory (EDAC) In-orbit reprogrammable OS-link between OBC and payload 5 Mb/s Payload has its own computer Datation: ±15 ms/UT (at 700 km altitude)
Communications S band	TX link: CCSDS and coding RX link: CCSDS and coding Emitter (QPSK modulation) THALES Receiver (QPSK demodulation) 2 antennas (SHELTON)	Error Bit Rate: 10^{-10} Error Bit Rate: 10^{-10} 10 or 400–600 kb/s – cold redundant 20 kb/s – hot redundant opposite sides, omnidirectional coverage
Payload data downlink with X band	Option: X band emitter for payload (*Thales Alenia Space*)	18 Mb/s–80 Mb/ more than 100 Mb/s in development
Payload management and data storage	Performed by dedicated payload electronic computer with microprocessor, solid state memory (*STEEL*)	8, 16, 32 Gbits mass memory included in payload electronics

3 In Flight Missions Highlights

Up to now:

- 2 MYRIADE CNES missions (DEMETER and PARASOL) are operating in flight.
- 4 other MYRIADE satellites part of the ESSAIM French Defence mission are also operating in flight.

DEMETER was the first MYRIADE in flight. Launch occurred on DNIEPR on June 29th 2004 (see Fig. 2).

Fig. 2 DEMETER accommodation on DNIEPR with various nanosats

DEMETER (Fig. 3) is a scientific mission proposed by CNRS agency (LPCE laboratory was the orderer). It is dedicated to the detection and the characterisation of electro magnetic waves signals associated with telluric activities (earthquakes, volcanic) or issued from human activities (Power lines, VLF, HF broadcasting).

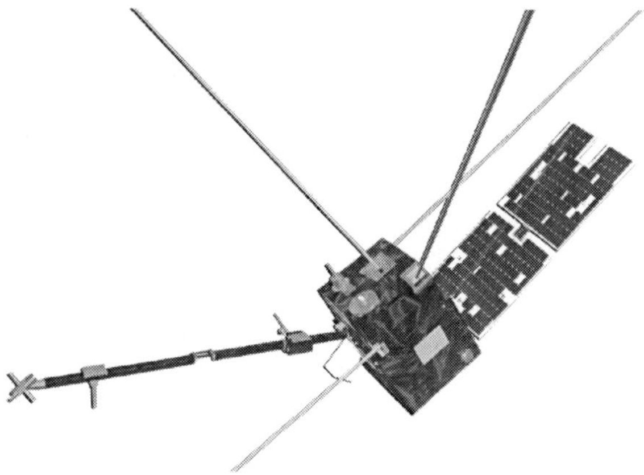

Fig. 3 DEMETER satellite layout

These signals disturb the ionosphere and high atmosphere. These perturbations should occur between a few tens of minutes and a few hours prior to the seism. The scientific payload is made of very sensitive magnetic and electric sensors. To limit bus perturbations, the magnetic sensors are set at the end of a 2 m length deployable boom, while the 4 electric sensors are set at the extremity of 4 m length expendable booms.

The main challenge of DEMETER bus was the necessity to procure to payload the lowest possible electric and magnetic perturbation levels (1,000 Hz range).

The magnetic satellite mission is limited to –24 dBpT from 1 to 7 kHz (some streaks are over but there are few). This has been done by specific wirings, active filtering, magnetic shields on wheels, connectors, ITO on solar generator . . .

In flight operations confirmed the good platform design. Just a sensitivity of reaction wheels electronics to SEU requested some SW modifications so as to reinitialize autonomously communications on these events. An other concern was high sensitivity of the star tracker to moon in the field of view which required operational modifications.

After almost 4 years in flight, more than 1,500 seisms have been registered and some maps as depicted on Fig. 4 show some correlations in electrical field near the geographical location a few hours before.

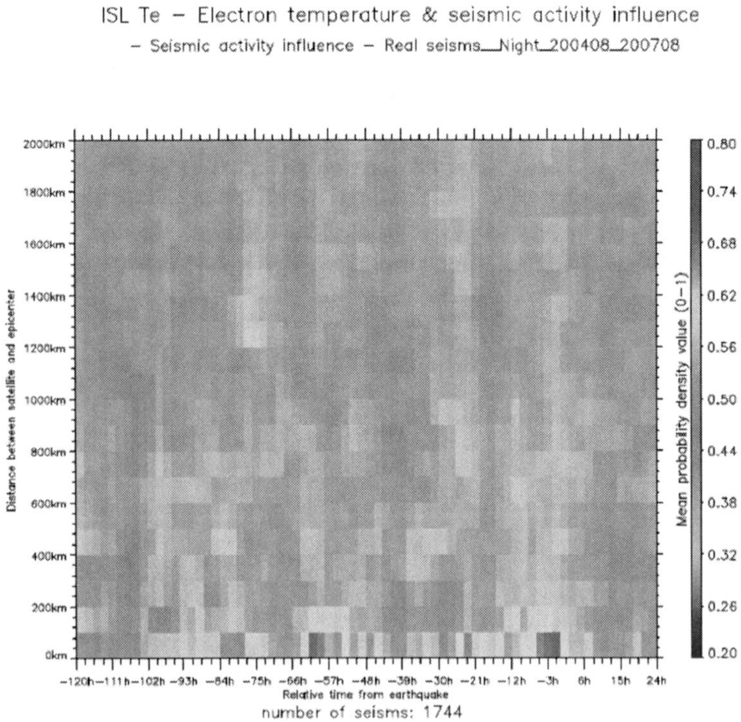

Fig. 4 DEMETER science Results

Besides the primary mission, an autonomous orbit control was implemented inside the satellite SW as a technological demonstrator. This demo closes the loop between on line orbit restitution on board through GPS and thrusters burns to fulfil ground programmed criteria: position on orbit, ground track,

Figure 5 highlights the good behaviour evidenced in flight on more than 150 manoeuvers.

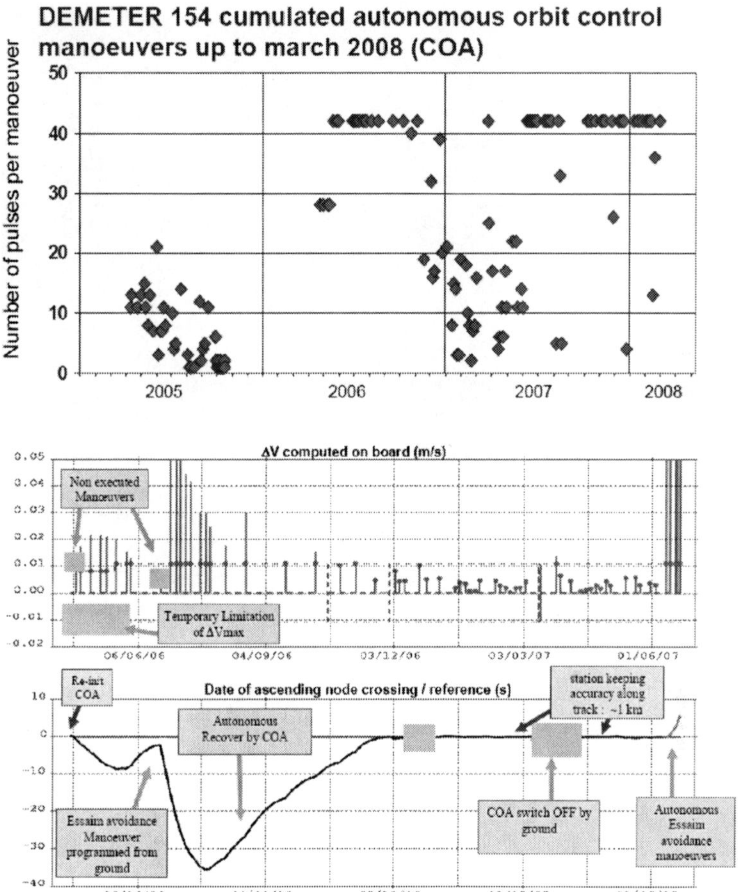

Fig. 5 DEMETER autonomous orbit control

Overall experiment availability is close to 90%. Mission has been extended until end 2009.

An in flight experimentation session is foreseen during first semester 2010, in order to enrich the knowledge of equipment performances after 4.5 years of in flight operation, before deorbitation and passivation operations.

PARASOL is the second mission developed by CNES within the frame of MYRIADE. It has been proposed by CNRS agency (LOA laboratory was the orderer), as being a part of the AQUA train composed by CALIPSO, AQUA, OCO, AURA and CLOUDSAT.

PARASOL addresses climatology topics, in particular measurements of clouds and aerosols properties, and radiative budget interaction (contribution to the global warming).

PARASOL bus design (Fig. 6) is almost similar to DEMETER/MYRIADE one.

Fig. 6 PARASOL satellite layout

Minor modifications concern:

- The solar generator cant angle has been set to 0° to avoid disturbances not compliant with the pointing stability requirements of the mission (< 0.01°/s over 5 s) while the solar generator is in rotation.
- A yaw steering capacity has been implemented in order to compensate for the earth rotation when the payload is taking images.
- Concerning the payload, mass memory capacity was extended from 8 to\break 16 Gbits.

PARASOL is the first demonstration of recurring mission developed within less of 3 years with a low recurring cost and with a very reduced time gap with previous DEMETER mission: 6 months: launch occurred on December 2004, aboard the ARIANE 5 ASAP together with the 4 other MYRIADE used for French Defence ESSAIM mission: see Fig. 7.

Overall mission availability is close to 90%. The star tracker concerns found on DEMETER were exacerbated because of higher mean temperature levels. SW

Fig. 7 PARASOL on Ariane 5 ASAP with companions

Fig. 8 PARASOL and ESSAIM satellites

modifications have allowed to retrieve correct tracking mode statistics. Mission has been extended till mid 2010.

In addition of these CNES missions, the MYRIADE standard has been selected by ASTRIUM to provide French Defence with 4 ESSAIM satellites (Figs. 7, 8, and 9) in orbit at time being.

4 CNES Missions Under Development

Regarding CNES missions, 3 MYRIADE satellites are under development (PICARD, TARANIS, MICRO-SCOPE), planned to be launched respectively in 2009, 2011 and 2013. Eight others satellites have been ordered to prime contractors by other customers.

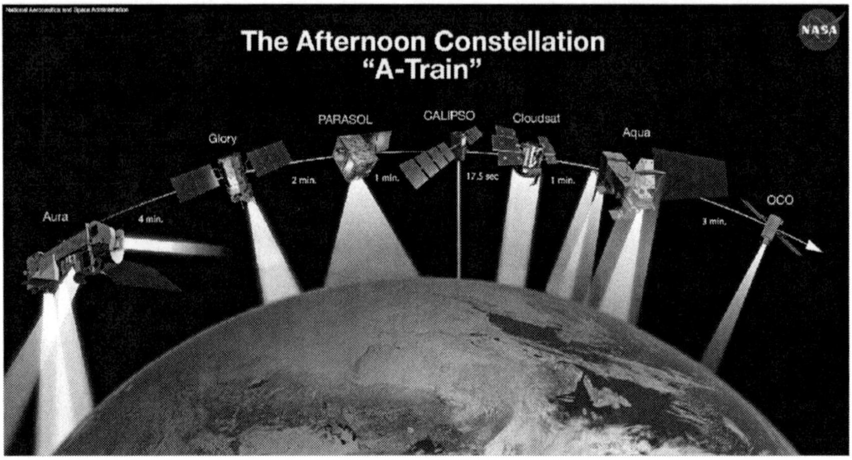

Fig. 9 PARASOL in the AQUA train constellation

PICARD is the next CNES MYRIADE mission scheduled for a November 2009 launch together with Swedish PRISMA mission on DNIEPR. Launch date is constrained by the beginning of the next solar cycle (Fig. 10).

This scientific mission has been proposed by CNRS agency, Belgium and Switzerland laboratories. Service Aéronomie /CNRS is the prime for the main payload. It is dedicated to the sun observation (diameter, irradiation, differential rotation, UV) in relation with the earth climatology, together with solar physics measurements (oscillations, UV).

The payload is composed of three instruments (total 60 kg, 60 W):

Fig. 10 The PICARD satellite

- a high resolution and high thermomechanical stability telescope SODISM (SA/CNRS) with CCD detector, for sun diameter measurement at high precision: a few milliarcseconds
- a radiometer SOVAP (IRMB) for total irradiation measurement
- a UV radiometer PREMOS (WRC/PMOD)

Platform has already been integrated and Payload mating is scheduled for July 2008.

The payload main sizing requirement is linked to bus pointing: the need is absolute sun pointing precision of less than $\pm$ 0.01° and pointing stability less than 0.01°/s. For that purpose, symmetrically from COROT, a fine sun sensor included in the payload telescope has been put in the AOCS loop through a specific mode.

Notice that this mission does not require the propulsion module nor the X band emitter (2 Gbits/day). On the opposite, AOCS mixes Payload sensor for sun 2 axes precise pointing. Third axis is provided by one of the 2 star trackers fitted in different directions.

The *TARANIS* mission is dedicated to study the magnetosphere-ionosphere-atmosphere coupling via transient processes. The mission will address unanswered questions relating to the recently discovered discharges in the stratosphere and mesosphere above thunderstorms, the so-called "Transient Luminous Events" (TLEs), and their relationship with the Terrestrial Gamma-ray Flashes (TGFs) of radiation from relativistic electrons accelerated in the atmosphere above thunderstorms.

The study of these phenomena is extended to the transient LEP (Lightning Induced Electron Precipitation) and accelerations of energetic electrons.

According to the mission objectives, the necessary instruments to fulfill the mission are:

- optical sensors,
- X and γ detectors,
- electron detectors,
- electric antennas,
- magnetic antennas.

This leads to an overall satellite mass close to 160 kg, which requires structural stiffening, but as a side effect, allows to set a new standard above the original MYRIADE 130 kg one, particularly fitted to the 200 kg ASAP capacities developed in the frame of SOYUZ.

Figures 11 and 12 depict satellite and mission.

Deployed antennae are close to the ones used on DEMETER. As well, stringent EMC requirements apply to the satellite, for which similar methods than DEMETER will be applied: filters at the output of the SA and SADM, between OBC and SST, reaction wheels covered by magnetic shielding ...

Satellite is currently in B phase and launch is scheduled for beginning of 2012.

Fig. 11 The TARANIS satellite

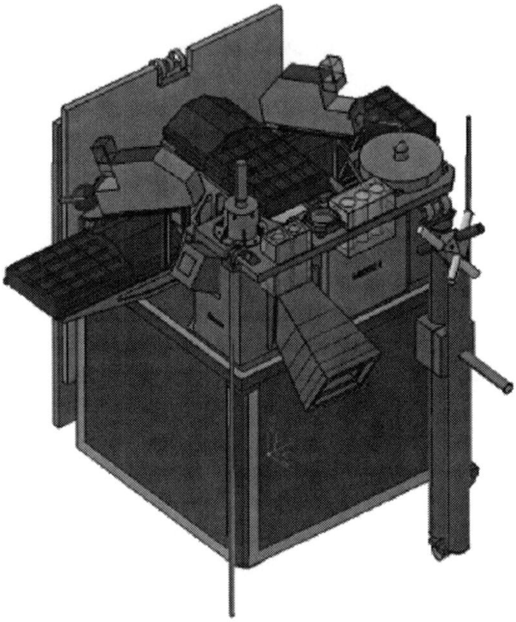

Fig. 12 TARANIS mission artist view

MICROSCOPE is a scientific mission proposed by ONERA who is also payload prime. MICROSCOPE is dedicated to the demonstration of the EP (Equivalence Principle). This principle states that the gravitational mass and the inertial masses should be equal (see Fig. 13).

The payload is based on two high performance (10^{-12} m.s^{-2} precision) differential accelerometers developed by ONERA. Detection of an acceleration on the second accelerometer, using a proof mass with a different density, will bring evidence of a violation or not of the equivalence principle at a 10^{-15} precision level.

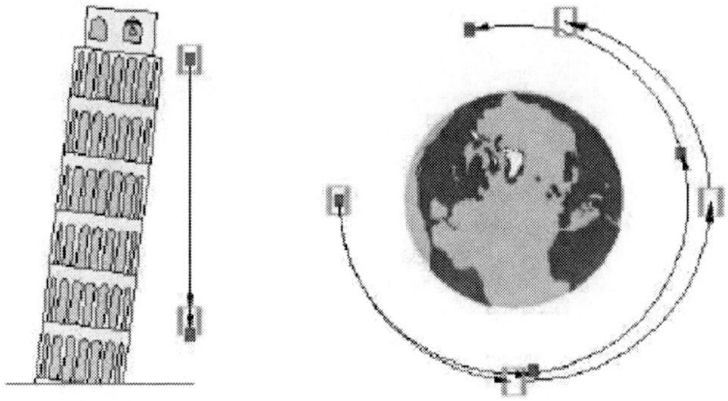

Fig. 13 The equivalence principle

Fig. 14 The MICROSCOPE satellite

In order to measure such differential accelerations, the accelerometers have to operate in proper conditions:

- the satellite (Fig. 14) has to be controlled drag free with residual linear acceleration less than 3.10^{-10} m.s^{-2}.Hz$^{-1/2}$.
- the satellite has to be spinned at 10^{-3} Hz.
- MICROSCOPE requires important adaptations of the basic MYRIADE design.
- payload accelerometers have to be located precisely at the satellite CoG, with very high thermal decoupling from spacecraft structure.
- dedicated AOCS modes have to be developed, using low thrust electric thrusters (Field Emission Electric Propulsion thrusters) as actuators. These thrusters are developed under ESA responsibility. A trade off is currently performed to also evaluate less complex cold gas proportional thrusters.
- bus architecture is specific; a new solar generator (symmetric) has been designed.

Main mission characteristics are:

Orbit	700 km, SSO 6 am/18 pm
Pointing	spin axis ⊥ to orbit plane
AOCS	new modes for drag free at 3.10^{-10}
Power	2 deployable then fix solar panels
Propulsion	12 FEEP thrusters on 4 pods 150 μ N thrust each
Weight	Total: 190 kg

The program is in phase B. Launch is scheduled in 2013. This mission is an example of the high flexibility of MYRIADE design.

In addition of these CNES missions, the MYRIADE standard has been selected by ASTRIUM and THALES to provide French Defence with 6 new satellites: 2 SPIRALE and 4 ELISA satellites under development. The MYRIADE standard is also the basis for imaging systems in development at ASTRIUM: 3 satellites for 2 export missions.

5 CNES Missions Under Study

The positive experience acquired in previous MYRIADE projects encourages the CNES engineering teams to envisage such hypothesis for the various missions submitted by CNES Program Directorate. Typically, even if preliminary system requirements lead to payload in the class 200–500 kg, an analysis is conducted in order to identify the possibility to support a part of the mission, or reduced performances with a MYRIADE solution. Depending on the programmatic context, this analysis can also highlight the interest for a demonstrator satellite based on MYRIADE. The following missions are under evaluation:

SMAR: Monitoring of maritime traffic. MYRIADE satellites could be involved in reception of AIS messages and transmission to ground processing systems. The need for less than 3 h delays in the operational chain could lead to a constellation of TBD satellites.

Reactive capacity: Imaging system (Pa+4XS), complementary to Pleiades system. The system aims to provide a short delay (3–4 h) to access to any point. Preliminary studies conducted at Astrium and Thales show that with a high resolution (1.5 m) payload can be installed on MYRIADE platform but may difficulties appear with an orbit at 450 km

Mistigri: Evaluation of ground temperature through observation in thermal infrared bands, representative of ground dryness. A MYRIADE satellite could accept a payload providing measurement in 4 bands with a 60 m resolution in TIR and 30 m in VNIR from a 720 km orbit. The system is designed for observation of test sites in South of Europe every 2 days.

6 Return of Experience

Obviously, development of lines of product has allowed CNES to propose for missions an efficient tool on a high variety of satellites configurations. Two of them are already in flight and at least 3 others are scheduled.

Several common determinant factors may be highlighted from the line of product approach:

- after the first mission, the system is well known and it is possible to register a drastic decrease of major anomalies. Of particular interest are all the design documentation, the SW, the validation benches and operation simulators, SDB, AIT GSEs, procedures
- using a generic ground segment with well known interfaces allows to concentrate new developments to the specific mission center; limited adaptations ease the operational teams efficiency and synergies among missions
- well known interfaces between platform and payload allow to specify payload without settling from the beginning the satellite team
- return of experience in flight allows mutual enrichment to continuously increase robustness; moreover, true performances are well known and ease the adequacy check for new missions without considering high margins
- procurement in batch of equipment units guarantees their industrial availability and allows drastic cost decrease, typically more than 30%; moreover, spares are naturally available
- this line of product approach is also of interest for prime contractors: after MYRIADE DEMETER launch, CNES decided to create a partnership with French prime contractors Astrium and THALES, allowing them to benefit from the design, to duplicate benches, to have common LTA with subcontractors. This has proved very efficient since now more than 12 MYRIADE have been or will be used on non CNES missions.

7 2009 A Decisive Year for MYRIADE Product Line

In March 2009, a prospective seminar will be with the same objective as those of Arcachon 1998 and Saint Malo 1994. Decision related to new missions are expected, and in particular MYRIADE implementation will be studies.

The development of a new mechanical structure will allow to take benefits from Soyuz ASAP capabilities: launch mass 200 kg for a payload mass at 95 kg. TARANIS project will be the first user of this structure.

On the other hand, progress are expected on ISIS program (Initiative for Space Innovative Services) aiming to support payloads in the range 100–600 kg. For MYRIADE product line, sharing industrial sources with ISIS program for components or pieces of equipment will allow to take benefits from new technologies, to deal with obsolescence problems and improve cost by common batch in equipment purchase.

8 Conclusion

Evidence of the gains brought by MYRIADE satellites lines of products has been highlighted. Commonalities and synergies are constantly evidenced through the various application missions using them. It is proposed to perpetuate and still strengthen this approach for next generation in the decade to come. This relates to a wide perimeter encompassing all the common services that allow to operate payloads in orbit. Increased flexibility has to be looked for, with respect to the size of the missions to accommodate. Commonality with ISIS new product line will be searched for cost efficiency and perpetuity. The broader the community using these standards will be the larger efficiency should be.

Acknowledgments Acknowledgments to all the CNES and industrial teams, as well as all cooperants who have permitted all the missions using MYRIADE and PROTEUS lines of product to be successful.

Further Reading

1. CNES Small Satellites Earth Observation and Scientific future Missions Acta Astronautica, Vol. 46, 2000.
2. DEMETER and PARASOL after 2 years in orbit: where are we, 4S 2006 Symposium, Ch B. Thiry.
3. TARANIS, a MYRIADE Mission dedicated to sprites characterization, 4S 2006 Symposium, Ch B. Thiry.
4. MICROSCOPE, a femto-g accelerometry mission, 4S 2006 Symposium, JB Dubois.
5. MYRIADE CNES Small satellites programme, Performances, Missions, Tool for education and cooperation, 4S 2006 Symposium, MA Clair.
6. Overview on CNES Micro and Mini satellites missions, in flight, under development and next: 4S Symposium, may 2008, Rhodes, Greece, Ph. Landiech.

Small Earth Observing Satellites Flying with Large Satellites in the A-Train

Angelita C. Kelly, Adam Loverro, Warren F. Case, Nadège Quéruel, Chistophe Maréchal, and Thérèse Barroso

Abstract This chapter presents Examples of the benefits of flying small satellites with other satellites, large and small. Scientists worldwide are beginning to take advantage of the opportunities for coincidental science offered by the international Earth observing constellation known as the A-Train. The A-Train comprises two large satellites (EOS-Aqua and EOS-Aura) flying with 4 small satellites (CloudSat, CALIPSO, PARASOL, and Glory). By flying in a constellation, each mission benefits from coincidental observations from instruments on the other satellites. The A-Train can be envisioned as a single, virtual science platform with mulTiple instruments. This chapter describes the challenges of operating an international constellation of independent satellites from the US and Europe to maximize the coincidental science opportunities while at the same time minimizing the operational interactions required between the teams. Constellation operations must be safe, but the team members must be able to conduct their respective science missions independently. The A-Train mission teams have also demonstrated that flying as members of a constellation does not take away the flexibility to accommodate new requirements.

1 Introduction

Flying small satellites with other satellites provides a lot of benefits. Typically, most small satellites fly payloads consisting of one or two instruments and fly in orbits that are independent from that of other satellites. The science data from these satellites are used alone or correlated with instrument data from other satellites. Data correlation with other satellites is greatly improved when the measurements of the same air mass or the same point on the ground are taken at approximately the same time. This chapter presents real-life Examples demonstrating the benefits of flying small satellites with other satellites, large or small, and vice versa.

A.C. Kelly (✉)
NASA Goddard Space Flight Center, Code 428 Earth Science Mission Operations (ESMO) Project, Mission Validation and Operations Branch, Greenbelt, MD, 20771, USA
e-mail: angelita.c.kelly@nasa.gov

R. Sandau et al. (eds.), *Small Satellite Missions for Earth Observation*,
DOI 10.1007/978-3-642-03501-2_2, © Springer-Verlag Berlin Heidelberg 2010

2 A-Train Description

Scientists worldwide are beginning to take advantage of the opportunities for improved data correlation, or coincidental science, offered by the international Earth Observing Constellation known as the A-Train, also referred to as the Afternoon Constellation (Fig. 1).

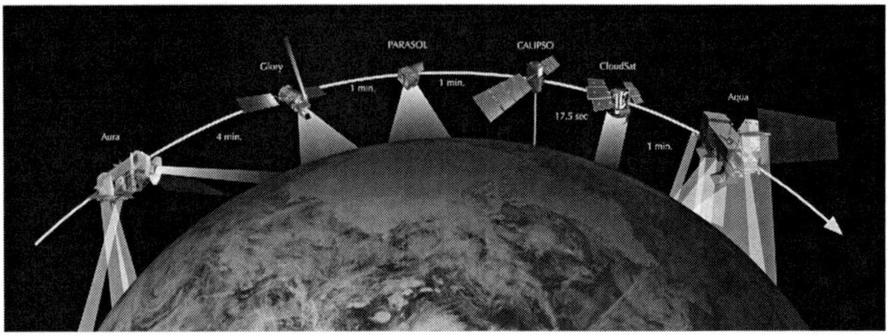

Fig. 1 Artist's depiction of the afternoon constellation (A-Train) in orbit

Most of the A-Train satellites (Table 1) are small. The A-Train is anchored by two large NASA satellites (Earth Observing System [EOS]-Aqua and EOS-Aura), but consists also of 4 small satellites (CloudSat, Cloud-Aerosol Lidar and Infrared Pathfinder Satellite Observations [CALIPSO], Polarization and Anisotropy of Réflectances for Atmospheric Sciences coupled with Observations from a Lidar [PARASOL], and Glory [to join in 2010]). Another satellite, JAXA's Global Change Observation Mission-W1 (GCOM-W1) recently agreed to join the A-Train (launch planned for 2012).

Table 1 A-Train member missions (both current and planned)

Mission	Launch	Agencies
EOS-Aqua	May 4, 2002	NASA
EOS-Aura	July 15, 2004	NASA
PARASOL	December 18, 2004	CNES
CALIPSO	April 28, 2006	CNES and NASA
CloudSat	April 28, 2006	NASA and USAF
Glory	2010	NASA
GCOM-W1	2012	JAXA

2.1 Aqua

The Aqua mission is a part of the international Earth Observing System (EOS) Program. Aqua was formerly named EOS PM, signifying its afternoon equatorial crossing time. Aqua was launched on May 4, 2002, into a near polar, sun-synchronous orbit with a period of approximately 100 min. The spacecraft repeats its ground track every 16 days to provide atmospheric measurements over virtually every point on the Earth in a repeatable pattern, permitting assessment of atmospheric phenomena changes in the same geographic locations throughout the life of the mission. Aqua has six Earth-observing instruments on board, collecting a variety of global data sets.

Aqua, Latin for water, is a NASA Earth science satellite mission named for the large amount of information it collects about the Earth's water cycle, including evaporation from the oceans, water vapor in the atmosphere, clouds, precipitation, soil moisture, sea ice, land ice, and snow cover on the land and ice. Aqua's instruments also measure additional variables including radiative energy fluxes, aerosols, vegetation cover on the land, phytoplankton and dissolved organic matter in the oceans, and air, land, and water temperatures.

2.2 Aura

Aura was launched two years after Aqua and is positioned at the back of the A-Train, 8 minutes away (along-track) from Aqua. Aura studies the Earth's ozone, air quality, and climate. It is designed exclusively to conduct research on the composition, chemistry, and dynamics of the Earth's atmosphere. Its instruments provide unique and complementary capabilities that enable daily global observations of Earth's atmospheric ozone layer, air quality, and key climate parameters. Aura measures five of the six "criteria pollutants" as identified by the US Environmental Protection Agency. Aura is providing the first-ever daily, direct global measurements of low altitude (tropospheric) ozone and many other pollutants that affect our air quality.

2.3 Parasol

PARASOL was launched for a nominal two-year mission and has almost doubled its nominal lifespan. PARASOL is the second micro-satellite in the MYRIADE family developed by Centre National d'Etudes Spatiales (CNES).

PARASOL is carrying a wide-field imaging radiometer/polarimeter called POLDER (Polarization and Directionality of the Earth's Reflectances), designed in partnership with the LOA (Atmospheric Optics Laboratory of CNRS-USTL) in Lille, France. POLDER is designed to improve our knowledge of the radiative and microphysical properties of clouds and aerosols by measuring the directionality and polarization of light reflected by the Earth's atmosphere.

Data collected by PARASOL have enabled the science community to establish the quantity and size distribution of aerosols over ocean regions, as well as their turbidity index over land surfaces, and to evaluate radiative forcing from solar radiation. The data are also used to help detect clouds, determine their thermodynamic phase and altitude, and estimate reflected solar flux. The PARASOL mission characterizes the clouds and aerosols radiative and microphysical properties by exploiting its complementarity with the other instruments in the A-Train constellation, which include the Clouds and the Earth's Radiant Energy System (CERES) and Moderate Resolution Imaging Spectroradiometer (MODIS) radiometers on the Aqua satellite, the CALIPSO lidar, and the CloudSat radar.

2.4 *Calipso*

CALIPSO was launched for a nominal three-year mission. It is based on a partnership between NASA and CNES, in association with the French national scientific research center CNRS/IPSL (Institut Pierre-Simon Laplace). Built around a PROTEUS platform, this mini-satellite has a payload composed of a backscattering lidar main instrument equipped with a 1 m diameter telescope provided by NASA, a visible Wide-Field Camera provided by NASA, and an Infrared Imager Radiometer provided by CNES.

CALIPSO is studying the clouds and aerosols radiative impacts. It will supply (for the first time) a data set of atmosphere vertical profiles as measured by a lidar on-board a satellite. The lidar data are used for characterization of the vertical distributions of the multi-layer clouds.

CALIPSO flies in tandem with the CloudSat radar satellite; together, they are dedicated to the observation of clouds, aerosols, and the water cycle.

2.5 *CloudSat*

CloudSat is a NASA Earth Systems Science Pathfinder (ESSP) mission, first proposed in the spring of 1998. CloudSat's Cloud Profiling Radar (CPR) is the first spaceborne Earth Observing sensor to peer into the structure of clouds, and measure the vertical profiles of liquid and water ice content within. Using this powerful tool, CloudSat explores the relationship between cloud structure and content, with the radiative properties of clouds, an understanding that is key to unraveling the effects and progression of global climate change. CloudSat's measurements also evaluate and validate the representation of clouds and cloud processes in global atmospheric models to better predict tomorrow's weather and global climate future. The CloudSat mission is a collaboration between NASA, the Canadian Space Agency, the Cooperative Institute for Research in the Atmosphere (CIRA) at Colorado State University (CSU), and the US Air Force. The project is managed by the Jet Propulsion Laboratory in Pasadena, CA.

3 Constellation Advantages

A-Train satellites fly at 705 km in sun-synchronous orbits. The mean local times at their equator crossings are within seconds to a few minutes of each other. By flying in the A-Train, each mission benefits from coincidental observations with instruments on the other satellites in the constellation. From an instrument observation point of view, the A-Train can be envisioned as a single, virtual science platform with mulTiple instruments. This is the primary benefit of constellations – data from small satellites can be enhanced with the data from the other satellites' instruments.

Specifically, the A-Train enables scientists to use numerous instruments for observing the land, atmosphere, oceans, and clouds, using data from passive radiometers, active lidar, and radar sounders. The scientists have a unique opportunity to gather a wide range of data and to combine for the first time vertical and horizontal views of the atmosphere. For Example, they are able to compare vertical profiles supplied by the CALIPSO lidar and the CloudSat radar with reflectance over a very wide spectral range provided by Aqua's MODIS instrument, while at the same time taking advantage of information on reflected light polarization from PARASOL's Polder instrument. First, the CALIPSO lidar and the CloudSat radar each provide complementary measurements for classifying cloud layers and determining cloud water content. Secondly, the lidar and the radar images are improved by integrating information provided by imaging radiometers. At the same time, the vertical profiles enable inversions obtained from passive observations along the ground track to be validated and improved, which then increase the quality of wide field observations (Fig. 2).

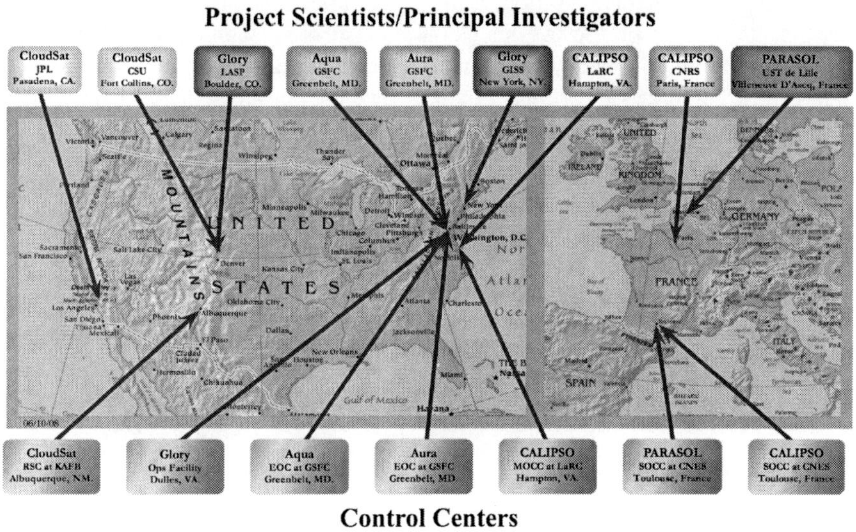

Fig. 2 Global distribution of A-Train member organizations and operation centers

A-Train mission teams also benefit from the increased safety, efficiency, and lowered costs through the coordination and centralized monitoring of all satellites in the constellation. For safety, an orbital configuration monitoring system provides Warnings of possible close approach/collision between member missions or with orbital debris. The information provides the missions enough time to consider what action may be needed. As all the satellites fly in the same portion of space, all debris or other spacecraft that could collide with a satellite of the constellation can be watched at the same time, providing surveillance for the entire constellation. This operational activity saves money and time. Moreover, orbit specialists of each team work together, which can be very valuable in case of crisis management. This is the case for Example for PARASOL (a fully French mission), which, thanks to the participation to the A-Train, gets a much appreciated support from NASA each time a potential collision has to be avoided. PARASOL has already made a collision avoidance maneuver once, thanks to the constellation monitoring.

4 A-Train Concurrent Imaging Examples

CloudSat's partnership and science collaboration with the missions in the A-Train is quite unique. First, by flying in close formation with the CALIPSO mission spacecraft, CloudSat has been able to co-register more than 90% of its CPR cloud profiles with the aerosol profiles from the CALIPSO lidar instrument. CloudSat's CPR can penetrate deep into the cloud layer, often all the way through to the ground, but it can miss the structure of thin clouds above of a thick cloud layer. CALIPSO's lidar by contrast measures the fine structural detail of the atmosphere above and between the clouds, but is attenuated very near the tops of thick cloud layers. Together, these instruments provide profiles of our global atmosphere, spanning the width of the troposphere more completely than either could individually. Figure 3 provides an Example of combined radar/lidar products by showing areas of radar only, lidar only, and radar and lidar combined observations.

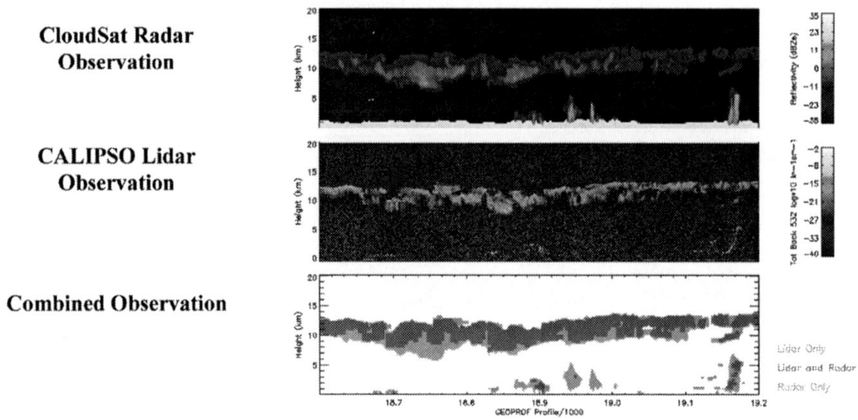

Fig. 3 Combined radar/lidar observations from CloudSat and CALIPSO (Credit: CloudSat Data Processing Center [DPC])

CloudSat's science collaboration in the A-train extends far beyond its relationship with CALIPSO. At the leading edge of the A-Train, the MODIS instrument on the Aqua satellite sweeps out a wide cross-track swath of multispectral imagery which are concurrent with CloudSat's cloud observations.

Some early Results of this collaboration are illustrated in Fig. 4, which shows a CloudSat quick-look display from an entire orbit overlaid on Aqua MODIS infrared imagery. The CPR profile for that entire orbit is shown immediately below, and one segment of the orbit is shown complete with reference topography and MODIS IR imagery.

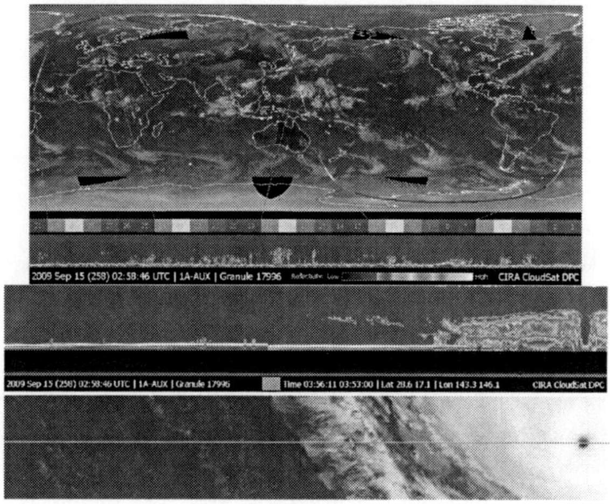

Fig. 4 CloudSat quick-look products overlay radar vertical profiles over MODIS IR imagery (Credit: CloudSat DPC)

5 Coordinated Constellation Operations

Each project or agency usually determines what satellites to launch into a specific orbit. The orbit requirements are primarily driven by the science requirements. Science can benefit greatly if the mission scientist and the mission planners spend some time during the mission concept and design stage to determine if the science can be enhanced by coordinating their location with other satellites that are already flying in the same orbit. If the answer is yes, then the mission planners should work with the scientists to understand where the new satellite should be placed relative to the on-orbit satellites to maximize the benefits of co-flying. The next step would be to coordinate the chosen location with the other satellite teams (in the case of the A-Train, that would be with the Mission Operations Working Group).

Coordinated constellation operations in the A-Train did not come about automatically. A concerted effort was undertaken 3 years before the launch of CALIPSO and CloudSat to ensure that the entry of the two satellites would not pose a safety risk for the EOS Aqua and Aura satellites. Through the cooperation among the member mission teams (Aqua, Aura, CloudSat, CALIPSO, PARASOL, OCO, and Glory), the orbital configuration, operations coordination requirements, processes, and Procedures were established and agreed to by all member missions and their management. These agreements are documented and kept under strict configuration control. Key provisions in these documents include the following: First, constellation operations must be safe so coordination is a must. Second, constellation operations must not place an undue operational burden on the mission teams. Third, the mission teams must be able to conduct their primary science missions independently. Finally, a mechanism has been set up to facilitate data sharing and enable the data users to obtain orbital data and status. Member missions perform their constellation operations in accordance with these documents. New missions who want to join the constellation are also required to adhere to the baselined agreements, processes, operating guidelines, and procedures.

6 A-Train Flexibility to Accommodate Recent Changes

Since the launch of CALIPSO and CloudSat in 2006, the A-Train teams have been able to demonstrate that flying as an international constellation does not take away the flexibility to accommodate new requirements. The regular communication and collaboration of the A-train team members has fostered an environment where each mission can adapt quickly and efficiently to configuration changes. A good Example is the change requested by the CALIPSO science team in May 2007. The CALIPSO science and instrument teams determined that a change in the pitch of the lidar 3° forward would dramatically improve data quality. While it seemed like a simple change at first look, it turned out that the change would adversely affect the data overlap between CloudSat and CALIPSO (running at ~95%). Through the framework of the A-Train working group, the CloudSat and CALIPSO teams, with the approval of the other A-Train members, were able to implement a change in the A-Train orbital configuration to maintain footprint overlap, wherein CALIPSO's pitch would be shifted forward as requested, and CloudSat would move forward relative to CALIPSO. The transition was so well-executed that the two satellites together lost only a few hours of combined data during the implementation period. In addition, the change was coordinated with the A-Train science community. The CloudSat science processing team, as well as CALIPSO, had to modify its processing algorithm. A minor change was also made to the constellation monitoring system. This Example demonstrated that if there is thorough coordination between the science users and the mission operations teams, a "simple" change could be implemented properly with all impacts taken into account.

Another change that was successfully negotiated and implemented was the repositioning of the Aura satellite closer to Aqua in May 2008. In all cases, while safety is the first priority, science return is enhanced, and operational flexibility is retained to the maximum extent possible.

7 Summary

The content of this chapter is based on several years of A-Train experience and should be helpful to missions considering constellation flying. The A-Train's approach to Constellation management and coordination has proven beneficial by enabling mulTiple missions to collect coincident measurements. Straightforward and frequent coordination mechanisms have helped to make the operations both lightweight and robust. Even with well-designed interfaces and processes, missions have the flexibility to change configuration to respond to changing science requirements and desirements. The future of the A-Train depends both on the investment in large spacecraft, and an ever-growing population of small satellites.

Acknowledgements The authors thank the Afternoon Constellation mission and science teams for their contributions to the report, and their continuing support of the A-Train constellation.

Additional Information

Visit the following websites for additional mission and instrument descriptions:

- Aqua – http://aqua.nasa.gov/
- Aura – http://aura.gsfc.nasa.gov/
- PARASOL – http://smsc.cnes.fr/PARASOL/
- CloudSat – http://cloudsat.atmos.colostate.edu/
- CALIPSO – http://smsc.cnes.fr/CALIPSO/ (CNES)/http://www-calipso.larc.nasa.gov (NASA)
- Glory – http://glory.gsfc.nasa.gov/

Acronyms

CALIPSO	Cloud-Aerosol Lidar and Infrared Pathfinder Satellite Observations
CERES	Clouds and the Earth's Radiant Energy System
CIRA	Cooperative Institute for Research in the Atmosphere
CNES	Centre National d'Études Spatiales
CNRS	Centre National De La Recherche Scientifique
CPR	Cloud Profiling Radar
CSU	Colorado State University
DPC	Data Processing Center
EOC	EOS Operations Center
EOS	Earth Observing System
ESMO	Earth Science Mission Operations
ESSP	Earth Systems Science Pathfinder
GCOM	Global Change Observation Mission
GISS	Goddard Institute for Space Studies
GSFC	Goddard Space Flight Center
IPSL	Institut Pierre-Simon Laplace
JPL	Jet Propulsion Laboratory
KAFB	Kirtland Air Force Base
LaRC	Langley Research Center

LASP	Laboratory for Atmospheric and Space Physics
LOA	Atmospheric Optics Laboratory
MOCC	Mission Operations Control Center
MODIS	Moderate Resolution Imaging Spectroradiometer
NASA	National Aeronautics and Space Administration
POLDER	Polarization and Directionality of the Earth's Reflectances
PARASOL	Polarization and Anisotropy of Réflectances for Atmospheric Sciences coupled with Observations from a Lidar
RSC	Research, Development, Test & Evaluation (RDT&E) Support Complex
SOCC	Satellite Operations Control Center
UST	Université des Sciences et Technologies

INTAμSat-1 First Earth Observation Mission

Manuel Angulo, Laura Seoane, Elisa Molina, Manuel Prieto, Oscar Rodriguez, Segundo Esteban, Jordi Palau, and Stefania Cornara

Abstract This chapter is a description of the INTAμSat-1 first Earth observation mission objectives and its development status. With a mass around 100 kg and a 60 × 60 × 90 cm size compatible as an auxiliary payload for VEGA, Soyuz-ST and Dnepr, the project is running now the Phase-C and will be ready for launch by the middle of 2011. This new INTA small satellite programme initiated in October 2005, belongs to the well known μSAT class (up to 150 kg) as a further step after the Nanosat programme success: first Nanosat-1 with a 19 kg mass and still working OK in orbit that was launched by Ariane-5 in December 2004; and Nanosat-1B with 23.5 kg set for launch in a DNEPR by July 2009 from Baikonur. In fact both programmes will be running in parallel at INTA with next Nanosat-2 tentatively set for launch by 2013. It will be an evolved 20–50 kg s/c with improved service module resources and a separated payload modular design. After the preliminary progress on the different platform subsystems design, it was decided by the middle of Phase-B in October 2007 that the first μSAT demonstration flight will be devoted to an experimental Earth observation payload with several instruments, together with some technological experiments. Both the Service Module (SVM) and Payload Module (PLM) are fully developed at INTA, with contributions from several Universities and few small Spanish companies. INTAμSat-1 PLM will carry the following instruments: (1) CINCLUS, a 30 m GSD pushbroom TDI camera with a wide swath, that will try to collect information on the water quality over more that 150 reservoirs spread out over the Spanish territory, with a repetition cycle of 9 days; (2) MS-WAC, a 10 m GSD wide swath MS camera with 4–5 channels that will provide full coverage of Spain also in 9 days; PAU, a combined reflectometer–radiometer that using GPS reflected signals will try to measure the sea surface height variations and salinity. All this PL data will be transmitted down to Maspalomas ground receiving station in X-band at 40–80 Mbps, while the Control Centre (CC) will use the same Nanosat CC facilities for S-band TT&C available at INTA headquarters in Torrejón de Ardoz, Madrid.

M. Angulo (✉)
INTA, Dto. de Programas Espaciales y Ciencias del Espacio, Torrejon de Ardoz 28850, Madrid, Spain
e-mail: angulom@inta.es

R. Sandau et al. (eds.), *Small Satellite Missions for Earth Observation*, DOI 10.1007/978-3-642-03501-2_3, © Springer-Verlag Berlin Heidelberg 2010

1 Introduction

This chapter contains a description of the first INTAμSat Earth observation mission devoted to routine wide swath 10 m resolution multi-spectral imaging, together with other R&D remote sensing payloads. Previous IAA paper on the μSat programme [1] (see IAA-B6-016 from the same authors presented at the 6th IAA Symposium) was a general presentation of this new INTA small satellite programme, in line with the previous Nanosat initiative. Here we will take a look on the mission requirements, mission analysis, short description of the different payloads, and review of present status of the Space and Ground segments development. During the last 2 years the satellite design, and especially the Service Module (SVM), has evolved as expected with some improvements over the original requirements established at the end of Phase-A.

From a programmatic point of view at this moment we are in the middle of Phase-C with an envisaged end by January next year. The SVM CDR is planned for February 2010 followed by each particular payload CDR. These important milestones will clear the way for the final INTAμSat-1 FM manufacturing, integration and verification during all 2010 and most probably the first half of 2011. Apart from the different challenges that a new 100 kg class spacecraft (S/C) programme must face in terms of design objectives fulfilment, we had some delay impact over the original planning due to the in house running second Nanosat mission, supported by the same project team (always the limited resources problem . . .).

The SVM subsystems design progresses as expected including the platform resources that will be available for the different experiments (i.e. Mass Memory Unit, high speed data links, etc.). On the other hand, to find attractive R&D Payload (P/L) candidates compatible with the time constraints of an Earth observation mission opportunity is not an easy task. At this stage of the programme some of the proposed payloads are in line with the planning, but others are not. This is quite common everywhere and it is clear that we should face a final selection decision before the end of this year 2009, based on the development status, maturity of the proposed experiments, and budget funding support from other national financing sources.

The strategic aim of the Nanosat programme [2] originally established in 1997 after the in orbit commitment of first Minisat-01, was to have ready for launch a mission every 4 years. Up to now we have succeeded on this objective and the Nanosat-1B second mission is now ready, around 4 years after the Ariane-5 launch of the first Nanosat-1 satellite from Kourou in December 2004. The planned launch date for Nano-1B by a Dnepr from Yasny (Russian Federation near the border of the Republic of Kazakhstan, 500 km up north from the Baikonur Kosmodrome), was by the end of last year 2008. Some problems encountered by Kosmotras with the official approval of the first stage dropping zone over Turkmenistan and the final decision to move the launcher to Baikonur with a manifest plenty of launches, has delayed the final launch up to next July 2009. It seems finally that the integration campaign for the 6 small satellite passengers will start in June, so good luck for everybody on board this time!!! (Fig. 1)

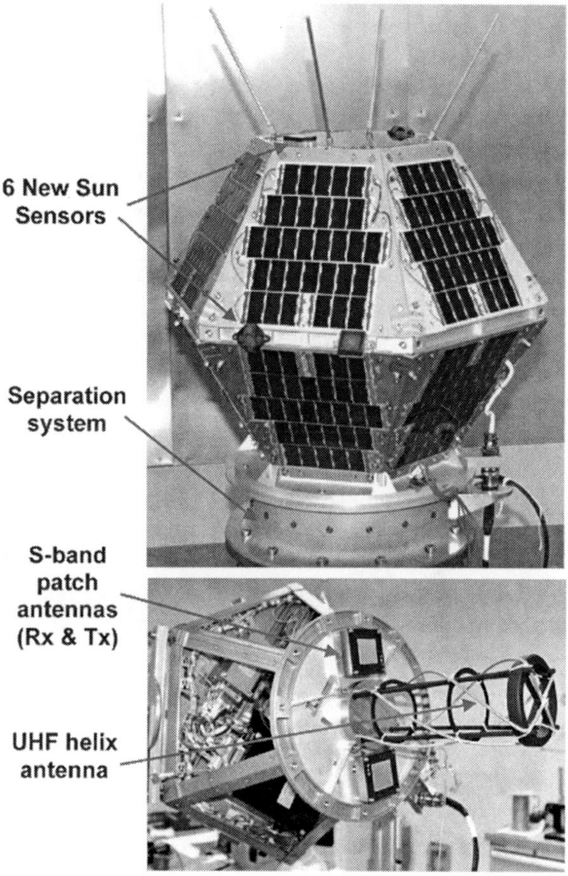

Fig. 1 Nanosat-1B FM during the vibration tests, and the new antennas under the separation ring (not shown above)

To have the same 4 years periodicity objective for the μSat missions together with the Nanos (i.e. alternatively each 2 years) is something really demanding from a budgetary point of view. At this moment this is not yet clear, but we will see what the future brings considering the present economic restrictions.

2 Nanosat and MicroSat Mutual Relationship Benefits

It is important to say that both programmes belong to a R&D initiative almost fully funded by the INTA internal budget. For this reason, we use to develop almost all the units and subsystems for the Space and Ground Segments. Practically speaking this philosophy implies a systematic but risky approach for the satellite bus and the payload designs. All new developments must be carefully verified and run safely in orbit as far as possible. It is then a must the acquisition by our team of the associated know-how and new technologies. Having two small satellite programmes running in parallel is of mutual interest, because a more than 20 kg "Fat Nanosatellite" is a cheap and perfect platform to test in orbit new designs. In a 100 kg class S/C the involved costs are 5–10 times higher together with bigger mission success expectations, and as a consequence, we need that the associated system and subsystems (S/S) design risks be reduced as far as possible. The following paragraphs explain in a more detailed way the very important synergies between both programmes.

Safe communications: A satellite in orbit is completely depending on its TTC Subsystem to communicate in both ways with the ground Control Centre (CC). If you are not able to contact in a routine and safely way your satellite at any time and attitude orientation, forget about having success in the space practice. That was the way with our first Nanosat-1 at the beginning [3], because although we suffered during the firsts 4–5 months some minor problems with the up-link communication channel (the on board receivers were not able to listen well to the uploaded TCs), in general we have had until now good access to the satellite. This open door let us improve the Flight SW in several ways and the automatic control and mission exploitation Procedures. Of course your on board computer also needs to be able to accept new uploaded Flight SW versions!!!

One important in orbit verification step is that Nanosat-1B carries as a technological P/L, a new S-Band TTC transceiver that will be used on all INTA missions from now on, if it works well in accordance to the expected performances. The critical point is a new digital Modem with transmission and reception lower speeds fixed on purpose at 100–200 Kbps depending of the selected FEC by TC (1/2–7/8), in order to be able to test the new FPGA design. Another reason is that we can use a small 2.5 m ground dish antenna for both up and downlinks placed over the roof of the Control Centre (CC) in Madrid. Although it can run up to 2 Mbps with the same HW resources on this unit, this would have required at least a 5 m parabolic antenna placed quite far away from our CC (Maspalomas in the Canary Islands).

Next Fig. 2 shows the brand new S-band Transceiver for TTC (100 Kbps) and TM (up to 2 Mbps) inside a Box with 106 × 166 × 90 mm and a mass of 1.6 kg, that will be used in the future INTA satellites.

On top of that, the on board OBDH CPU based on a 68332 Motorola microcontroller chip has processing speed limitations and is just able to handle this throughput attending at the same time other real time tasks on board. One of the system functional objectives was to share the CPU baseband operations at 200 Kbps having the capability to use both communication front-ends (UHF Store and Forward mission, and the S-band new development transceiver) when selected from ground with a dedicated TC. At the CC we also have full UHF and S-band front-end chains. In fact this is an important redundancy at system level because in case of failure in one of the chains, the other will take care of the TTC functions.

ACS subsystem: Any small satellite mission requires at least coarse Sun and Magnetic sensors for the initial acquisition and survival modes. We have already tested in orbit on board Nanosat-1 a new 3-axis magnetic sensor using a commercial available Magneto-resistance from Honeywell. Based on the good Results in orbit (robustness, accuracy and low power consumption) the same design is ready to fly in the Nano-1B follow-on mission, and has been selected for future INTAμSats and

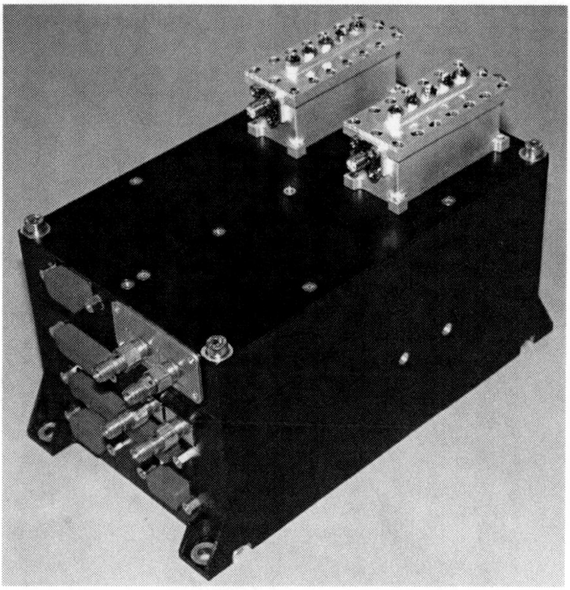

Fig. 2 The new S-band TTC fully redundant transceiver that will be tested very soon in orbit (July 2009)

second generation Nanos (Nanosat-2). Next Fig. 3 shows a view of the sensor head assembly.

For the Coarse Solar Sensors (CSS) we are taking advantage of the two different designs flying on Nanosat-1 (2 sets with 3 space qualified GaAs, together with 2 more Nanosensors with 3 experimental porous Si cells each) and still running in orbit after 4.5 years (normal degradation on the Si cells is in accordance with the expected values). We have introduced a new very small design with 5 GaAs cells that provide 0.2° accuracy in a quasi hemispherical angular aperture. As Nanosat-1B is an Earth pointing S/C (due to the UHF medium gain helix antenna), we need three SS placed in ±Xs and +Zs directions. INTAμSat-1 spacecraft will make use of the same 3 SS with a very similar orientation configuration (±Ys and –Xs). The ACS sensor driver and S/C acquisition SW needed to control both satellites is almost identical. So we will be able to test very soon in orbit both new HW and SW designs.

In parallel with the nominal SS we have 3 more experimental μsensors called VectorSol that have been developed by UPC (Barcelona) together with the US (Sevilla), and qualified at INTA facilities.

All the 3 × 4 Si cells analogical signals are sent to the Sensor Box electronics after a pre-amplification inside each VectorSol (using operational amplifiers and a differential lines I/F). Once mulTiplexed, they are then sent to the OBDH CPU where the ACS flight SW makes good use. Sensors A & C dimensions are just 30 × 30 × 14 mm and one of them (sensor B) that is a little bit bigger: 30 × 50 × 14 mm, holds inside a PIC μprocessor including the A/D conversion and all the S/W calculations to provide through a SPI bus channel, both zenith and elevation Sun vector directions to the ACS SW (Figs. 4 and 5).

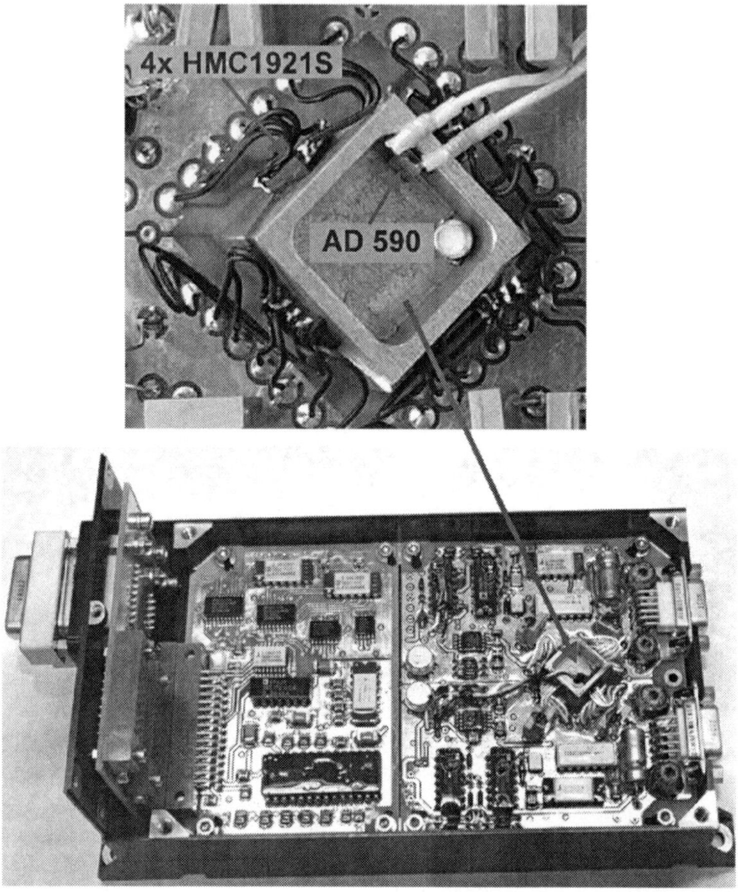

Fig. 3 Triaxial magnetic sensor head (14 mm side cube with 4 single axis chips at 45° with a hot redundant configuration), and its board inside the Sensors Box unit, that will be also used in the μSat missions

As an evolution and using the same micro-technology manufacturing facilities the UPC have in Barcelona, they are now developing upon request from INTA a Fine Sun Sensor (μFS) for the MicroSat and Nanosat-2 future programmes. This promising sensor will be tested in orbit on board INTAμSat-1, where it can be used once fully tested as a back-up fine SS in case of need. Several missions could make use of this μFS in order to save the cost of a Star Tracker. For this reason it is also important to see the Results in orbit of their first design VectorSol after the launch of Nanosat-1B.

For a Remote Sensing mission like the first INTAμSat-1, it is interesting to have an ACS intermediate mode with a coarse (2–3°) Earth pointing capability after a transition from a pure inertial Sun pointing Survival mode (we have selected the –Xs zenith solar panel direction) that can be achieved using the 3-axis Magnetic

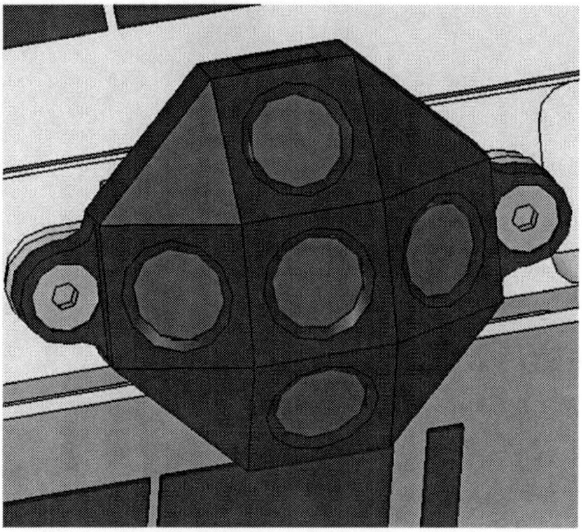

Fig. 4 The new 5 GaAs cells Sun Sensor (30 × 30 mm and 0.2° in 120°)

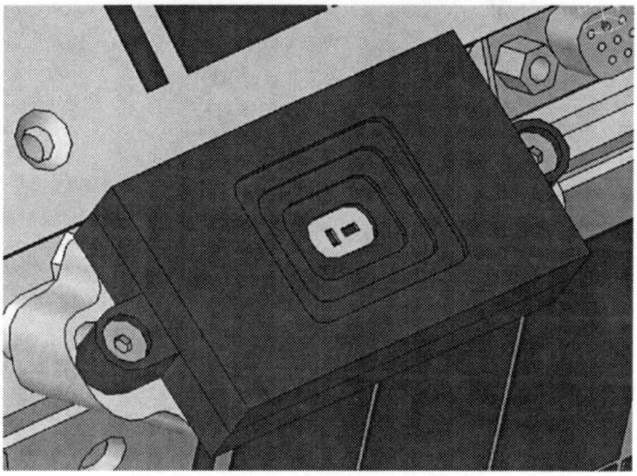

Fig. 5 VectorSol is an experimental 4 cells μSun Sensor (30 × 50 mm and 0.1° in 120°) including a PIC processor and its flight SW

sensor and CSS. After this coarse Earth mode is secured, we will be able to use the FSS to increase the pointing accuracy and switch to the Nominal mode.

Power S/S: A new 5.8 Ah 24 V Li-ion battery from SAFT will be flying on Nanosat-1B. This is a new cells combination design specifically qualified for this mission, but with the same approach used for the ESA's PROBA-2 S/C battery.

In this case the Nano-1B takes benefit in advance from the same battery that will be used in the INTAμSat mission.

The same applies for the solar panels, because after a dedicated assessment made together with Selex Galileo in Italy for the μSat best solution (triple junction cells over 20 mm Al honeycomb core with carbon fibre skins), we are going to use the

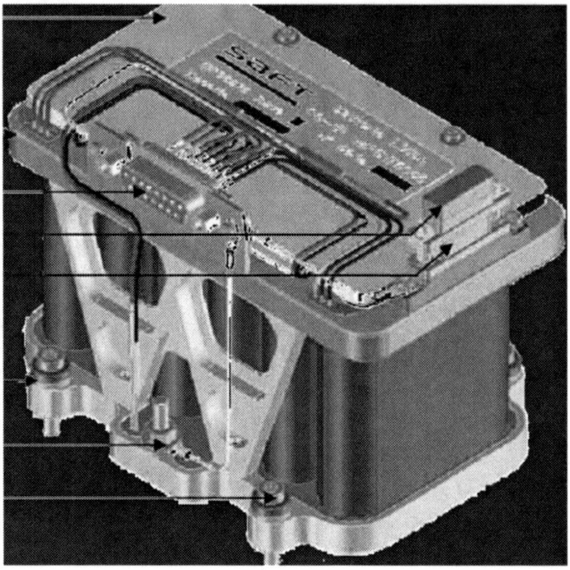

Fig. 6 Nanosat-1B Li-ion SAFT battery

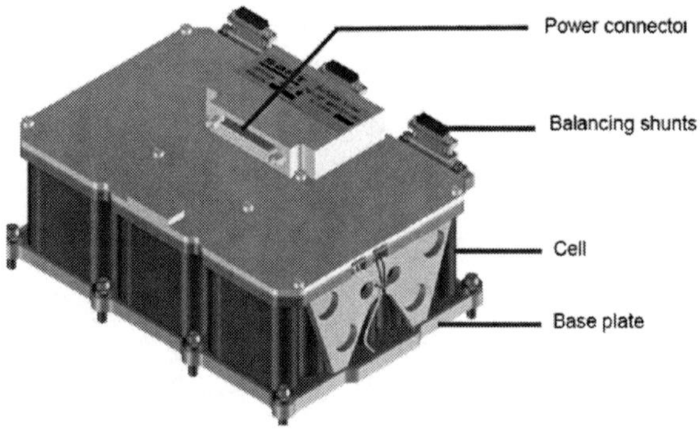

Fig. 7 INTAμSat-1 Li-ion SAFT battery

same design (12 mm) in the future Nanosat-2 solar panels. This together with a change in the external shape to a traditional Box, will increase the solar panels sunlit power from 23 W in Nano-1 up to 50 W, which is remarkable for such a small size S/C (Figs. 6 and 7).

OBDH S/S: As mentioned above, we are using for the first Nanosat generation a Mil-883 68332 processor previously tested for radiation tolerance, together with a latch-up protection circuit that avoids destruction in case of a latch-up build-up. For the INTAμSat missions we have already run a new CPU based on the well known TSC-695 from Atmel, which is a radiation hardened ERC-32 chip device. The CPU is a single string design and for this reason, we need to rely on proven and space qualified components (the same applies for the SRAM, EEPROM, FPGAs, buffers, drivers and so on). Also in this case, the future Nanosat-2 missions will take advantage of carrying the same but simplified design. The quite open CPU philosophy approach we are following in μSat allows us to upgrade in the next future to the LEON-3 SPARC-V8 installed in a RTAX FPGA from Actel, with not so big PCB lay-out impact.

3 INTAμSat Mission Requirements

The next table is a summary of the main mission requirements selected for this first R&D Earth observation mission associated to the satellite and each payload.

Manufactured by: INTA – Spain *Destination*: R&D in remote sensing	*Model/bus*: INTAμSat-1
Mass: • Total mass at launch: <110 kg • Mass of satellite in target orbit: <110 kg	*Dimensions*: • Stowed for launch: • 620 × 620 × 950 mm • Deployed in orbit: the same
FINAL ORBIT: • 7,081 × 7,089 km; i = 98.2°; w = 66°; RAAN = 156° (around 700 km 22:30 LTAN SSO orbit)	LIFETIME: • Target = 3 years, • Extended = 5 years
EXPERIMENTAL PAYLOADS: • CINCLUS: a camera with 4 narrow MS bands of 30 m GSD for in-land reservoirs water quality control, with a 10–15 days revisit time (developed at INTA) • MS-WAC: two cameras with 4–5 MS bands of 10 m GSD for Remote sensing of the Spanish territory, with 9–10 days revisit time (developed at INTA) • PAU: an experimental radiometer and GPS reflectometer to study the sea salinity and surface state mainly at night (developed by UPC & AD Telecom in Barcelona)	
TECHNOLOGICAL PAYLOADS: • μFS: a Fine Sun Sensor (0.01–0.02°), developed by UPC & US (Sevilla) based on VectorSol • GPS Rx: an experimental GPS L1 receiver using COTS RF front-end components, developed by INTA, AD Telecom and SRG-UAH (Alcala de Henares, Madrid) • LASER downlink: an optical link communication experiment between a LEO S/C and the OGS in Tenerife (Canary Islands) at 100 Mbps, (UPC, ICFO, AD Telecom, INTA, IAC, ROA)	

Since the middle of Phase-B it was decided to make a first qualification flight with a Remote Sensing mission. The basic idea is to reduce the complexity of the S/C as much as possible and one of the possible ways is to keep always the Earth pointing mode. This implies for any payload on board, that in order to cope with the selected revisit time coverage, the cameras or instruments should deal with a predefined swath. The main optical cameras (MS-WAC and CINCLUS) design driver is the ability to provide both the required resolution and swath for the selected orbit altitude.

Figure 8 shows a map of Spain with more than 130 water reservoirs spread over the peninsula that must be revisited each 10–15 days. For the observation task of the

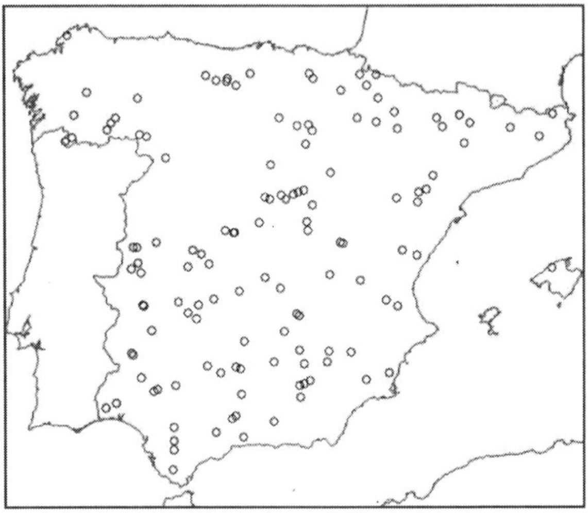

Fig. 8 Main reservoirs map in Spain

CINCLUS payload, Deimos Space have made detailed analysis to see if a slewing satellite with a 30 km swath camera was able to cope with the stringent coverage requirements. As it is easy to imagine and assuming the high number of targets to check in such a short revisit time, the only way was to keep the S/C nadir looking, and to increase the swath of the camera or to use two identical units to cover the 280 km distance between two orbits in 9 days. This was exactly the decision taken for the MS-WAC camera. Figures 9 and 10 show in a simplified way the mission analysis logic behind the orbit selection process.

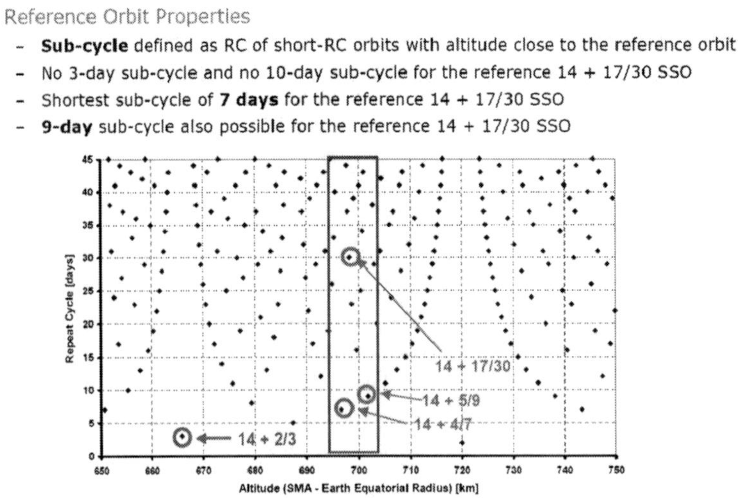

Fig. 9 Orbit selection

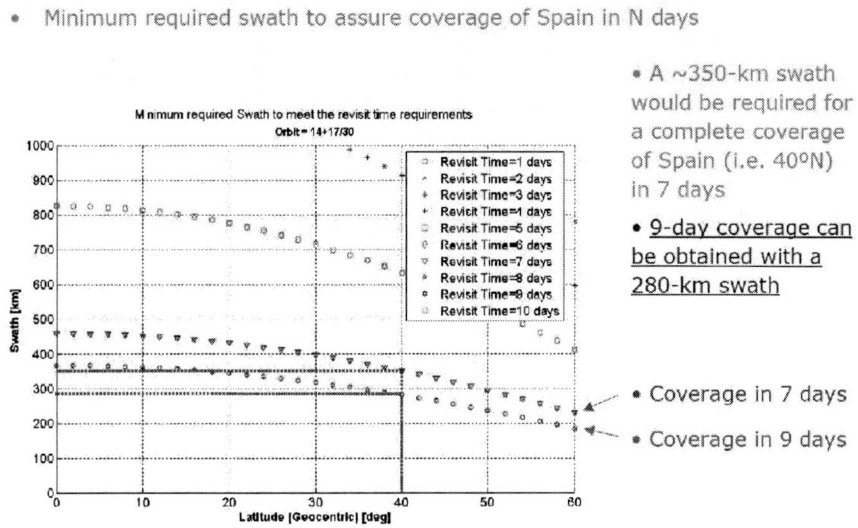

Fig. 10 Coverage alternatives

The reference SSO orbit has a 696 km height and is very sensitive to decay drifts. For this reason it has been recommended the use of a propulsion S/S to correct the injection errors and orbit drift with time.

PAU payload sensor uses an array antenna with 4 or 7 patches (TBC) to have at the same time a radiometer to look for the sea salinity, together with a GPS reflectometer that will try to measure the sea surface state. This is accomplished comparing the reflected and dispersed signal from the sea surface to the direct GPS reception. Due to this requirement the P/L also needs another patch antenna on the S/C zenith face. It is optimised for the sea observation at night time to avoid the noise coming from the Sun that could bother the radiometer, and for this reason it is a good complementary P/L for the optical cameras mission that should work at day time and mainly over the Spanish territory. From an operational point of view the only conflict will be the on board archiving of the retrieved data, for this reason we will need to optimise the use of the on board Mass Memory Unit (MMU). This point depends to a great extent on the potential use of the Svalbard or Kiruna receiving stations, to dump down in a more efficient way the data from the MMU.

4 MicroSat Project Team

The project working group at this stage of the Phase-C is the following:

- *SRG-UAH*: The Space Research Group belongs to the Universidad de Alcalá near Madrid. They are working with INTA in the design and development of the SW and HW for the OBDH CPU and MMU.

- *RM-EUITA-UPM*: The Structural Group at the Escuela Universitaria de Ingenieros Técnicos Aeronáuticos (EUITA) belongs to the Universidad Politécnica de Madrid (UPM), and they are doing the NASTRAN analysis for INTAμSat.
- *UCM*: The Automatic Control Research Group from the Universidad Complutense de Madrid, have previous experience in control laws design and implementation. They are integrated in the AOCS design and testing group, with INTA as leader.
- *AD Telecom*: is a small private company in Barcelona responsible for all the communication developments in collaboration with INTA.
- *TTI Norte*: is another little company in Santander, which was responsible for the specific communications protocol development for Nanosat-01. They are now working on the INTAμSat TTC protocol that is compatible with ESA CCSDS recommendations.
- *DEIMOS*: is a well known private Spanish small company in the GN&C activities for ESA and other European programmes. Within INTAμSat they are participating in the mission analysis and supporting the AOCS design and testing group.

There are other participating universities and research organisms engaged in the development of the experimental payloads (see the mission requirements table for further details).

5 System Design Status

The baseline design for this first mission is a S/C configuration with fixed solar panels optimised for an average power production all along the sunlit portion of the orbit.

Figure 11 shows the flight configuration and P/L instruments accommodation within the PLM. On both Earth and zenith sides there are a number of antennas needed to fulfil all communication or detection tasks.

Figure 12 contains the present block diagram for the SVM and Fig. 13 for the PLM. At this stage of the design this is showing almost the final selected configuration for all the S/S and P/L units I/F connections.

Taking a look we have to point out the simplified I/F lines between the PLM and the SVM due to the extensive use of standard buses and a distributed processing design approach. This is achieved thanks to the Remote Terminal Units (RTU) placed where a CAN Bus service is needed, and the SpW point to point high speed lines. This applies for the Telemetry, house-keeping (H/K) information, and for power distribution for the P/L users by mean of a dedicated P/L Power Unit, controlled by the main satellite PDU (Power Distribution Unit).

One (or two, TBC) Star Tracker sensor head is placed near the cameras structural wall in order to keep the thermo-structural misalignments to a minimum. As it can

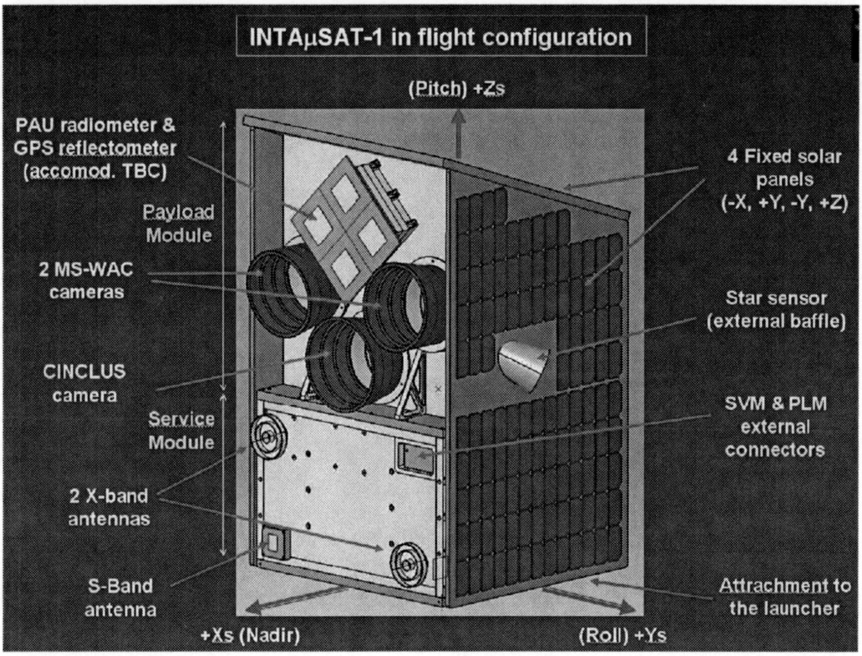

Fig. 11 INTAμSat-1 S/C description

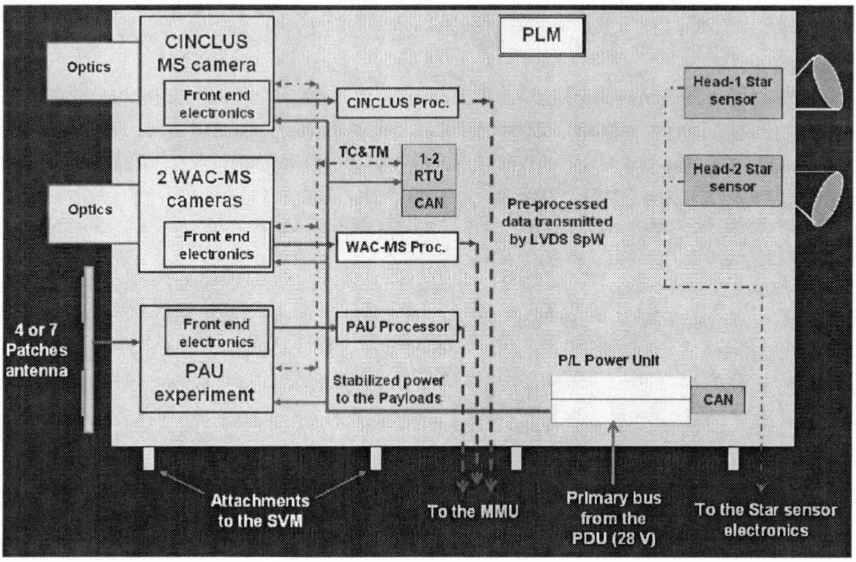

Fig. 12 SVM block diagram

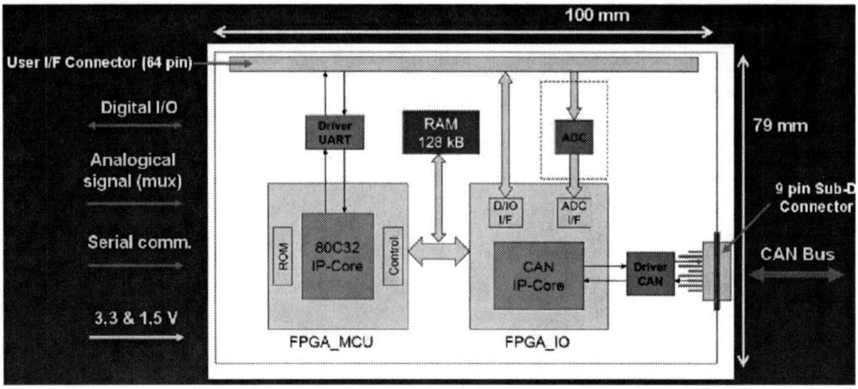

Fig. 13 PLM block diagram

be seen, the final accommodation of the PAU antenna is open, as there is another feasible place on the lower +Xs wall of the SVM.

6 OBDH S/S

The OBDH Box is based on a double size Europa board (200 × 160 mm) organised in three floors. The CPU PCB occupies the lower base of the unit while the first and second are taken by the Mass Memory Unit (MMU) boards. There is also a back panel to interconnect the 3 PCBs and the required external connectors in the front side.

A good connectivity between P/L instruments, all SVM S/S units and the central processor, is mandatory in view of performance, flexibility and modularity. The CAN bus concept is used for H/K, TC distribution, and for ACS sensors and actuators communications in real time with the CPU.

Dedicated Remote Terminal Units (RTU) with a local 80C32 microcontroller embedded in a RTAX 250 FPGA, are used to interface with the CAN bus at <500 Kbps. Higher speed (up to 100 Mbps) lines needed by the P/L instruments for the data I/F with the MMU, and also during downloads from the MMU to the S and X Band modems, are based on a LVDS physical layer and ESA's SpaceWire (SpW) protocol. AD Telecom and the SRG-UAH teams are still working on the development of a first functional model of the SpW node installed in both S and X Modem FPGAs, nearing completion.

Figure 14 shows the block diagram of the standard RTU with open and flexible connections to a generic user. Some particular FPGA configurations can be adapted to very specific users (i.e. the communication modem).

The first MMU breadboard is at this moment in its final design stage and it will have 80 Gbits capacity that could be easily enlarged (Figs. 15 and 16).

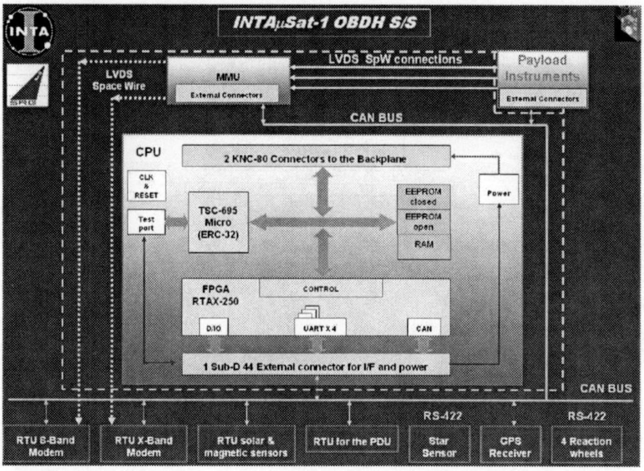

Fig. 14 RTU block diagram

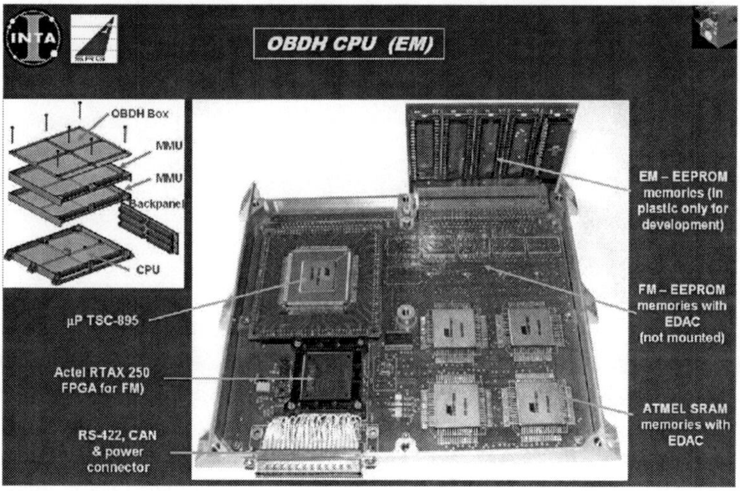

Fig. 15 OBDH block diagram

7 Communications S/S

During the last 2 years a lot of advance has been made on this really critical S/S taking into account the number of cameras and Remote Sensing P/L on board this first mission. Our INTA communication and antennas expert team in cooperation with AD Telecom already fixed the baseline design. Next Fig. 16 depicts the block diagram of the complete S/S.

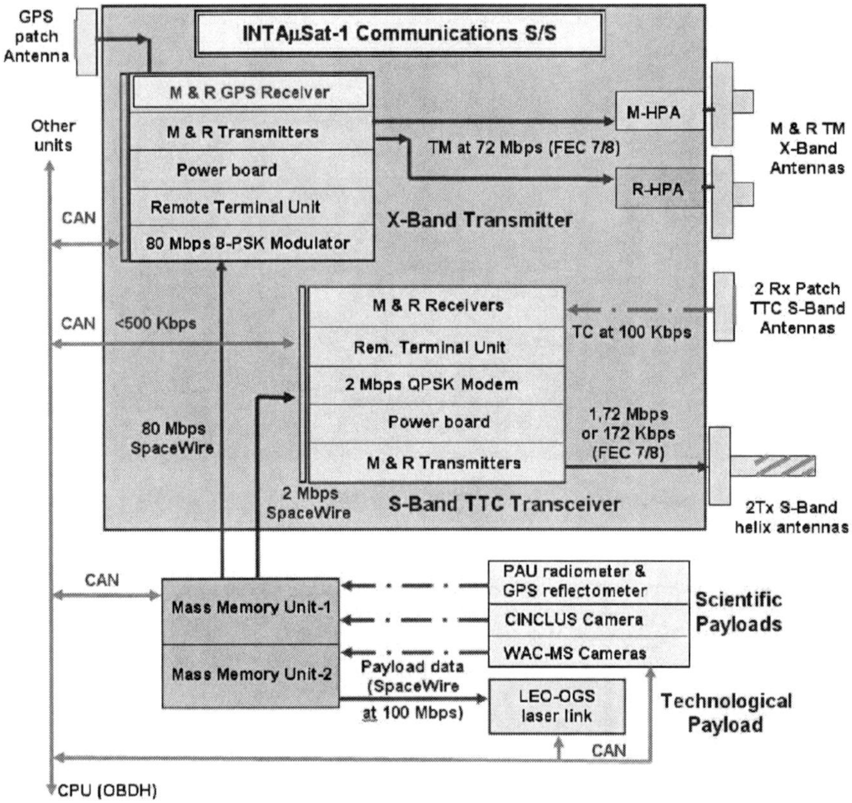

Fig. 16 CPU engineering model

Part of the spent development effort has been focused on the new digital Modem, to try to reach the required transmission speed. Originally the ceiling was set to 20–40 Mbps, but recently we have been able to test in the Lab at 80 Mbps. This is a remarkable reuse of the RTAX 2000 FPGA using almost the same HW resources. The demonstration was also made using the X-band front-end transmitter without any problem. The manufacturing and testing of the 12 w RF output Solid State Amplifier (SSA or HPA) is now near completion.

Other important devices are of course the S and X band antennas. Rx patch antenna for S-band will fly very soon in the Nanosat-1B and we are working on a very light design of a new quad helix antenna. This shaped diagram transmission antenna is critical for the S-band downlink channel. The same can be said about X-band LEO optimised shaped antennas that are now in the manufacturing process. They are developed at INTA where we have long time experience and nice testing facilities.

Finally, we are working in a very interesting optical communication terminal proposal able to downlink the on board TM data directly from the MMU at 100 Mbps, from a LEO μSat to the ESA Optical Ground Station (OGS) placed in Tenerife, the Canary Islands. The baseline is to use a 1 mm IR laser beam aimed at the OGS by means of a two degrees of freedom small mirror. The mirror mechanism on board shall aim precisely at the OGS during a pass, from a 10° elevation angle. The link budget calculations with some extra margin will in principle guarantee a communication free of errors (10^{-6}) from 20°. In this way we will test the feasibility and problems at lower inclinations. At this moment there are two scenarios: with or without a laser beacon from the OGS. The beacon from Earth will allow a much better pointing in a closed loop arrangement, but you need then an error sensor on board placed in the same optical path.

8 Conclusions

1. On November 2005 we started the MicroSat programme, a 100–150 kg microsatellite after the Nanosat programme success. This will follow the same principles and development rules of Nanosat, together with the acquired know-how and lessons learned up to now.
2. At the present stage we are in the middle of Phase-C with an expected completion date by the end of 2009. CDR meetings will be organised along February 2010 that will give the green light to the Phase D.
3. The first MicroSat mission will carry a Mid Resolution Multispectral Camera (10 m GSD) with 4–5 channels, another 4 narrow band MS cameras using TDI to detect contamination on in-land reservoirs over Spain. PAU is a very interesting Remote Sensing radiometer and GPS reflectometer to watch the sea surface.
4. Other interesting R&D technological payloads are also envisaged to check in orbit novel ideas.

Acknowledgements 1. The Nanosat and MicroSat programmes are running thanks to several funding from the National Space Plan, either Scientific (PNE) or Technical (CDTI), and from our internal budget.

2. Special thanks must be given to all the Nano and MicroSat project team members for their outstanding and encouraging work made during the last 10 years.

References

1. M. Angulo, J.M. Mi, P. de Vicente, M. Prieto, O. Rodriguez, E. de la Fuente, J. Palau. *Development of the MicroSat programme at INTA*. 6th IAA Symposium, Berlin 2007.
2. M. Angulo, M.R. Canchal, J.M. Mi, P. de Vicente. *Development and qualification of the Nanosat programme at INTA*. 57th IAC Valencia 2006.
3. M. Angulo, L. Seoane, R. Blanco, J. Iglesias, J. R. de Mingo. *Nanosat-1 three years in orbit experience*, 4S Symposium Rhodes, 26–30 May 2008.

VENμS (Vegetation and Environment Monitoring on a New Micro Satellite)

Philippe Crebassol, Pierric Ferrier, Gérard Dedieu, Olivier Hagolle, Bertrand Fougnie, Francesc Tinto, Yoram Yaniv, and Jacob Herscovitz

Abstract VENμS (Vegetation and Environment on a New Micro Satellite) is an Earth observation demonstration mission developed in cooperation by FRANCE and ISRAEL in the framework of the European Global Monitoring for Environment and Security Program (GMES).

This program gathers both a scientific mission (earth images at high resolution and high repetitivity) and a technological mission (low orbit control by IHET – Israeli Hall Effect Thruster). It is due to fly at mid of 2011, for an overall duration of 3.5 years.

This chapter presents all the aspects of the VENμS mission: missions' description, missions' implementation, missions' concurrence, ground image processing, ground control segment, satellite and payloads descriptions. It will also detail the cooperation aspects.

1 Introduction

The main goal of the VENμS *scientific mission* is to acquire data over land in order to improve the understanding and modeling of land surface and vegetation processes, and to develop new applications such as water balance, crop yield and carbon fluxes assessments. The VENμS data should also prove to be useful for a variety of studies, such as water color monitoring or the mapping of glacier velocity fields.

The satellite also flies a *technological mission* which is aimed at qualifying an electric propulsion technology (IHET – ISRAEL Hall Effect Thruster) and at demonstrating its profitable contribution for low altitude missions and LEO (Low Earth Orbit) to LEO orbit transfers.

The VENμS missions, combined scientific and technological, will be implemented on a 265 kg satellite. The satellite will be injected into a polar sun-synchronous orbit, will cross the equator of the earth on its southbound path when the local time of day below the spacecraft is 10:30 a.m.

P. Crebassol (✉)
Centre National d'Etudes Spatiales, 31401, Toulouse, France
e-mail: philippe.crebassol@cnes.fr

R. Sandau et al. (eds.), *Small Satellite Missions for Earth Observation*,
DOI 10.1007/978-3-642-03501-2_4, © Springer-Verlag Berlin Heidelberg 2010

2 The VENμS Missions

2.1 Scientific Mission

2.1.1 Mission Description

Global models of biosphere processes and dynamics have been developed in the past few years and are used for carbon cycle studies. Today they are mainly based on ground measurements provided by networks of experimental stations. On the other hand, the water cycle is driving important research activities on the interface between land and atmosphere for improving atmospheric models. A major limitation of these approaches is the lack of consistent ground network and measurements and the models' imperfection. The development of Earth observation data assimilation techniques and aggregation or desegregation procedures is leading to more promising integrated approaches.

The main scientific objectives of the Earth observation mission for terrestrial environment monitoring are:

- to monitor and analyze surface functioning under the influence of environmental factors (climate, topography, soils etc.) as well as human activities,
- to study processes and their interactions with natural and human factors,
- to develop and validate natural and cultivated ecosystem functioning models and to improve and validate global carbon cycle models,
- to develop remote sensing data assimilation techniques within vegetation and surface flux models,
- to define theoretical and practical methods for scale transfer, i.e. up and down-scaling,
- to interpret data collected by those low spatial resolution sensors,
- to develop, validate and implement integrated modeling platforms.

The uncommon feature of the VENμS scientific mission is the conjunction of the 4 following characteristics: high revisit capability (2 days), high resolution (5.3 m), spectral richness (12 bands), and constant viewing angles from the satellite at constant sun lighting angles (excluding sun seasonal variations). Advantages of this unique combination will be profitably exploited by newly developed processing methods.

About 100 selected sites of interest all around the world are currently being selected. They will be scanned throughout the duration of the scientific mission. It must be pointed out that *VENμS will image the very same pre-selected sites throughout the scientific mission's duration.*

VENμS is also intended to demonstrate the profitability of such observation capabilities for the GMES program.

The scientific mission is called VM1 (VENμS Mission 1). Imaging has priority during this period of time, although the technological mission will be carried out during some limited periods of time. The hydrazine budget allows both the on-orbit

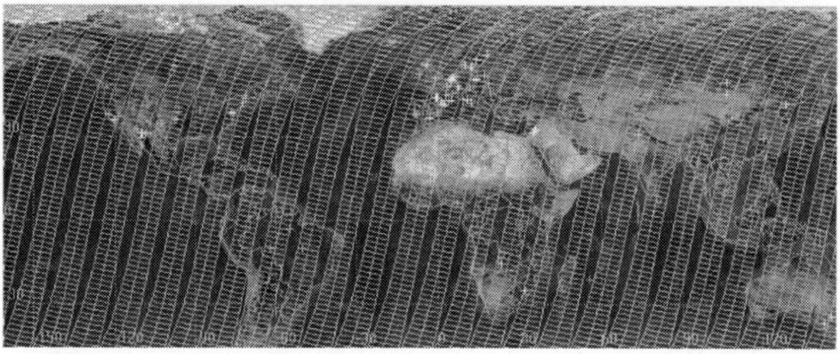

Fig. 1 VENμS orbits and accessible areas

injection errors correction and the orbit maintenance throughout VM1. In other words, although the missions are concomitant, the scientific mission is not dependent on unforeseen turn of events of the technological mission (and vice versa). In order to image two full season cycles in both hemispheres, VM1 duration will be 2.5 years, providing measurements over land surfaces.

2.1.2 VM1 Orbit

The VENμS orbit is the following:

- Semi major axis: 7,105 km
- Eccentricity: 0.0005
- Inclination: 98.28°
- Local time of the descending node: 10:30 a.m.

With such a sun-synchronous orbit, the repeat cycle is 2 days with 29 orbits per cycle. The instrument swath allows 27 × 27 km individual images at nadir (images can be taken in sequence along the track, to allow large scenes imaging).

The entire satellite can be tilted across track and along track. This feature provides flexibility when selecting the scientific sites among the proposed ones. For image quality purpose, sites requiring more than 30° tilt angles are not selected. The satellite yaw steering capability will be used in order to limit the smearing and thus, to globally improve the MTF.

2.1.3 Image Quality Requirements

Radiometry

Table 1 shows the spectral bands characteristics and the required signal to noise ratio (SNR) for the minimum radiance (Lmin) defined for the mission. These requirements are based on radiative transfer simulation done to retrieve leaf index area (LAI) and chlorophyll concentration (Cab) accuracies required for scientific needs.

Table 1 The 12 bands description

Band	Equiv Central wavelength (μm)	Band width	Lmin (W/m²/sr/μ)	SNR at 10 m ground sampling distance
B1	0.420	40	40	50
B2	0.443	40	50	80
B3	0.490	40	30	100
B4	0.555	40	30	100
B5	0.620	40	20	100
B6	0.620	40	20	100
B7	0.667	30	20	100
B8	0.702	24	15	100
B9	0.742	16	20	100
B10	0.782	16	30	100
B11	0.865	40	30	100
B12	0.910	20	30	50

Most of the bands (565, 620, 670, 702, 742, 782, and 865 nm) are designed to characterize different parts of the chlorophyll spectrum: absorption features, red edge. Some bands are dedicated to atmospheric corrections: 910 nm (water vapor absorption), 420, 443, 490, and 620 nm (aerosol characterization), while some bands may be used for water color studies in coastal or lake environments. Lastly, the 620 nm band has been duplicated with a slight viewing angle difference (1.5°) so as to produce Digital Elevation Models for image location and to detect clouds through their altitude.

As level 2 and 3 products will be provided to users at 10 m resolution, the required SNR will be better than 80 at Lmin radiance for the bands dedicated to the land mission.

The accurate knowledge of the band equivalent wavelength is very important as an error of 1 nm can induce a few percent errors on the reflectance, especially in the blue part (atmospheric scattering) and the red part of the spectrum (vegetation red edge). Therefore the equivalent wavelength is required to be known with an uncertainty below 1 nm and with a stability within 1 nm throughout the instrument life in orbit and for all the bands.

Getting a physical value (radiance or reflectance) from the numerical output provided by the camera requires the knowledge of the camera sensitivity. Any error on the absolute calibration measurement directly affects the accuracy of this physical value. That's why a maximum 5% absolute calibration knowledge uncertainty was required for the mission, with an objective of 3%. In the same way, the cross-band and multi-temporal calibration knowledge accuracy were set to respectively 3 and 2%. Moreover the camera non linearity will be known with an accuracy better than 0.5% and will have to be stable enough so that the detector non uniformity can be calibrated at one radiance level in flight.

Finally the system MTF (Modulation Transfer Function) including the telescope, the detector and the smearing will be better than 0.1 at the NYQUIST frequency.

Fig. 2 Bands

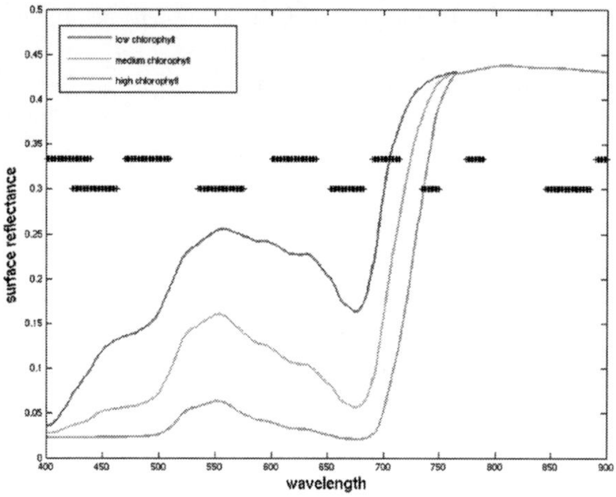

Geometry

The geometric image quality requirements correspond to two types of requirements, and are summarized in Table 2:

- programming requirements to guarantee the satellite pointing and steering accuracy over the sites to be viewed
- image quality requirements to be met by the level 1 products using the auxiliary data provided by the satellite and the data and models either calibrated in-flight or before the launch.

The accuracy of the image location, without ground control points, is not a high constraint for the system as it is only used to predict the pixels position in the image used as a reference (see level 1 processing) for the geometric model calibration.

The multi-temporal and multi-spectral registration performances require knowledge of the non linear temporal component of the attitude better than 1 μrad/s; we

Table 2 Geometric image quality requirements

Geometric image quality performance (pixel size = 5.3 m)	Ground processing
A priori accuracy of image location: 1,000 m (max)	(1)
Accuracy of image location: 100 m	(2)
Accuracy of image location: 10 m	(3)
Multitemporal registration: 3 m (rms)	
Multispectral registration for any couple of spectral bands: 2 m (rms)	
Sampling regularity during the sampling time (0.850 msec): 1 m (max)	(1)

(1) No processing
(2) After image processing without control points
(3) After image processing with control points

assume here that the ground processing will be able to calibrate any linear temporal variation of the attitude using a reference image for each site, providing ground control points.

The sampling regularity refers to the capacity of the system to sample the landscape according to a regular theoretical grid. A good sampling regularity is important for the internal image consistency and constrains both the satellite (stability) and the camera (detector regularity) to 1.4 μrad during the sampling time.

2.2 Technological Mission

2.2.1 Mission Description

The technological mission is split in two phases: in space verification where electric propulsion is tested and qualified, and in space validation which demonstrates how mission operations can be enhanced by the use of electric propulsion:

- VM1: electric propulsion verification though orbit maintenance and short/extended experiments. Concomitantly with scientific mission. It has to be emphasized here that, thanks to the hydrazine fuel budget, the scientific mission per se (VM1) is not dependent on the IHET availability. The baseline is to use the IHET for orbit correction after injection and for orbit maintenance. But if it cannot be the case, the hydrazine propulsion system will be used for those operations
- VM2: validation phase: transfer to a lower altitude orbit
- VM3: validation phase: low altitude orbit keeping, compensating drag, while resuming imaging as a secondary mission

The technological mission has priority during VM2 and VM3, although the scientific mission will be resumed concomitantly during VM3.

VM2 is expected to be 6 months and VM3 around one year (depending on the xenon remaining quantity).

Current target for IHET utilization (2 thrusters):

- 2,000 h cumulated functioning time
- 4,900 cumulated activations.

2.2.2 IHET Verification

The verification of the IHET will start immediately after launch, in the IOT (In Orbit Tests) period. A series of experiments are planned to test the IHET at various operating conditions (constant and variable power levels). The verification goals are to test the proper operation of the thrusters and the whole electrical propulsion system in space, and to check the performance (thrust, specific impulse, efficiency).

In addition, the lifetime of the IHET will be checked out according to the behavior of these parameters as EOL (End of Life) is approaching.

Every month, a two day period – sometimes a four day period – will be devoted to short experiments. At the end of each experiment, IHET thrusters will be used to bring the satellite back to its nominal orbit.

Each year, a whole month will be dedicated to perform multiple experiments and to accumulate activation time.

All IHET experiments and operations will be analyzed and processed by the TMC (Technological Mission Center, in Israel). Each IHET activation will result in an orbit change. These changes are monitored and recorded continuously, by the GPS (Ground Positioning System) equipment of the satellite. The TMC will fit the best estimations of the instantaneous thrust that match the orbit change. Other vital parameters, such as operating currents and propellant flow will be downloaded by the satellite telemetry, and analyzed in TMC along with the orbit data.

2.2.3 IHET Validation

IHET advantages will be demonstrated during VM2 and VM3 phases.

Since the most noticeable profit of an electrical propulsion system is its high specific impulse (Isp), compared to conventional monopropellant propulsion systems usually used in small satellites, the IHET challenge will be the demonstration of its capabilities in this field.

As said in the scientific mission description, one of the unique features of this mission is perpetual imaging with a constant viewing angle, at a constant sun angle lighting on the targets. This requirement is very stringent, and calls for a very delicate and accurate orbit control system using the HET thrusters.

In VM1, in addition to the month long extended experiments period, the IHET was used to control the orbit and bring back the satellite to nominal orbit.

In VM2, the IHET's efficiency will be demonstrated by a LEO to LEO transfer to VM3 orbit. This major orbit change, of about 300 km in altitude, 1.2° in inclination angle, and controlling the RAAN and equatorial crossing local time (performing a maneuver of about 350 m/s) will need about 7.5 kg of Xenon.

Finally, in VM3, VENμS will keep on imaging at a circular sun-synchronous orbit while maintaining the 2-day revisit ground track. Since VENμS satellite was not designed for a very low altitude, its cross-section area is quite large and the atmospheric drag at this altitude plays a significant role. In VM3, the thrusters will have to be used intensively, to keep the imaging requirements accuracy. This task is hardly achievable by a comparable traditional chemical propulsion system.

In this phase, the images resolution increases and its swath decreases with respect to VM1 mission phase.

The challenge of VM3 is to remain imaging as many sites as possible among the sites already imaged in VM1, while maintaining the orbit with electric propulsion.

2.2.4 VM3 Orbit

Once VM2 is completed, the final orbit will still be Sun-synchronous with a revisit time of 2 days (31 orbits) and an altitude of 410 km:

Semi-major axis: 6788.76 km
Eccentricity: 0
Inclination: 97.07°
Local time of the descending node: 10:30 a.m.

Ground definition will be then 3 m.

3 The VENμS System

The VENμS system is comprised of:

- The Satellite
- The Ground Control Station
- The Data Processing Ground Segment
- The KIRUNA Receiving Station
- The Technological Mission Center
- The Launch Vehicle

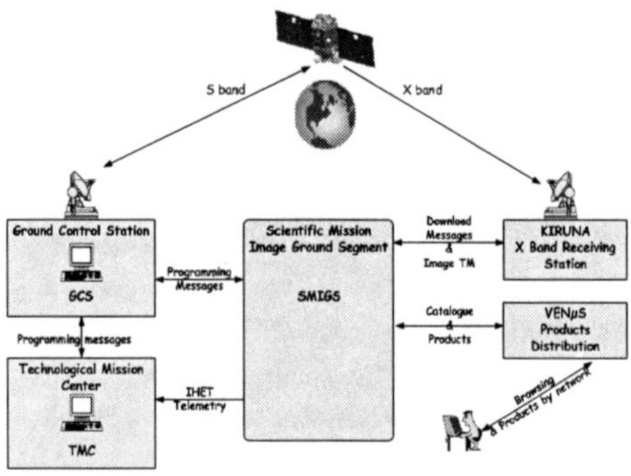

Fig. 3 VENμS system

4 Cooperation Basis

VENμS is a demonstration mission developed in cooperation by ISRAEL (ISA – ISRAEL Space Agency – IAI – ISRAEL Aerospace Industry – and RAFAEL – ISRAEL) and FRANCE (CNES – Centre National d'Etudes Spatiales).

CNES is responsible for the provision of the camera and the image ground segment. ISA is in charge of the platform, the IHET, the satellite control center and the satellite operations. Launcher provision is a cost-shared task between the parties. It is the first cooperation of this type between our two countries.

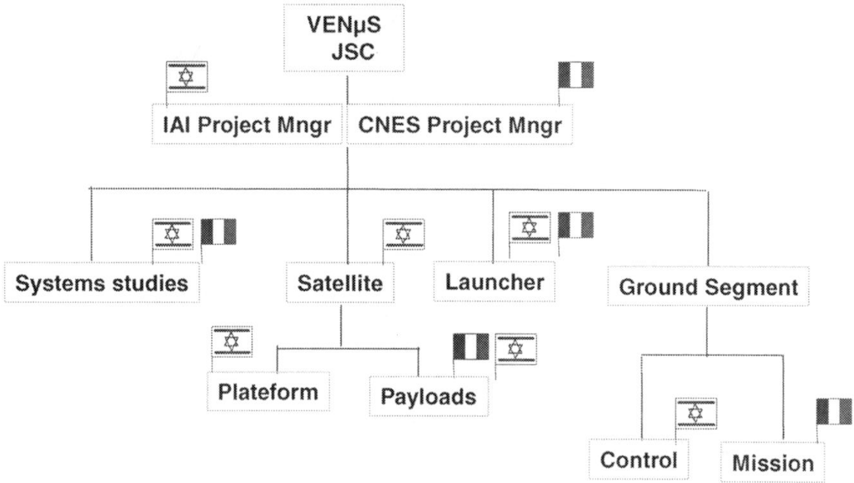

Fig. 4 Tasks sharing among the two space

5 Satellite Description

5.1 Overall Presentation

VENμS is implemented on a modified IMPS bus (Improved Multi Purpose Satellite). The modifications concern mainly the base plate – to accommodate the 2 IHET engines and the tanks – and the solar panels – to provide the amount of power needed for the IHET. As it contains two different propulsion systems, it has one 7 kg Hydrazine tank to feed 4X 1-Newton redunded thrusters and one 16 kg Xenon tank for one redunded IHET.

The power management system size can fulfill the high energy demand of the IHET system (operating range up to 600 W). Two RF links (X-band and S-band) are implemented respectively to flow the image data and the housekeeping data. The AOCS houses Star Trackers GPS and Reaction Wheels, giving very high pointing accuracy performances while imaging. It also allows the high agility required for imaging a large number of sites in tight sequences (angular speeds of 1°/s and more). In order to prevent pollution of the optic on the ground and during the launch phase, the telescope tube is fitted with a cover door, deployed once in orbit (one shot deployment). As far as on-board memory is concerned, two 15 GB units are available for a total amount of 30 GB.

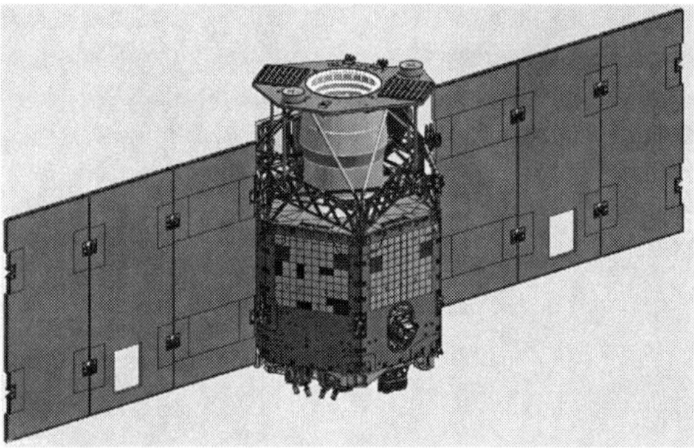

Fig. 5 VENμS in deployed

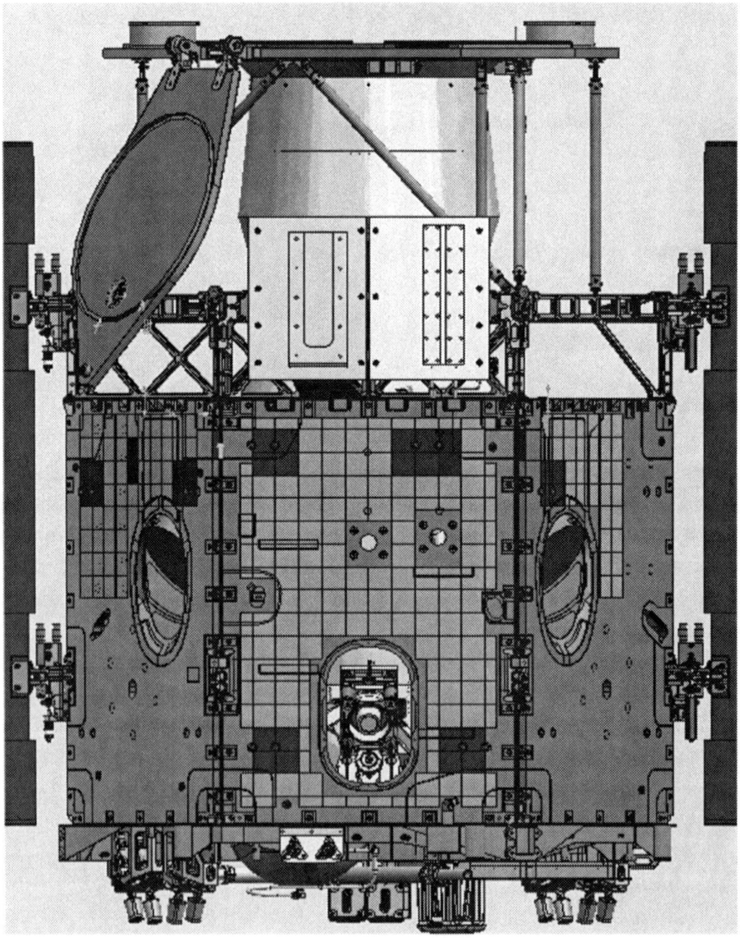

Fig. 6 VENμS in stowed configuration

5.2 *Scientific Mission Payload*

5.2.1 Camera Description

The VENμS Super Spectral Camera (VSSC) is specified by CNES and built by El-Op (Elbit Systems Electro-Optics ISRAEL). It comprises a catadioptric objective, a focal plane assembly with narrow band filters and 4 detector units with 3 separate CCD-TDI (Time Delay Integration) arrays each, operating electronics, operational control and thermal control (operated by the satellite). The 12 spectral bands are scanned in pushbroom mode by the 12 parallel detector arrays in the focal plane. The pixel pitch is 13 μm which gives a 5.3 m resolution at nadir on the ground at 720 km.

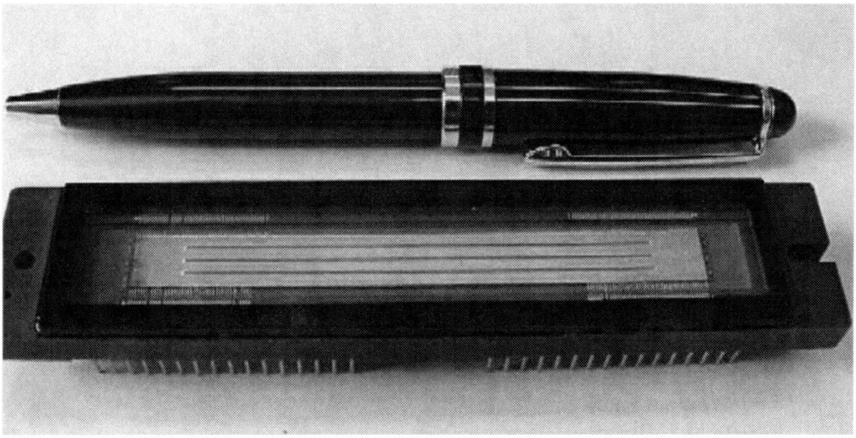

Fig. 7 A TriTDIdetector, without the spectral window (ELOP)

The objective is a Ritchey-Chrétien type with auxiliary lenses for field correction and telecentricity. There is no refocus mechanism but properties are maintained thanks to the careful choice of materials and processes and a close thermal control loop which maintains constant temperature of the camera.

The VSSC focal length is 1.75 m, with an objective diameter of 0.25 m.

As there is a 1.5° angle between the first band and the last band, the scanning start for each detector will be delayed until the first line of the scene reaches the first band on each detector. That will allow an efficient on-board memory management, by minimizing the amount of non useful recorded data.

The camera total mass is 43.5 kg.

Particular attention was paid to stray light since it has a serious impact on the radiometric performances. The sources of stray light are diverse, either scattering due to mirror roughness and contamination or ghost images due to reflection on the optics surfaces. The camera baffles' design was done in order to minimize the stray light and the particle contamination level on optical surfaces is limited to 300 ppm EOL.

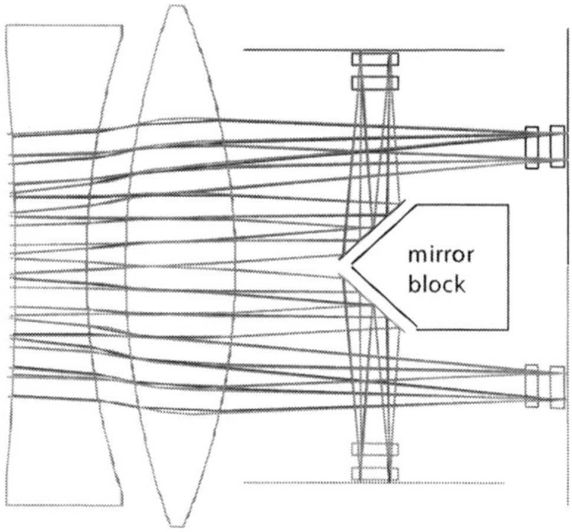

Fig. 8 Focal plane layout (ELOP)

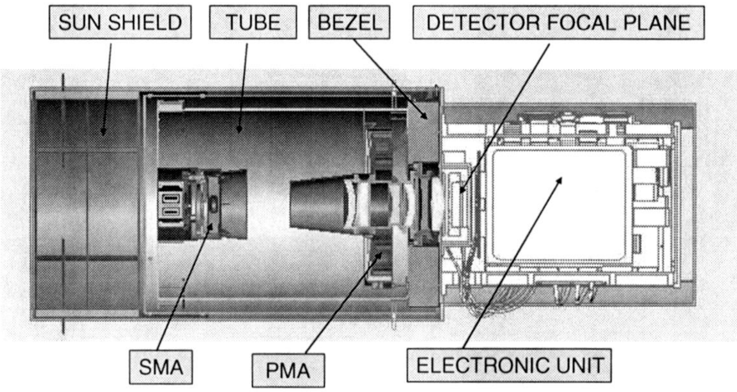

Fig. 9 Camera longitudinal section (ELOP)

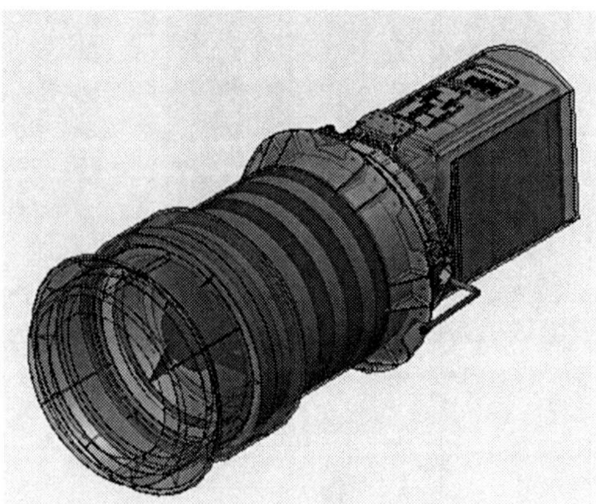

Fig. 10 Camera layout (ELOP)

5.3 Technological Mission Payload

5.3.1 HET-300 Thruster

The heart of the technological payload, the IHET, is the Hall Effect Thruster (HET), codenamed HET-300. It operates on Xenon propellant, which is ionized by electrons emitted from the cathode and accelerated as plasma using a high electrical field.

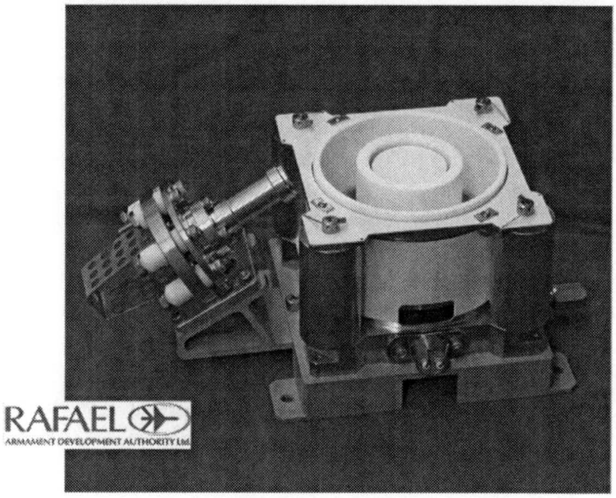

Fig. 11 The IHET engine

This thruster is specifically designed for use onboard micro satellites, operating nominally at only 300 W anode power. However, its useful range of operation is between 250 and 600 W. Thus, it can use the instantaneous available power from the satellite. HET-300 performance at 300 W is listed here below in Table 3.

Table 3 I HET-300 main characteristics

Thrust (@300 W)	>15 mN
Specific impulse (Isp @ 300 W)	>1,300 s
Nominal anodic power	300 W
Power operation range	250–600 W
Operating life	>1,000 h
Number of operations	>2,000

HET-300 mass is about 1.5 kg and its dimensions are 170 × 120 × 90 mm.

6 Ground Segments

6.1 Image Ground Segment

6.1.1 Top Level Functions

The SMIGS (Satellite Image Ground Segment) is in charge of performing:

- Mission planning
- Image processing
- Image products distribution

6.1.2 Mission Planning

As previously said, VENμS is not intended to provide operational imaging services with a dynamic programming loop (such as SPOT etc.. . .). As the targeted sites are always the same, the image acquisition plans are basically limited to a regular set and two calibration sets.

6.1.3 Image Products

The ground segment provides with 3 product levels:

- The VENμS level 1 will provide geolocated top of atmosphere reflectance with a subpixel multi-date registration (requirement 3 m); a cloud mask at a coarse resolution will be associated to the product. The ground sampling distance of level 1 product will be 5 m. VENμS level 1 products will be equivalent to SPOT level 3 products.
- Level 2 provides surface reflectance ("Top of Canopy"). Its aim is to suppress as much as possible the atmospheric contribution to the signal. During this process, a high quality cloud and cloud shadow mask is generated. The algorithms used for level 2 processing take advantage of the 2 days revisit period with constant observation angle: most short term variations of TOA (Top Of the Atmosphere) reflectance are due to atmosphere variations.
- The level 3 product is a 10 day composite of level 2 products. The aim is to provide the users with cloud-free, easy to use images of surface reflectance, every 10 days.

The ground sampling distance of level 2 and 3 products will be 10 m, whereas level 1 products are delivered at 5 m resolution.

These images will be available on a web server, to registered members which will get access to the data after a simple (and free) authorization Procedure.

6.1.4 Level 1 Processing

During level 1 processing, radiometric and geometric calibration is performed. In order to produce the best possible calibration parameters, the VENμS Image Quality software will be run intensively during the IOT phase (3 months duration) and once a month in routine.

The radiometric corrections deal with:

- normalization (dark signal and detector relative sensitivity correction)
- defective pixels

- image restoration combining deconvolution to correct the camera MFT (spectral processing) and denoising based on a wavelet processing; if the denoising is not necessary, it will be automatically deactivated through a noise criteria.

The goal of the geometric correction is to perform the temporal and spectral registration of all images taken over a given site. To reach this objective, the geometric model, which associates a viewing direction to any detector, has to be refined. This is done using the duplicated bands B5 and B6, and a reference image for each site, with the associated DEM (Digital Elevation model). As the 2 bands are located at the 2 extremities of the split focal plane, it allows to temporally cover the viewing of the site by the twelve bands. The refinement uses the homologous points provided by the correlation matching between B5, B6 images and the reference image. This refined geometrical model is then propagated to all the bands and used to project them on a geographical or cartographical reference frame. This reference frame will be chosen by the user.

The reference image is a monospectral VENμS image which geometric model has been refined through DEM correlation. It requires a VENμS DEM produced by stereo matching between the duplicated bands B5 and B6 and a well located DEM. The small Base to Elevation ratio (0.026) between B5 and B6 is not a problem as a very efficient CNES correlation algorithm is used. Any image which geometric model has been refined with regard to this reference can become a new reference; this allows updating the reference according to the landscape seasonal evolution [4].

6.1.5 Level 2 and 3 Processing

The level 2 and 3 algorithms for cloud detection and atmospheric correction intensively use the VENμS unique characteristics: very frequent observations at high resolution with constant viewing angles. Thanks to these features, the evolutions between two successive acquisitions are very likely to be related to changes in the atmosphere rather than in the surface. This is used to detect clouds and aerosols before atmospheric correction. The atmosphericcorrection is detailed in [5].

6.2 Ground Control Station (GCS)

The satellite will be controlled via the Ground Control Station located in IAI in ISRAEL. It will interface with the two system components in charge of each mission planning:

- the scientific mission planning located in CNES TOULOUSE, FRANCE,
- the technological mission planning located in the TMC (Technological mission Center), RAFAEL, HAIFA, ISRAEL.

6.2.1 Top Level Functions

They are the following:

- Checking that the mission plans provided with by the mission centers are compliant with the satellite resources,
- Command Generation,
- Camera operation,
- IHET operation,
- Interfacing (External Systems),
- Telemetry (Spacecraft Monitoring),
- Simulator.

6.2.2 Satellite Operations

The GCS will receive the mission plans (originated by the scientific and technological centers), check their feasibility, create downlink plans for the KIRUNA X-band receiving station, create command files to be uploaded (from the IAI S-band station) to the satellite and send upload reports to the two Mission Planning Centers. Additionally, GCS is in charge of the general health of the satellite sub-systems.

Fig. 12 Simulation of level 1 product (using FORMOSAT 2 images)

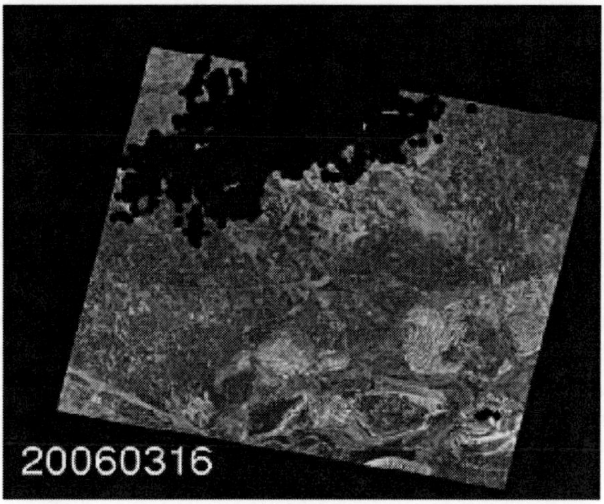

Fig. 13 Simulation of level 2 product

Fig. 14 Simulation of level 3 product (using FORMOSAT 2 images)

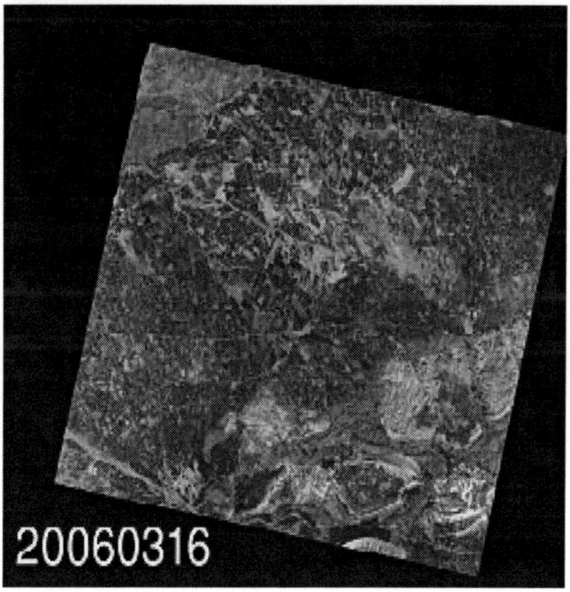

6.3 Technological Mission Center

The TMC performs evaluation and analysis of the data related to the technological payload (the IHET) and to the technological mission.

TMC receives data from the satellite via X-band telemetry, through Kiruna.

It performs several functions, such as preparation of updated operation plans for the technological mission, and analysis and diagnostics of the performance of the electrical thrusters. TMC will evaluate the IHET performance, in verification and validation phases. It will estimate the relationship between the thrust and the input electric power, as well as between specific impulse and input power. The goal is to model the IHET behavior in space, during various conditions as well as aging information. Another task is Orbit Managing and Performance Analysis. It will be accomplished by performing system-level analysis of the orbit control function performance during all technological mission operation phases. TMC will also track and record the interaction between IHET and the satellite subsystems, to complete and validate the ground-based assessments and analyses of IHET plume influence on satellite external subsystems (such as solar panels) and also the electrical and magnetic influence (such as the effect of the IHET on the magnetometers). All TMC functions will be reported and distributed. Technological mission data gathered during VENμS mission (whether raw or processed) will be archived for future and extensive research.

7 Launcher

VENμS should be launched from INDIA. Launch services would be provided by ANTRIX, using a PSLV rocket in a dual launch configuration.

8 Project Status (at Beginning of 2009)

VENμS has successfully achieved its C-phase. The camera CDR has been successfully held in 2007 and the satellite and system CDR at fall 2008. VENμS is planned to enter into service mid 2011.

Following a very successful research announcement and call for paper concerning the sites selection, around 100 sites will be finally selected, half of them being located in Western Europe and in ISRAEL.

VENμS web site: http://smsc/VENUS/Fr

Further Reading

1. G. Dedieu, A. Karnieli, O. Hagolle, H. Jeanjean, F. Cabot, P. Ferrier, Y. Yaniv. VENμS: a joint French Israeli Earth observation mission with high spatial and temporal resolution capabilities, RAQRS II conference.
2. J. Topaz, F. Tinto, O. Hagolle. The VENμS super-spectral camera." In Sensors, Systems, and Next-Generation Satellites X, 6361:63611E-8 presented at the Sensors, Systems, and Next-Generation Satellites X. Stockholm, Sweden: SPIE, 2006. http://link.aip.org/link/?PSI/6361/63611E/1
3. F. Cabot, O. Hagolle, C. Ruffel, P. Henry. Remote sensing data repository for in-flight calibration of optical sensors over terrestrial targets, SPIE proceedings, Vol. 3750, Earth Observing Systems IV, Denver, July 1999.

4. O. Hagolle, Ph. Goloub, P. Y. Deschamps, H. Cosnefroy, X. Briottet, T. Bailleul, J.-M. Nicolas, F. Parol, B. Lafrance, M. Herman. Results of POLDER in-flight calibration, IEEE transactions on geoscience and remote sensing, Vol. 37, No. 3, May 1999.
5. Hagolle, O., G. Dedieu, B. Mougenot, V. Debaecker, B. Duchemin, A. Meygret. Correction of aerosol effects on multi temporal images acquired with constant viewing angles: Application to Formosat-2 images.Remote Sensing of Environment, Vol. 112, No. 4, April 15 (2008): 1689–1701.
6. J. Inglada, H. Vadon. Fine registration of SPOT5 and Envisat/ASAR images and ortho-image production: A fully automatic approach, IGARSS'05, Seoul, North Corea, July2005.
7. A. Meygret, M. Dinguirard, P. Henry, O. Hagolle. Calibration of SPOT4 HRVIR and VEGETATION cameras over the Rayleigh scattering, SPIE, San Diego, August 2000.
8. A. Meygret. Absolute calibration: from SPOT1 to SPOT5, SPIE, San Diego, August 2005.
9. P. Ferrier. VENμS, Vegetation and Environment on a new micro satellite, Rhodes, May 2008.

The “Ulingo” Mid-Low Latitudes Observation (MILO) Mission

Carlo Ulivieri, Giovanni Laneve, and Emiliano Ortore

Abstract This chapter aims at proposing a small mission devoted to observe the mid-low latitude regions. The selected orbit, called multi-sun-synchronous (MSS), represents an innovation with respect to the classical sun-synchronous orbit. The advantage offered by this kind of orbit, more suitable for observing tropical regions, will be evaluated and described. The satellite will be equipped with three optical sensors: a medium-high spatial resolution VIS-NIR multi-spectral sensor, allowing the surface monitoring and land-use and land-cover studies; a medium spatial-resolution 3-band thermal (MIR-TIR) sensor allowing the surface temperature (LST, SST) estimate and hot-spot (fires, volcanic eruption, etc.) detection; a panchromatic VIS-NIR camera for night-time observation able to reveal artificial and natural lights. Further, such an orbit allows the observation of the same region of the Earth at different local-times. In this way, the diurnal cycle of surface temperatures can be reconstructed with a 2.4-hour local-time interval. The amount of data acquired by an equatorial station has been estimated. Orbital perturbations effects have been taken into account in order to verify the feasibility of the mission; the Results have been reported in another chapter.

1 Introduction

This study is devoted to describe a small mission for monitoring the mid-low latitude regions. The mission has been called “Ulingo” because oriented to African tropical belt and “Ulingo” is the Swahili word for “surveillance”. The mission objective is in line with several international initiatives, which focus their attention on tropical areas (lacking in both monitoring and natural resources management), as TIGER (ESA), GMES (the European contribution to GEO – Group on Earth Observations) and other projects concerning the Kyoto Protocol. Furthermore it aims at providing

G. Laneve (✉)
CRPSM (Centro di Ricerca Progetto S. Marco) – Sapienza University of Rome, Via Salaria 851, 00138, Rome, Italy
e-mail: laneve@psm.uniroma1.it

R. Sandau et al. (eds.), *Small Satellite Missions for Earth Observation*,
DOI 10.1007/978-3-642-03501-2_5, © Springer-Verlag Berlin Heidelberg 2010

an effective and technologically advanced tool to enhance the sustainable development of such areas and the prevention of crisis conditions, related to environmental changes, resources scarcity and natural disasters. In fact, the data provided by such a mission would allow carrying out a great deal of studies regarding geology, agronomy, utilization of soil, water quality, etc.

The orbit selected for these applications is a direct and low Multi-Sun-Synchronous Orbit (MSSO), that allows, for a given area, a repetitivity interval of the same solar illumination conditions which is multiple of the observation cycle (revisit frequency) [1, 2]. With respect to a Sun Synchronous Orbit (SSO) the drawback of this solution is represented by the necessity to take into account the different illumination conditions of the scene, due to the different local times at which the successive observations of the same area are performed. The advantage of MSSO lies in the fact of having a greater flexibility in the orbit inclination choice (SSO are retrograde and usually quasi-polar and that determines a limitation in the employable launching sites). Moreover, the MSSO represents an optimal solution to the problem of having a high observation frequency for the regions of interest (it allows a sampling frequency of the tropical areas at least double with respect to a quasi-polar orbit). The daytime cycle is sampled in a uniform way in a relatively brief period (1–2 months). A global longitudinal coverage is easily obtainable by suitably selecting the values of the semi-major axis a and the inclination i.

The proposed payload includes a multi-spectral sensor that allows to distinguish Earth surface properties, a thermal sensor to monitor surface temperature, identify possible hot spots (fires, volcanic eruptions, etc.) and assess cloud-contaminated pixels, a panchromatic sensor VIS/NIR operating at night-time, able to detect artificial lights presence (night-time illumination) or natural (fires, etc.).

2 Mission Design

2.1 Orbit Characteristics

Taking into account only the J_2 perturbative effects, for an MSSO, the same illumination conditions occur after n nodal days, where n is a multiple of the revisit period m ($n = Nm$, with N integer; m is expressed in nodal days). Since the orbit is periodic, every m days the sub-satellite point crosses the equator in the same point, but at a different local time; every n days (after N observation cycles, each of m days), the satellite crosses the equator at the same initial local time. In such a way, the satellite observes each zone N times in n days, and every time in a different solar illumination (important requirement for observations in the thermal infrared region) condition. The characteristic equation of the MSSO is the following:

$$\left[n \cdot D_n = \frac{2\pi}{\dot{\Omega}_S - \dot{\Omega}} \right] \tag{1}$$

where $\dot{\Omega}_S$ is the Sun angular velocity in the apparent motion with respect to the Earth and $\dot{\Omega} = -\frac{3}{2}J_2\frac{R^2}{a^{7/2}}\sqrt{\mu}\left(1-e^2\right)^{-2}\cos i$ represents the variation of the Right Ascension of the Ascending Node, $R = 6371$ km is the equatorial mean radius of the Earth, $\mu = 398600.44$ Km^3/s^2 is the Earth's gravitational constant, e is the orbit eccentricity. By replacing in Eq. (1), the expression of the nodal day D_n (time necessary for the Earth to re-encounter the orbital node):

$$D_n = \frac{2\pi}{\omega_E - \dot{\Omega}}$$

where ω_E is the Earth's angular velocity, the following relationship can be obtained:

$$n = \frac{\omega_E - \dot{\Omega}}{\left|\dot{\Omega}_S - \dot{\Omega}\right|} \Rightarrow a^{3.5} = -K \cdot \frac{n \mp 1}{n \cdot \dot{\Omega}_S \mp \omega_E} \cdot \cos i \,(-\text{ for } \dot{\Omega} < \dot{\Omega}_S, + \text{ for } \dot{\Omega} > \dot{\Omega}_S) \tag{2}$$

with $K = \frac{3}{2}J_2R^2\sqrt{\mu}$.

The possible solutions of Eq. (2) are shown in Fig. 1.

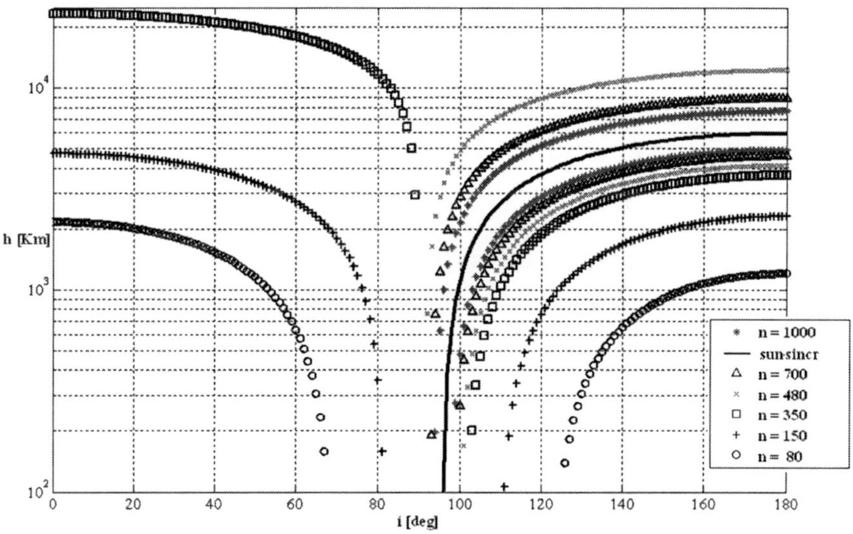

Fig. 1 Semi-logarithmic representation of possible solutions of periodic MSS0

By combining Eq. (2) with the periodicity condition $m \cdot D_n = N_t \cdot T_n$, where N_t represents the number (integer) of ground tracks covered by the satellite during the repetitivity period and T_n is the nodal period of the satellite (interval between two consecutive co-rotating ascending equator crossings), we derive the polynomial relationship given in Eq. (3):

$$C_1 \cdot a^7 + C_2 \cdot a^2 + C_3 \cdot a^{0.5} + C_4 = 0 \tag{3}$$

where:

$$C_1 = \frac{4}{\mu K} \cdot \left(\frac{n \cdot \dot{\Omega}_S - \omega_E}{n-1}\right)^2, \; C_2 = -\frac{1}{\sqrt{\mu}},$$

$$C_3 = -\frac{m}{N_t} \cdot \frac{n-1}{n \cdot (\dot{\Omega}_S - \omega_E)}, \; C_4 = -\frac{K}{\mu}$$

Once the values of n and m are fixed, the one and only physically acceptable solution of the equation (3) provides the semi-major axis of a periodic MSSO.

Table 1 shows a series of possible solutions of Eq. (3): $k = N_t - mN_i$, where N_i is the integer number of orbits covered by the satellite in a nodal day; S_m = distance, measured along the equator, between two adjacent satellite ground tracks. The selected orbit, evidenced in Table 1, is a direct, LEO (Low Earth Orbit) and periodic MSSO. A given region is observed every 5 days ($m = 5$) and the same illumination conditions are re-obtained after 50 days ($n = 50$). This means that the same area during the 50 days is observed in 10 different illumination conditions. The selected circular orbit has the following characteristics: altitude $h = 547.68$ km, $i = 31.59$ deg, $N_t = 74$, $S_m = 541.55$ km ($S_m = Swath/\sin i$).

Table 1 Possible solutions of periodic MSS0 with m = 5 days

m (days)	n (days)	K	N_t	N_i	h (km)	i (deg)	Sm (km)
5	45	4	74	14.80	542.28	16.06	541.55
5	50	4	74	14.80	547.68	31.59	541.55
5	50	3	73	14.60	611.78	28.39	548.95
5	50	2	72	14.40	677.42	24.64	556.58
5	55	4	74	14.80	552.56	40.31	541.55
5	55	3	73	14.60	616.53	38.05	548.95
5	55	1	71	14.20	749.08	32.76	564.42
5	60	4	74	14.80	556.93	46.51	541.55
5	60	3	73	14.60	620.77	44.71	548.95
5	60	2	72	14.40	686.19	42.76	556.58
5	65	3	73	14.60	624.66	49.78	548.95
5	65	2	72	14.40	689.97	48.16	556.58
5	65	1	71	14.20	756.84	46.42	564.42
5	70	2	72	14.40	693.39	52.45	556.58

Orbital perturbations effects have been taken into account in order to verify the feasibility of the mission; the Results have been reported in another paper [2].

2.2 Payload Description

The mission would be accomplished by using a small satellite (200 kg or less), carrying on board remote sensing instrumentation requiring an accurate pointing and a three-axis attitude system: a hyper-spectral (multi-spectral) sensor VIS/NIR/SWIR, a thermal sensor (TIR), a panchromatic night-time sensor VIS/NIR. The mission is based on the utilization of a combination of sensors representing an evolution of already existing systems [3, 4].

The hyper-spectral sensor we are thinking about should have number of bands similar to the CHRIS/PROBA (63 bands) sensor, and should operate in the spectral interval 0.4–2.5 μm rather than 0.4–1.05 μm with a slightly inferior spatial resolution (possibly around 50 m). In order to guarantee the required high revisit frequency (5 days), the proposed instrument is characterized by a swath greater than CHRIS (14 km).

The TIR sensor should be characterized by 3 bands operating around 3.9, 10.5 and 11.5 μm with a spatial resolution of about 60 m.

The night-time sensor is a panchromatic VIS/NIR camera sensitive enough to allow its utilization in night time hours (if possible, higher than the one foreseen for the future night time sensor DNB, Day-Night Band, that will fly on the satellites NOAA/NPOESS). At the present time, few sensors are working in such modality (OLS on board of the satellite DMSP and HSTC on board of the satellite SAC-C). Data provided by similar instruments, thanks to their high sensitivity, have been already used by the authors in order to estimate the extension of areas hit from natural or man-made disasters and assess the amount of affected population. Therefore, the main applications of a night-time sensor concern problems linked to security.

The thermal and night-time sensors should work in a synergic way in order to mask cloud-contaminated pixels by and distinguish "ephemeral" lights (forest fires, bonfires, etc.) from permanent lights (artificial illumination).

2.3 Ground Segment

Based on the orbit characteristics, it is possible to use as ground station the BSC (Broglio Space Centre) of Malindi, already equipped with 2 CRPSM-operated (Centre of Research S. Marco Project) antennas able to acquire band X data. In particular, the station is equipped with an antenna X-Y of 6.1 m, and a 10 m TT&C antenna capable of operating in S, L and X bands. Thus, the same station could carry out the operational management of the satellite. Taking into account a satellite minimum elevation angle of 5°, in m days the number of ground tracks inside the Malindi acquisition circle is equal to 32 (satellite-station contacts). The mean contact time is close to 8 min while the total visibility time in m days is about 4 h.

3 Conclusions

In this chapter a satellite mission devoted to mid-low latitudes observation, by using a small satellite in a low, direct and periodic MSSO (particularly suitable for monitoring these regions), has been proposed and described.

The synergic utilization of hyper-spectral, thermal and panchromatic (VIS/NIR) night-time sensors with appropriate spatial and temporal resolutions, would allows a great deal of studies concerning geology/environment (desertification, fire detection, soil and vegetated areas utilization, etc.), atmospheric and soil (water, vegetation, etc.) pollution assessment, and security related applications.

References

1. M. Castronuovo, C. Ulivieri, G. Laneve. Multi-sunsynchronuos Constellations for Continuous Surveillance in Tropical Regions. *AAS 98–301*, 19–30 (1998).
2. G. Del Gaudio, G. Laneve, C. Ulivieri. A Small Satellite Mission Devoted to Mid-Low Latitudes Earth Observation, *International Astronautical Congress*, Glasgow (2008).
3. C.S. Amaral, A.M.V. Monteiro, G. Camara, J.A. Quintanilha. DMSP/OLS Night-time Light Imagery for Urban Population Estimates in the Brazilian Amazon. *International Journal of Remote Sensing* **27**, 855–870 (2006).
4. D.A. Quattrocchi, J.C. Luvall. Thermal Infrared Remote Sensing for Analysis of Landscape Ecological Processes: Methods and Applications, Dordrecht, Kluwer Academic Publishers, **14**, 577–598 (1999).

The Italian Precursor of an Operational Hyperspectral Imaging Mission

Andrea Sacchetti, Andrea Cisbani, Gianni Babini, and Claudio Galeazzi

Abstract PRISMA, that stands for "PRecursore IperSpettrale della Missione Operativa" is an Earth Observation mission funded by Italian Space Agency (ASI) and is the follow-on of previous programmes, both national (HypSeo, phase A/B) and international (JHM phase A, in cooperation with Canada). The Mission will supply worldwide imaging access, with a specific area of interest covering the whole European and Mediterranean region, by providing with unique fusion of hyperspectral and panchromatic imagines. The program is conceived as a precursor of a future operational mission and, consequently, it is mainly driven by the needs of the Italian user community that, for the exploitation of the PRISMA data, will develop a set of dedicated applications. In fact, according to its "public good" nature, PRISMA will focus on the needs of the Italian institutional and research entities. Hyperspectral data, combined with panchromatic imagines, will help to efficiently address the selected applications, such as those related to quality and protection of the environment, sustainable development, climate change, strong economy. The quality of the images and products will be superior to previous and competing systems for what concerns geometry (also thanks to fusion of panchromatic and hyperspectral data), spectral and radiometry.

1 Overview of PRISMA Programme

PRISMA is an Italian small mission for Earth Observation having the following main objectives:

- in orbit demonstration and qualification of an Italian state-of-the-art hyperspectral camera;
- implementation of a pre-operative mission;
- validation of end-to-end data processing to be able to develop new application to manage the environmental risks and Earth observation based on high spectral resolution images.

A. Sacchetti (✉)
Carlo Gavazzi Space, Via Gallarate 150, 20151, Milano, Italy
e-mail: asacchetti@cgspace.it

R. Sandau et al. (eds.), *Small Satellite Missions for Earth Observation*,
DOI 10.1007/978-3-642-03501-2_6, © Springer-Verlag Berlin Heidelberg 2010

The mission is implemented by a single satellite, placed in a sun synchronous orbit, focusing primarily on the European area of interest, enabling the download of the data on two ground stations located in Italy. Once the hyperspectral images are downloaded on ground, they are archived and processed up to level 2.

PRISMA System is articulated in the following integrated segments:

- *Space Segment*, consisting in a single satellite placed in suitable LEO SSO orbit with an operational lifetime of at least 3+2 years.
- *Ground Segment*, comprising various centres located in Italy and including: [L1]
- Mission Control Centre (MCC), in charge of mission planning and management.
- Satellite Control Centre (SCC), in charge of satellite command and control.
- Instrument Data Handling System (IDHS), in charge P/L data reception, archiving, processing and interface with users. [L1]

A dedicated launcher will be used to directly inject the Satellite in its final orbit.

The baseline is VEGA launcher, as alternative options, other small launchers such as Dnepr, Eurockot, etc. will be considered in the satellite design.

2 System Description

2.1 Mission

The selected orbit for the area of interest (see Fig. 1) in the baseline configuration is sun synchronous with repeat cycle of 25 days (364 orbits)

- Orbit altitude: 702 km
- Inclination: 98.19°
- Local Time of Descending Node: 10.30.

Fig. 1 Area of interest

- Orbital period: 5,927 s (about 99 min)
- Average eclipse: 33.9 min

Considering the selected orbit, the daily accesses to the area of interest are usually 6. The average duration of the access to the area of interest is about 8.5 min. The daily imaging capability is of 88,600 km^2 (worst case) and 108,000 km^2 average.

The PRISMA satellite will communicate with two nominal Ground Stations, with a daily average contact duration of about 44 min:

- S-band (at Fucino), providing telemetry and telecommand links for satellite management.
- X band (at Matera), providing telemetry link for image data download at a maximum rate of 155 Mbps.

2.2 Satellite

The satellite is made up of:

- *Platform*, based on the Italian small satellites standard platform already used on MITA and AGILE missions
- *Payload*, consisting in a Hyperspectral instrument and PAN camera
- *Payload Data Handling and Transmission* (PDHT) recurrent from COSMO-SkyMed

The satellite configuration (Fig. 2) Results in a total wet mass of 500 kg with margins.

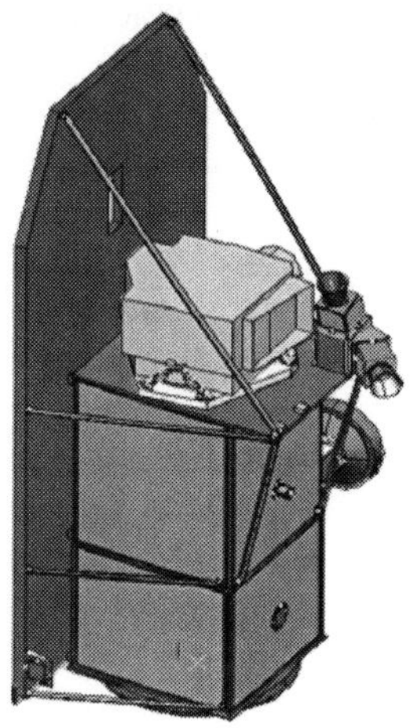

Fig.2 PRISMA satellite configuration

The average power consumption of the satellite is about 325 W with a peak of 720 W during acquisition and download.

2.2.1 Platform

The *PRISMA platform* represents a significant evolution of the MITA/AGILE platform. In particular it includes an enhanced attitude control system with pointing agility, a propulsion subsystem for orbit control and final disposal, and a power control electronics that makes maximum use of the competences in this domain at national level. The resulting key performances of the proposed PRISMA platform are in line with the current European state-of-art platforms used in first class remote sensing missions.

The on board operative and control functions of the satellite are performed by the On Board Data Handling (OBDH) Subsystem, which is in charge of on board telemetry and telecommand management, ground generated commands execution, AOCS, power management, monitoring of main platform H/K parameters, payload monitoring and image downloading during ground contacts.

The S band TM/TC subsystem carries out the telecommunications tasks. It is composed of two transmitters, two receivers and three antennas, managed in order to guarantee global coverage with every attitude condition.

The power to the Satellite is provided by fixed solar panels and by two Lithium-Ion battery packs during eclipse periods and during the attitude acquisition phase. The satellite attitude will be nominally kept Sun pointing while the fine nadir pointing attitude will be applied only during images acquisition and download. The electronics to convert and distribute the power to the satellite users is included in the Power Control and Distribution Unit (PCDU) and managed by the OBDH.

The PRISMA bus software will run on the OBDH and will be based on an Operating System kernel with Real Time and Multitasking features, so it will consists in a set of task working in parallel.

The sensors used to satisfy the payload needs are two star sensors, two GPS receivers, a set of gyros; two magnetometers, and a set of coarse Sun sensors. Four reaction wheels and three redundant magnetic-torquers constitute the actuators set of the ACS which is completed by ACS software running on OBDH.

The orbit control will be guaranteed by an hydrazine blow-down propulsion subsystem. Initial orbit correction and periodic orbit maintenance tasks will be calculated and planned by MCC, and uploaded to the satellite by the SCC. The satellite will then execute the required attitude manoeuvres and the thrusters firing according to the uploaded command sequence.

The satellite thermal control is mainly passive and it is realised by placing the most dissipating devices on radiating surfaces. Dedicated thermal control will be implemented for Li-Ion battery, for payload detectors and for the propulsion subsystem.

The satellite structure consists in a frame made of tubes and reinforcing side made up by Aluminium – Aluminium honeycombs and which supports internal subsystems hardware.

2.2.2 Instrument

The PRISMA Payload is composed of an imaging spectrometer (or hyperspectral imager), able to take images in a continuum of spectral bands ranging from 400 to 2,500 nm, and a panchromatic camera.

The Hyperspectral Instrument is based on a prism spectrometer concept. The main features, considering as reference a Sun-synchronous orbit with altitude of 702 km at the equator, are summarized in Table 1. The Instrument is composed of two units: the *Hyperspectral/PAN Optical Head* and the *Main Electronics* box. The only interface between them is electrical. The Instrument Optical Head is divided from the Main Electronics (ME) not to concentrate all the mass in the sensor head.

The Optical Head collects the radiation by a telescope, disperses the radiation by the two spectrometers, converts photons to electrons by means of appropriate detectors, amplifies the electrical signal and convert it into digital data stream. The Optical Head has mechanical and thermal interfaces with the S/C.

Table 1 Instrument main characteristics

Orbit altitude baseline	702 km
Lifetime	5 years
Swath/FOV	30 km/2.45°
GSD	Hyperspectral: 30 m PAN: 5 m
Spatial pixels	Hyperspectral: 1,000 PAN: 6,000
Spectral range	VNIR: 400–1,010 nm SWIR: 920–2,505 nm
Spectral resolution	≤10 nm
Radiometric quantization	12 bit
VNIR SNR	> 200:1 on 400–1,000 nm > 600:1 @ 650 nm
SWIR SNR	> 200:1 on 1,000–1,750 nm > 400:1 @ 1,550 nm > 100:1 on 1,950–2,350 nm > 200:1 @ 2,100 nm
PAN SNR	> 240:1
Absolute radiometric accuracy	Better than 5%
Aperture diameter	210 mm
MTF	VNIR @ Nyquist Frequency > 0.3 SWIR @ Nyquist Frequency > 0.3 PAN @ Nyquist Frequency > 0.2
Cooling system	Passive radiator

The Main Electronics box has the function to control the instrument and to handle, according to the agreed protocols, the bit stream representing the spectral images up to the interface with the S/C transmitter. It has mechanical, thermal and electrical interfaces with the S/C.

The cooling is obtained by means of a passive radiator facing the cold space. The Instrument head architecture main drivers are summarized hereafter:

- positioning of the cold zones (SWIR detector most critical) close to the radiator;
- positioning of the radiators towards the cold space;
- necessity to concentrate as much as possible the mass near the S/C mechanical interface to reduce the mechanical loads effect.

Optical Head

The optical system consists of a common telescope a double band spectrometer, operating in VNIR and SWIR bands, and a panchromatic camera. The telescope configuration consists of a Three Mirrors Anastigmatic (TMA) telescope that assures excellent optical quality with a minimum number of optical elements. This solution is very compact and without obstruction.

The spectrometer is a SELEX Galileo patented concept based on a prism solution. This prism system has high efficiency and low polarization sensitivity. The good efficiency allows to reduce the instrument dimension and mass with less demanding resources to the S/C and less criticalities for the optics design.

On the telescope focal plane it is located a slit common to VNIR and SWIR spectrometers, whose fields of view coincide. The separation between VNIR and SWIR is realized with a dichroic plate to improve co-registration features.

The entrance pupil diameter choice allows the system to meet high Signal-to-Noise Ratio considering an overall 10% degradation at EOL, while maintaining a F# high enough to improve the geometrical performances (in terms of smile, keystone, coregistration). Moreover, with this choice, the design of the system is less critical, for the reduced volume and mass, mirrors material costs, manufacturing times and calibration implementation.

Figure 3 shows the layout of the Instrument Optical Head.

Main Electronics

The Main Electronics has a redundant architecture: each sub-system is composed of a single board computer (SBC), a power distribution module (PDM) for complete system power supply, and the Scientific Data Acquisition Board (SDAB) that collects the signals coming from the PE, digitalize the detector output, compose the packets and interfaces the high speed serial link.

The Main Electronics has the following interfaces with the S/C:

- MIL-STD1553B standard, for Telecommands/Telemetries data packets exchange;

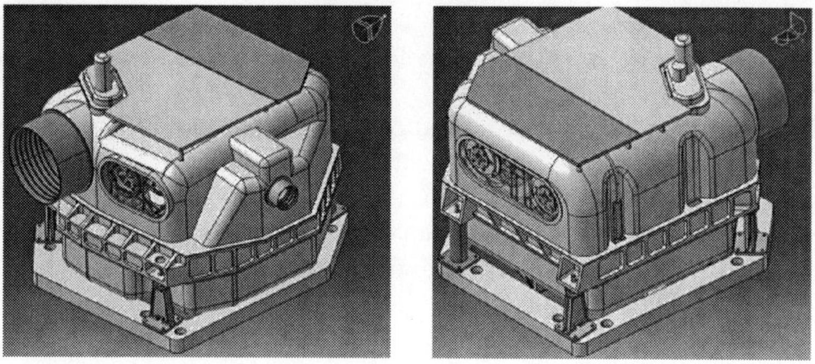

Fig. 3 Hyperspectral/PAN Optical Head layout

- High-speed serial data link (up to 1 Gbit/s), for sending image data packets to mass memory and relevant X-band transmitters;
- Bus power supply and discrete commands, for units power switching.

In-Flight Calibration

The in-flight calibration strategy is based on a dark calibration (close to the lower boundary of dynamic range), a Sun-calibration through a suitable diffuser (close to the higher boundary of dynamic range) and a spectral calibration. The calibration unit hardware is very compact, less expensive and with reduced degradation problems. The reliability of the system is so improved.

The calibration system uses a diffuser inserted in the optical path of the telescope at secondary mirror level. The lack of information on the primary mirror degradation can be overcame through a periodical vicarious calibration (e.g. view of Moon, Earth calibrated sites, etc.).

Dark calibration is implemented by means of a light-weight shutter placed in front of the optical slit.

The spectral calibration is implemented by the use of spectral lamps containing transparent filters with spectral signatures materials. The lamps can be oriented to illuminate the internal part of the cover in closed position or a solar diffuser directly.

2.3 Ground Segment

The PRISMA Ground Segment key aspects are:

- to develop the Italian capabilities and know-how for the handling and processing of the hyperspectral data and to promote the development of national applicative products based upon the hyperspectral remote sensing;

- to integrate EO data coming from different missions and based on different technologies so improving the national capabilities in remote sensing data exploitation;
- to benefit of the existing systems and operations know-how for ground control segment as well as user ground segment;
- to pursue the evolution and customization of Italian Space Agency's existing ground segment infrastructures developed in the frame of different programs to be enhanced, consolidated and re-used for different missions.

The Ground System is organized in a distributed architecture based on a network of centres connected by means of ASI's proprietary communication infrastructures (ASINET). ASINET, currently working and being used, among the others, in the frame of the AGILE mission. is a Customer Furnished Item (CFI) in the PRISMA contract.

3 Budgets and Performance

3.1 Revisit Time and Relook Time

The selected PRISMA orbit has been chosen to guarantee a repeating cycle shorter than 1 month, in particular the satellite pass over the same orbit every 25 days. The re-look time requirements are met with the in orbit pointing agility of the platform that can perform off-nadir pointing accordingly. A off-nadir (roll) manoeuvre of 15° guarantees the re-look of each point twice in 7 days as required.

3.2 Data Latency

For what concerns the "data acquisition latency" (i.e. maximum gap between the acceptance of a data order and the acquisition of these data) the performance of the system is below the requested limit of 9.5 days.

The overall response time is the sum of the "data acquisition latency" and the "data processing latency". The data processing latency time can be evaluated taking into account:

- the time for downloading an image just stored on board
- the time necessary to the ground station to archive the data downloaded from the spacecraft
- the time necessary to process data in order to generate the standard products
- the time necessary to acquire auxiliary data necessary to the data processing activity

The worst case for image download is the maximum gap between two access of the Matera ground station (about 11.8 h). The required period to perform the overall data processing activities is 4.5 days, therefore the ground segment dedicated to the archive, processing and delivery of the downloaded products will be designed in order to be able to execute these activities within 4 days from reception of the image data to the Matera ground station. So, the overall time response of the system is less than 14 days.

3.3 Imaging Daily Capability

The baseline configuration is sized to acquire, store and downlink up to 108,000 km^2 of hyperspectral/PAN images each day. Since the payload camera has a swath of 30 km, the total length of the daily images is 3,600 km. The satellite mass memory dedicated to the images storage is sized to guarantee the possibility to store the total daily image data considering a payload data generation rate of about 600 Mbit/s. The daily image capacity can be obtained as a single acquisition or in mulTiple acquisition according to the users needs. The image download can be performed using all the daily possible contacts with the Matera station, that are usually 4 or 5 for a total actual download period greater than 43 min. The X band data rate of 155 Mbps allows to download all the image data acquired in one day.

4 Programmatics

The prime contractor for the PRISMA mission is an industrial consortium formed by Carlo Gavazzi Space (CGS), Selex Galileo (GA) and Rheinmetall Italia (RHI), that form the industrial core team that coordinates the participation of other major Italian space companies, as subcontractors. The industrial team also includes major Italian companies, such as Thales Alenia Space, responsible for the PDHT, and Telespazio, responsible for the overall Ground Segment. The Kick off meeting has taken place in early January 2008 and the launch is planned in June 2011. After the Launch, a 3 month in orbit commissioning phase is foreseen. At completion of the commissioning phase, the system will be delivered to ASI.

Part II
Earth Environment Missions

Space System "Radiomet" for GLONASS/GPS Navigation Signal Radio Occultation Monitoring of Lower Atmosphere and Ionosphere Based on Super-Small Satellites

A. Romanov, A. Selivanov, V. Vishnyakov, A. Vinogradov, V. Selin, A. Pavelyev, O. Yakovlev, and S. Matyugov

Abstract Concept of new low-orbit system "Radiomet" is offered. This system is intended to global radio occultation monitoring and uses signals of radio navigation satellites passing through lower atmosphere and ionosphere. Authors of the concept – Russian organizations FSUE "RISDE" and IRE RAS.

"Radiomet", in comparison with existing system "Formosat-3" (COSMIC), should provide more than twice higher quantity of atmospheric parameters measurements per day, higher speed of final data deliver and some other significant improvements.

1 Now-Existing Space Systems for Radio Occultation Monitoring and Restrictions of Their Efficiency.

Radio occultation measurements of atmosphere are fulfilled when the emitting device (G) is located onboard GPS or "Glonass" satellite, and the receiver is installed onboard low-orbit satellite L. The altitude h of the investigated atmospheric area is determined by geographical location of point T – perigee of radio ray GTL, where the distance TD from the Earth's surface is minimal (Fig. 1).

The receiver of satellite L registers signals at two frequencies. The changes of phases and amplitudes of these signals contain the information about parameters of circumterraneous space along the radio ray GTL. An assumption of local spherical symmetry of the circumterraneous space makes it possible to consider that these changes are caused, mainly, by influence of those areas of ionosphere and atmosphere which are near to the ray perigee T. Further, using the measured changes of phases and amplitudes and known ephemeris of G and L satellites, it is possible to calculate altitude dependence of refraction angle of the ray GTL and then, by means of Abel transformation, to find altitude dependence of refraction parameter in atmosphere, and also electronic concentration in ionosphere (Fig. 2). Using the equations of ideal gas and hydrostatic balance we finally define parameters which are very

V. Vishnyakov (✉)
FSUE "RISDE", Russian Federation, Moscow, Russia
email: Vishnyakov@rniikp.ru

R. Sandau et al. (eds.), *Small Satellite Missions for Earth Observation*,
DOI 10.1007/978-3-642-03501-2_7, © Springer-Verlag Berlin Heidelberg 2010

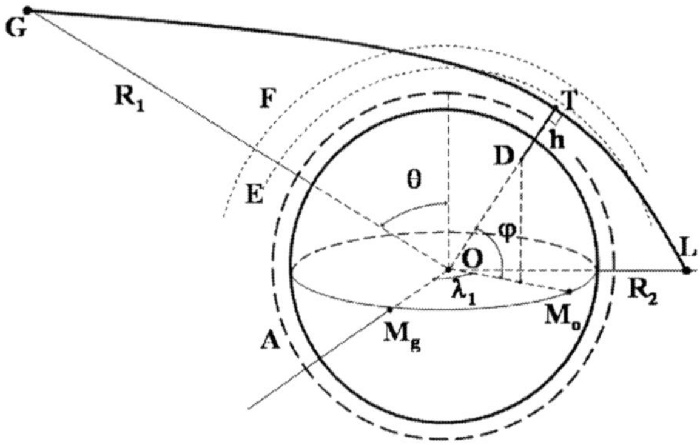

Fig. 1 Geometry of radio occultation measurements in atmosphere

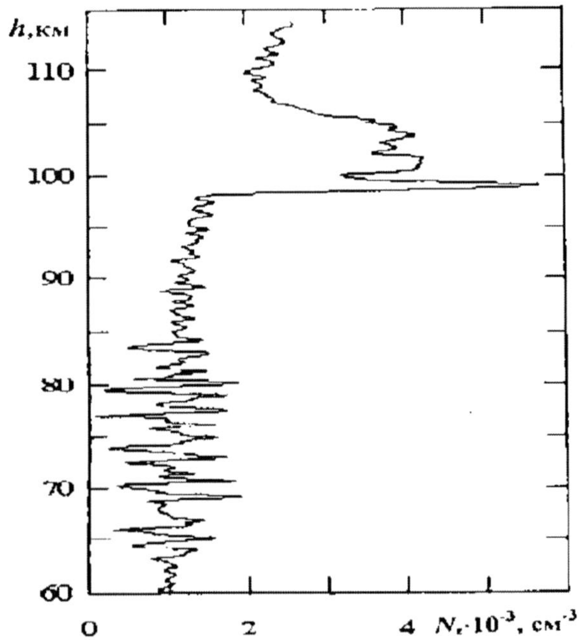

Fig. 2 Vertical profile of electronic concentration in ionosphere

important for meteorological monitoring – vertical profiles of density, atmospheric pressure, temperature (Fig. 3) etc.

Experimental research of radio occultation monitoring of the Earth have been made in Russia in 1990–1998 with use of "Mir" orbital station and two geostationary satellites [1], and in USA in 1995–1998 (GPS/MET experiment) with use of MICROLAB satellite which received two wave ($\lambda_1 = 19$ cm and $\lambda_2 = 24$ cm) signals of GPS navigating satellites in [2] and in other missions.

At a following stage (since 2001) several national space missions of global-scale radio occultation monitoring means have started. Nowadays both small satellites

Fig. 3 Vertical profile of temperature in atmosphere

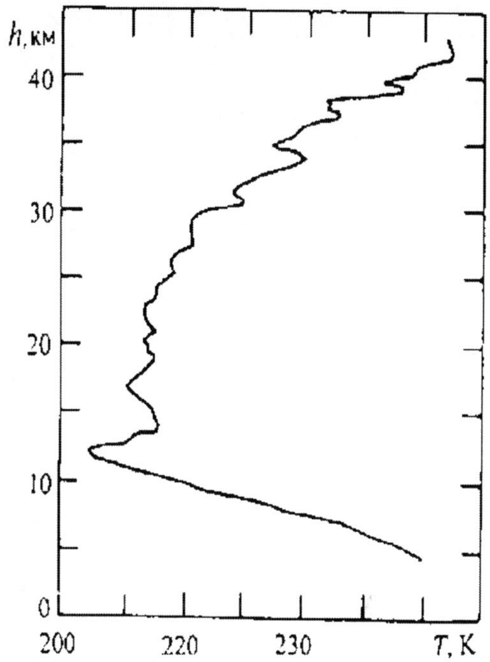

(GRACE–A, GRACE –B, “Champ”, “Formosat-3” (COSMIC) and large vehicles (MetOp-A, TerraSAR-X), intended mainly for other issues, are being used for radio occultation monitoring experiments. Six-satellite system “Formosat-3” [3], launched in 2006, may be considered as a prototype of future international system for regular radio occultation monitoring of atmosphere. “Formosat-3” satellites make ~2,500 soundings per day over various regions of the Earth, including remote regions (the seas, oceans, deserts, hilly terrains, etc.). It’s obvious that global monitoring possibilities of space monitoring systems are much higher than the possibilities of traditionally used meteorological stations and radio probes.

Basic features of some radio occultation monitoring missions which use GPS signals (L1 = 1575.42 MHz and L2 = 1227.6 MHz) for sounding of atmosphere, are resulted in Table 1.

It’s a common disadvantage of all missions (including “Formosat-3” (COSMIC) that they do not use signals of “Glonass” navigation system. That’s why they have limitations of the total number of atmosphere soundings per day by no more than 3,000. “Formosat-3” system has other restrictions of its efficiency. Orbit parameters (72° inclination) do not allow to spend frequent radio sounding in high-altitude and polar areas of an atmosphere (for Example, over subpolar territories and water areas of Russia and other countries). Besides, the “raw” data from satellites “Formosat-3” are passed through complicated and multistage Procedures in the ground segment – reception, preliminary and thematic processing, data dissemination, etc. As

Table 1 Basic features of some radio occultation monitoring missions which use GPS signals

Mission name and lifetime	Orbit parameters (altitude h, inclination i)	Measurement rate (Hz) of phase and amplitude	Periodicity of data deliver to ground stations
CHAMP 2001–2008	h = 470 km, i = 87°	0.1–50	10 MB, 2 times/day
SAC-C 2001–2002	Sun-synchronous orbit, h = 707 km	0.1–50	10 MB, 2 times/day
SunSat since 2001	Polar elliptical orbit, h = 650/800 km	50	2 times/day
GRACE-A *GRACE-B* since 2002	h = 500 km, i = 87°	0.1–50	10 MB, 2 times/day from each satellite
FORMOSAT-3/COSMIC (6 satellites) since April 2006	Circular orbit h = 400(initial) – 700 km (operational), i = 72°	100	5 MB/100 min from each satellite
MetOp-A since October 2006	h = 817 km, i = 98.7°	100	Once per day

a result, final data about physical parameters of lower atmosphere and ionosphere are delivered to consumers with a significant time delay (≥3 h). The exchange with consumers in «on-line» mode is not available.

2 Concept of "Radiomet" Low-Orbit Space System

The concept of the low-orbit radio occultation monitoring system "Radiomet" [4] has been developed by FSUE "RISDE" (Russian Institute of Space Device Engineering) and IRE RAS (Institute of Radio engineeing and Electronics of Russian Academy of Science).

FSUE "RISDE" (the integrator of "Radiomet" system) is the leading organization for development of Russian satellite navigation system "Glonass" and also the developer of the newest miniaturized onboard two-system (GPS/"Glonass") navigation receivers. IRE RAS develops new radio holographic methods for studying atmosphere by means of radio emission of satellites. IRE RAS participates in data processing of missions GPS/MET, "Champ", COSMIC and other systems of radio holographic sounding of atmosphere.

Figure 4 represents the structure of "Radiomet" system.

"Radiomet" system, in comparison with existing system "Formosat-3 (COSMIC)", should improve capabilities of atmosphere monitoring. First of all, the new radio occultation equipment that will be applied onboard "Radiomet" satellites will register radio occultation signals not only from GPS system, but also from

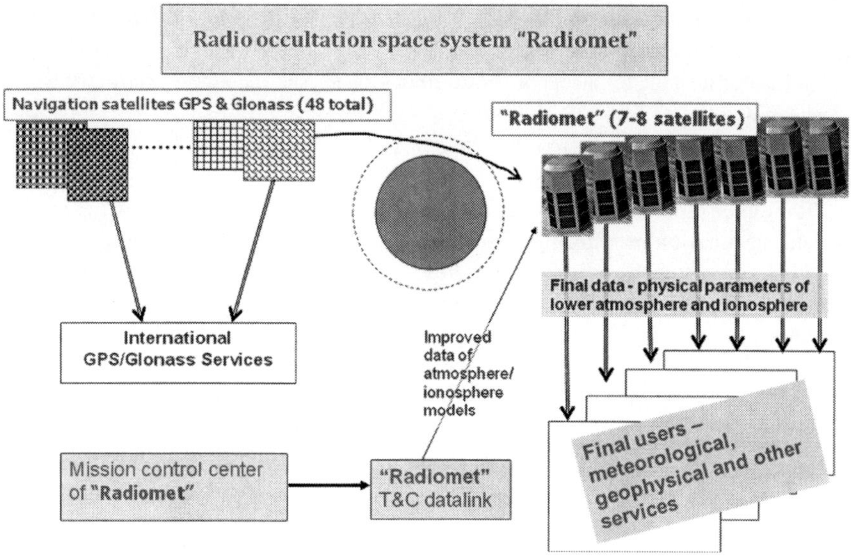

Fig. 4 Structure of "Radiomet" radio occultation monitoring system

"Glonass" system (and later – from "Glonass" and other global positioning systems). This will allow raising 2 and more times the daily number of soundings of atmosphere. Optimization of "Radiomet" satellites orbit parameters will allow to increase frequency of sounding sessions above any area of the Earth (including near-equatorial and subpolar areas). Two additional (compared to "Formosat-3") polar-orbit satellites may be included in the structure of "Radiomet" system for this purpose.

FSUE "RISDE" has developed the new two-system ("Glonass"/GPS) small-size receiver that will be a core of the radio occultation equipment of "Radiomet" satellites. Now the breadboard model of the radio occultation equipment is being prepared for ground tests in 2009–2010. Its flight test on board of "TNS-2" technological nanosatellite will start after 2010.

We develop now the special small-size receiving system for the radio occultation equipment of "Radiomet" satellites. This system should consist of 2 spiral cylinder aerials, each sized ca. 8 cm by diameter and ca. 40 cm by length. Such system provides gains large enough to compensate losses of radio occultation signals of L1 and L2 ranges due to attenuation in lower atmosphere within the field $70^{\circ} \times 30^{\circ}$.

The new methods offered by authors [5], will allow fulfilling the basic part of processing and the analysis of accepted radio occultation signals just onboard "Radiomet" satellites.

If necessary, the finalized data of atmosphere (parameter of refraction, density, temperature, pressure and humidity of a neutral atmosphere, electronic concentration of ionosphere) can be transferred from "Radiomet" satellites to ground stations of consumers. That significantly raises the speed of monitoring – up to "real-time" mode of exchange with consumers.

Multisatellite constellation of "Radiomet" system is developed on the basis of super-small satellites of "TNS-2"-type (weighing no more than 20 kg), thus minimizing the expenses for creation, launch and renewal of satellites. "Radiomet" system is expected to start in 2013–2014.

"Radiomet" space system is not considered as a replacement for existing system of radio occultation monitoring "Formosat-3" (COSMIC). On the contrary, the two specified systems can be operated simultaneously and supplement each other. As a result, the possibilities of global radio occultation monitoring of atmosphere should essentially extend (see Table 2).

Table 2 Possibilities of global radio occultation monitoring of atmosphere by "Formosat-3" and "Radiomet" space systems

Radio occultation space system	"Formosat-3" (COSMIC)	"Radiomet" + "Formosat-3"
Registered navigation satellite signals	GPS only	GPS, "Glonass", "Galileo"
Number of everyday observations	from 1,500 to 2,500	up to 8,000
Medium period of observations for each area	~once per day	3...4 times per day
Periodicity of polar zones observation	low (compared to other latitudes)	as high as for other latitudes
Final parameters of atmosphere and ionosphere are formed:	in the ground segment	mainly – on board of satellites
Time delay of data delivery to final users	≥ 3 hours	up to real-time

3 Modelling of Radio Occultation Monitoring System "Radiomet"

The mathematical model of multisatellite constellation of a radio occultation monitoring system is developed. The model provides calculation of relative movement parameters of low-orbit satellites and navigation satellites (during radio occultation sounding of an atmosphere). Implementation of the model allows optimizing structure and orbital parameters of "Radiomet" constellation.

During modelling those Navigation spacecrafts (NSC) are defined which are situated within "radio sounding cone" of a small spacecraft (SSC) "Radiomet". The vertex of "radio sounding cone" coincides with the center of SSC. The axis of this cone may be directed both forward – in a direction of SSC flight (then the NSC is in front of the SSC), and backward – against a direction of SSC flight (then the NSC is behind SSC).

The quantity of "Radiomet" SSCs, quantity of GPS/ "Glonass:"/ "Galileo" NSCs and their orbital parameters are the set of initial data for modelling. The sizes of "radio sounding cone" in horizontal and vertical planes and some other parameters may be varied as well. The bottom border of "radio sounding cone" in the orbital plane of SSC is defined by the lowest meaning of radio occultation ray perigee $HATM_{\min}$ (usually $HATM_{\min} = 0$) and the top border of "cone" is defined by the highest meaning of radio occultation ray perigee $HATM_{\max}$ (usually 150 km or more). The angular size of the cone in a plane of an orbit makes usually 35–40°; in a horizontal plane it can be varied (typically within 70–80°).

The above-mentioned model can provide:

- to determine those NSC which are now situated within "radio sounding cone" of SSC and duration of their staying within the cone;
- to calculate the total quantity of radio occultations (for each SSC) during the given time interval and duration of each radio occultation;
- to determine the trajectories in atmosphere of perigee T of radio ray GTL (altitude, longitude, height). In Fig. 6 you see the display of it in a two-dimensional mode (altitude, longitude).

It's shown below how the Results of modeling are displayed. The radio occultation constellation of 6 SSCs with 700 km height and 72° inclination of orbits is modeled. Figure 5 (above) shows the daily temporal diagrams of radio occultation signals received by one of the 6 spacecrafts from each of 24 GPS satellites. Each line of this diagram includes daily radio occultation sessions which are shown as small blue rectangles. Width of a rectangle is proportional to the duration of a session. Under the diagram the same data in a numerical form, as well as the direction of soundings (forward or backward off the SSC motion), are shown.

Figure 5 (below) shows the daily temporal diagrams of radio occultation signals, when not only GPS, but also Glonass system (totally 20 satellites by now) signals are received by SSC.

Two-dimensional daily diagrams of radio occultation ray paths in atmosphere are shown in Fig. 6 for the system of 6 SSCs, when GPS signals only (see above) and GPS & Glonass signals (below) are used. Such diagrams describe global coverage and density of radio occultation sessions in the atmosphere over any area of the Earth. The diagrams of Fig. 6 show a significant growth of global coverage and density of radio occultations when Glonass signals are additionally used.

4 Conclusion

Russian research organizations FSUE "RISDE" and IRE RAS have developed the concept of a new radio occultation monitoring space system – "Radiomet". Such a concept should overcome restrictions in efficiency of existing radio occultation monitoring systems ("Champ", "Formosat-3"/COSMIC) and provide the following advantages:

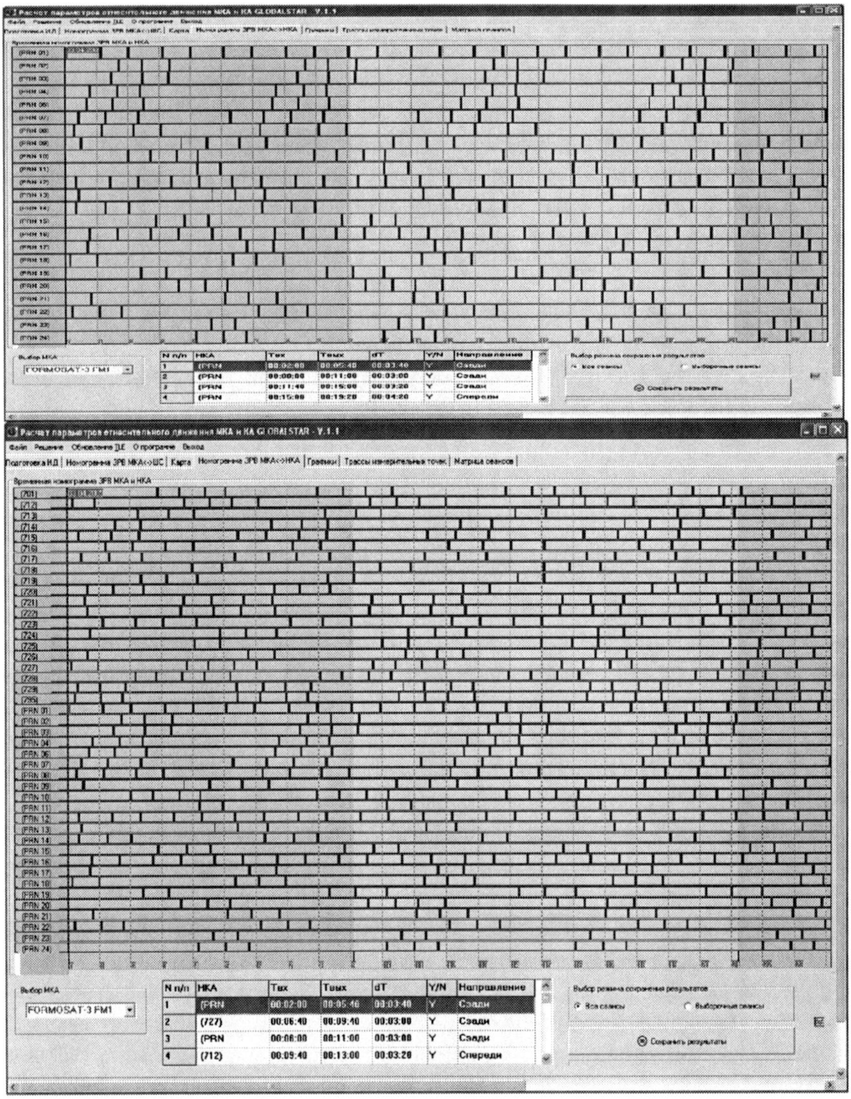

Fig. 5 Daily temporal diagrams of radio occultation soundings, made by one of spacecrafts of the constellation, when GPS signals only (see above) and GPS & Glonass signals (below) are used

- 2 and more times bigger everyday number of radio occultation soundings of atmosphere due to the new on-board receiving equipment which registers signals not only from GPS, but also from “Glonass” & “Galileo” navigation satellites;
- higher periodicity of measurements of atmospheric parameters over any area of the Earth (including near-equatorial and sub polar areas);
- a few times higher speed of monitoring due to:

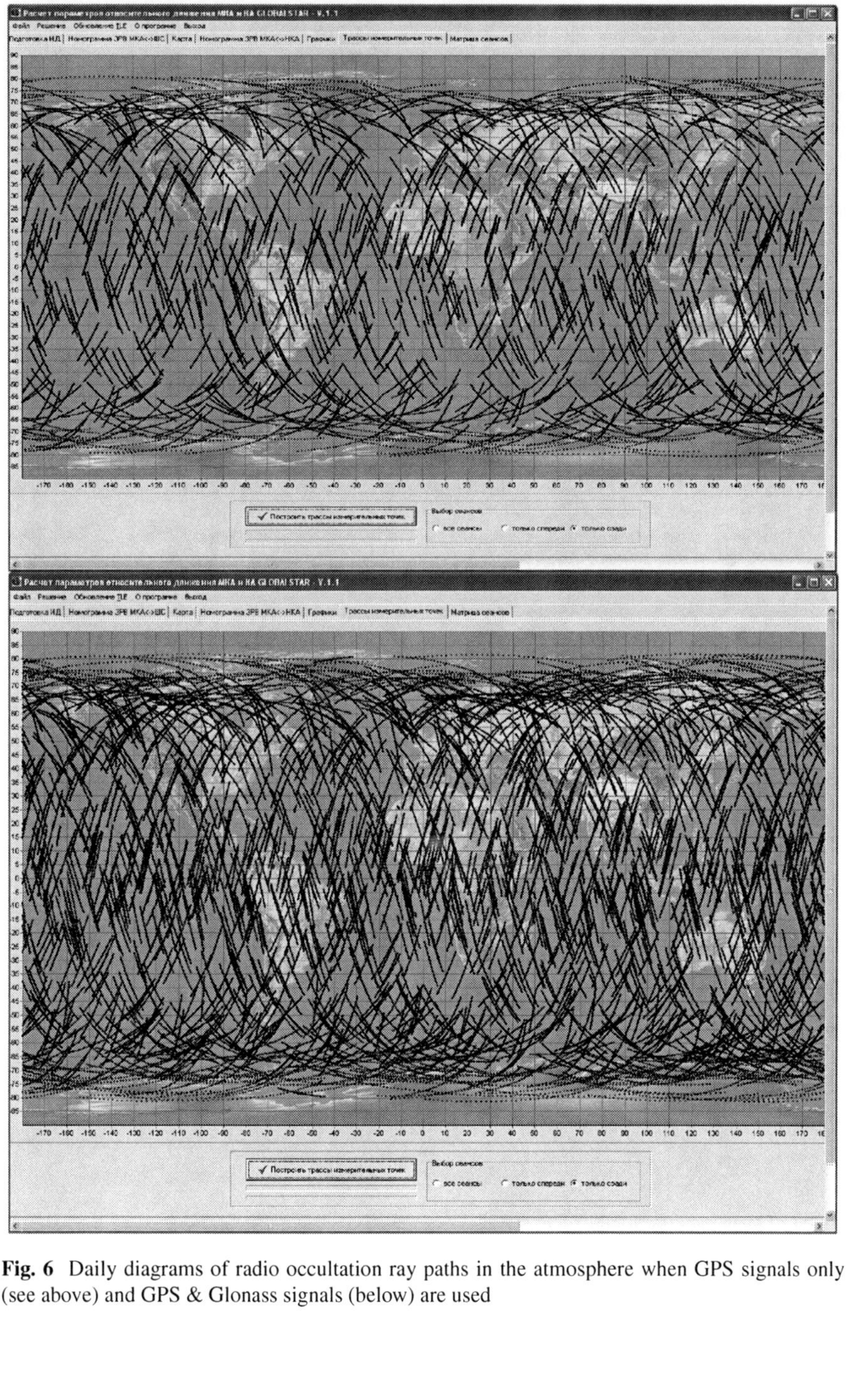

Fig. 6 Daily diagrams of radio occultation ray paths in the atmosphere when GPS signals only (see above) and GPS & Glonass signals (below) are used

- processing and analysis of occultation signals mainly onboard of "Radiomet" satellites;
- transmitting the final data on physical parameters of the lower atmosphere and ionosphere directly to the ground stations of consumers in real-time mode.

The presented work is supported by Russian fund of fundamental investigation (RFFI) (project N 08-02-13535).

References

1. O.I. Yakovlev, S.S. Matyugov, I.A. Vilkov. Attenuation and scintillation of radio waves Earth atmosphere in radio occultation experiments on the satellite-to-satellite link. Radio Science, 30(3), 591, 1995.
2. C. Rocken, R. Anthes, M. Exner, et al. Analysis and validation of GPS/MET data in the neutral atmosphere. Journal of Geophysical Research, 102, 29849, 1997.
3. C. Rocken, Y.-H. Kuo, W. Schreiner, et al. COSMIC system description. Journal of Terrestrial, Atmospheric and Oceanic Science. Special issue. 11(1), 21, 2000.
4. A. Selivanov, A. Romanov, V. Vishnyakov, A. Vinogradov, A. Pavelyev, O. Yakovlev. «Concept of space system for global radio occultation monitoring of lower atmosphere and ionosphere based on super-small satellites with GLONASS/GPS navigation signal receivers». «United Nations/Austria/ European Space Agency Symposium on Space Tools and Solutions for Monitoring the Atmosphere and Land Cover». Austria, Graz, 9–12 September 2008.
5. A.G. Pavelyev, K. Igarashi, K. Hocke. Patent of Japan P2001 – 18248. Int. Cl. G 0 1 S 13/89 5/14 G 03 H 5/00 (P2003 – 4844), 2003.

The Study of Electromagnetic Parameters of Space Weather, Micro-Satellite "Chibis-M"

Stanislav Klimov, Denis Novikov, Valeriy Korepanov, Andriy Marussenkov, Csaba Ferencz, Janos Lichtenberger, and Laszlo Bodnar

Abstract Plasma-wave experiment on the micro-satellite "Chibis-M" is aimed at the solution of fundamental problem – a study of the interrelation of the plasma-wave processes connected with the manifestation in the ionosphere of solar – magnetosphere – ionosphere – atmosphere connections and the parameters of space weather. Specific fundamental problem is the search for universal laws governing transformation and dissipation of plasma-wave energy in the magnetosphere-ionosphere system. The solution of this problem will be achieved employing the coordinated Procedure: (1) Study in situ of the fluctuations of electrical and magnetic field, the parameters of thermal and epithermal plasma in the ionosphere near layer F during different helio- and geomagnetic conditions. (2) Study of the geomagnetic and geophysical parameters on the ground-based observatories with the time scales from 10^{-1} to 10^{-3} s. (3) Study of the interrelation of electromagnetic phenomena (spectra of ULF/VLF- waves) in different regions of near-earth space by means of via the comparative analysis of the wave measurements of those carry out simultaneously on different spacecrafts and ground geophysical stations. Micro-satellite "Chibis-M" now designed in IKI. Total mass of "Chibis-M" with support systems construction and scientific instruments is 40 kg.

1 Introduction

Long-wave electromagnetic radiations is used for studies of plasma processes in space and it play the same role as seismic waves for studying of the Earth's structure. In comparison with the electromagnetic processes in other regions of physics waves in plasma possess a number of the specific characteristic features. The fundamental importance is a resonance effect which appears as a result of waves and particles interaction, conversion of modes, formation of resonators and waveguides. Because of resonance effect ultra-low frequencies waves transfer information about

S. Klimov (✉)
Space Research Institute (IKI) of RAS, 117997, Moscow, Russia
e-mail: sklimov@iki.rssi.ru

R. Sandau et al. (eds.), *Small Satellite Missions for Earth Observation*,
DOI 10.1007/978-3-642-03501-2_8, © Springer-Verlag Berlin Heidelberg 2010

the dynamic phenomena in the near-earth outer space and the upper atmosphere. In this case they reach sufficiently large amplitudes in order to have noticeable influence on the flows of plasma and to effectively accelerate electrons in the magnetosphere. It is recently shown that by the disturbance of plasma are accompanied not only magnetic storms but also the ground-based geophysical phenomena connected with the large ejection of energy – explosions, hurricanes and thunderstorm, earthquake etc. [1]

At present in a number of countries several space experiments directed toward the study of low-frequency wave activity in the ionosphere are planned and even it is located in the stage of realization. Widely known experiments on the satellite DEMETER (France, [2]), on the satellites COMPASS-2 (Russia, [3]) and "Sich-1 M" (Ukraine, experiment "Variant" [4]) can be attributed to them. The continuation of these studies will be the wave experiment planned on board the micro-satellite "Chibis-M" (Russia, [5, 6]). The launching is planned in 2010 (Fig. 1).

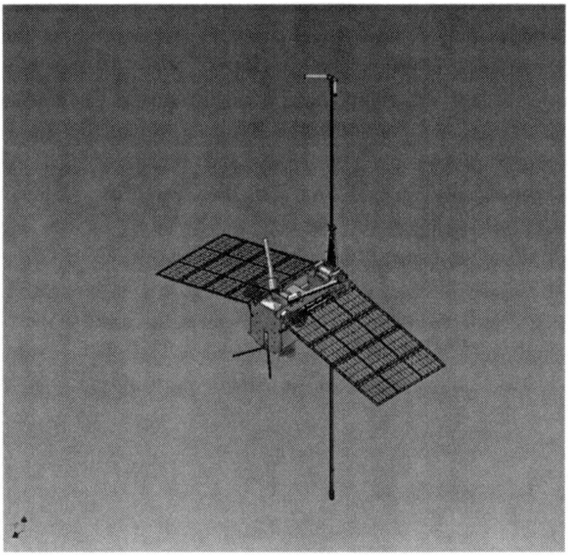

Fig. 1 General view of the micro-satellite "Chibis-M"

2 Wave Experiment on Board the Micro-Satellite "Chibis-M"

The basic scientific task solved by the complex of scientific instruments of satellite "Chibis-M" – KNA "Groza" – is the study of the process of the development of the stepped leader of high-altitude lightning in the high electric fields [6]. Magnetic-wave complex does not enter into the composition of the instruments which manufacture markers on the fact of the registration "of event" – stepped leader, but his onboard processor – PSA – can use the manufactured by BND-Ch strobe for purposes of the fixation of generation in this case electromagnetic radiations in the range 10^{-2} – 2∗104 Hz (ULF-VLF). It is known that the generation discharge electromagnetic radiations in the basic phase of the lightning called whistles are recorded in VLF frequency range and are observed on the satellites already long ago. Thus VLF – emissions are genetically connected with the lightning discharges that also explain the connection of their studies with the basic task of project "Chibis-M". The basic task, solved by magnetic-wave instruments in the process of the flight of micro-satellite "Chibis-M", is the study of the electromagnetic parameters in the range of frequencies 0.1–40 kHz. In this case magnetic-wave experiment on the "Chibis-M" it is aimed also at a study of the interrelation of the plasma-wave processes in the ionosphere, proceeding under the effect of the solar – magnetosphere – ionosphere – atmosphere connections and is of great interest for studying the parameters of space weather. It is expected that the systematic study of these connections will make it possible to find the universal laws, which regulate conversion and dissipation of plasma-wave energy in the magnetosphere-ionosphere system.

On the basis of the analysis of the extensive experimental material expected from the micro-satellite of "Chibis-M" and geophysical ground-based observatories, theoretical analysis and computer simulation will be developed the algorithms of wave and hydromagnetic diagnostics of magnetosphere- ionosphere system. It is assumed that these methods will make it possible to conduct the continuous quantitative monitoring of characteristic energies of particles, concentration and chemical composition of plasma in the magnetosphere in order to predict different forms of the geomagnetic disturbances created by the influence of different factors both "from above" (space), and "from below" (ground-based).

The solution of this problem will be achieved with the use of the coordinated Procedure of the studies:

(a) in situ of the fluctuations of electrical and magnetic field directly in the ionosphere layer F under different helio- and geomagnetic conditions;
(b) of the geomagnetic and geophysical parameters in the ground-based observatories with the time resolution from 10^{-1} to 10^{-3} s;
(c) of the interrelation of the electromagnetic phenomena of ULF-VLF ranges in different regions of near-earth outer space by means of the simultaneous comparative analysis of the wave measurements conducted on different spacecrafts and ground geophysical stations.

It is important to note that the signals recorded by magnetic- wave complex can be transmitted in the form "wave form" which makes it possible to conduct precision spectral- temporary processing on the Earth. Furthermore, they undergo onboard working by the processor of spectral analysis – PSA, which makes it possible to substantially press the volume of telemetry data. It should be noted that realized configuration of the arrangement of sensors makes it possible to conduct the vector measurements of the magnetic component of emissions (Fig. 2).

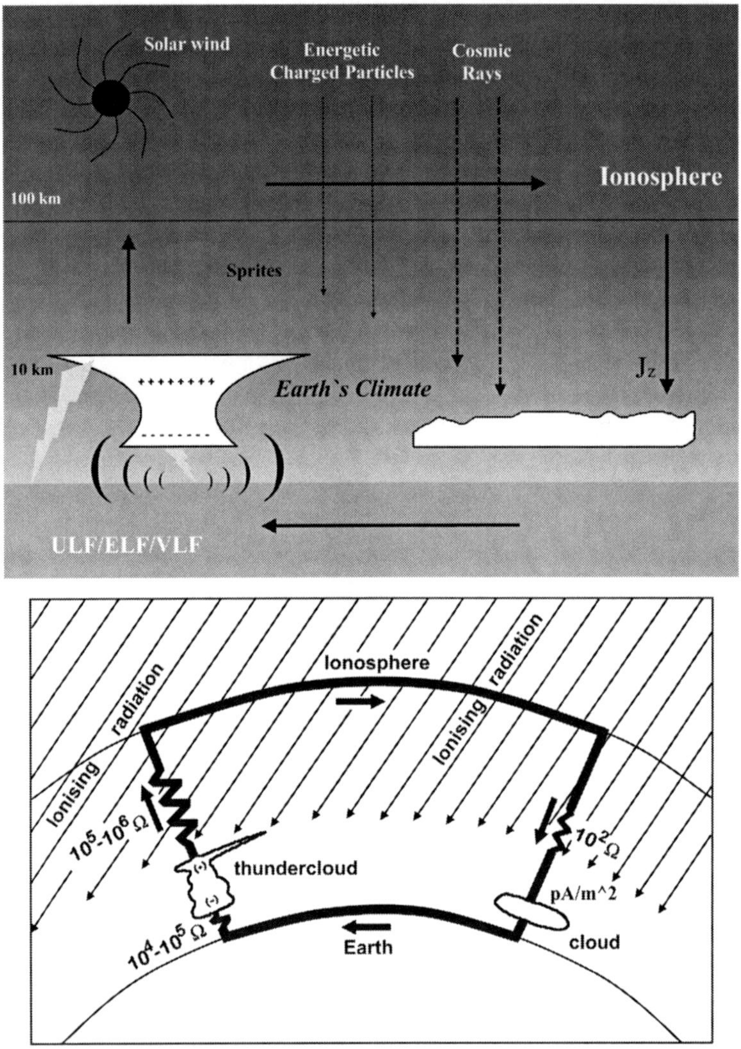

Fig. 2 Role of thunderstorms in global current system

Basic function of PSA is the onboard the processing of signals and the forming of spectra in the range (0.1–40 kHz). Furthermore, PSA must conduct the detection of event and flashes of the recorded modes as it realized on board other satellites (such as DEMETER or COMPAS-2).

The floating potential of the probe into the plasma is the integrated parameter, which characterizes space weather, since it is determined by the balance of the currents, which flow to the probe:

$$[Ie + Ii + Ip + Ih = 0] \tag{1}$$

where: Ie – current of electron; Ii – current of ion; Ip – photo-current; Ih – current of energetic particles.

On the micro-satellite "Chibis-K", "Chibis-M" is placed instruments for studying the parameters of the space weather:

Electrometer (DC – 40 kHz);
Fluxgate magnetometer (DC – 1 kHz);
Search coil (0.01–40 kHz);
Processor of spectral frequency analysis (0.01–40 kHz).

3 Registration of Event and the Modes of PSA Work

PSA is intended for the search of the electromagnetic signals considered as "the event" in the flow of data obtained by sensors and interesting from the point of view scientific tasks of "Chibis-M". The determination of event is accomplished by the integrated program module of PSA. In the case of electromagnetic event it is recorded in one of the counters depending on the type of event. In addition to this, the wave form which contains event remains in the memory of instrument with the programmed duration and then it is sent into telemetry system. Software of PSA can work in several modes. Mode switch and possible reprogramming of PSA is achieved on the telecommands whose enumeration is determined by the complex of stated problems.

Taking into account "Chibis-M" orbit inclination close to the inclination of the International Space Station (ISS) the important aspect of studies is the forming of the coordinated "two-point" observations. Contributory factor for these observations is the proximity of the parameters of the magnetic-wave sensors and the sensors of experiment "Obstanovka 1st stage" on board ISS [7]. Relatively low orbit inclination ISS and "Chibis-M" create a good basis for studying few investigated processes of space weather on the average and equatorial latitudes [8] where thunderstorm activity is also high. These regions are frequently subjected to the natural calamities whose reflection in the ionosphere will be investigated on two the satellites indicated as the component part of the program of prediction and diagnostics of space weather. Let us examine the special features of the work of wave complex in the operating modes accepted.

3.1 Basic Mode – "Mode A"

"Mode A" is basic mode for the evaluation of input signals. In this mode only one channel is processed. The actual configuration of channel (component of vector E or B, the speed of numbering) depends on the commands which entered in PSA.

Software of PSA conducts fast Fourier transform FFT and with the aid of the digital processor of signals calculates spectrum and average noise background of the selected channel. The detailed structural analysis of spectrum FFT, conducted by software of PSA, makes it possible to find the local peaks of energy in the spectrum and to estimate their following basic parameters:

- the level of peak (amplitude);
- the level of peak relative to noise background;
- the number of peaks;
- the width of peaks.

The given parameters are accumulated and are investigated in the selected sequence of frequency ranges. As a result carried out estimation signals are divided on the following categories:

1. Narrow frequency band (T_NB) – in signal is only 1–3 lines in the spectrum.
2. Wide frequency band (T_WB) – the strip of signal frequencies is located in the range 1. . .2 kHz.
3. Ultra wide frequency band (T_UWB) – signal is recorded in the larger part of the spectrum being investigated.
4. Multiple peaks (T_MP) – in the spectrum appears more than one narrow signal.

Reading data on the sequence of ranges it is possible to add to the categories mentioned above:

a) Time (duration of signal):
    1. 1 range;
    2. 2–10 ranges;
    3. prolonged.
b) Frequency (how signal frequency it changes in the time):
    1. the fixed frequency;
    2. the being risen frequency;
    3. the reduced frequency.

As a result the application of existing software at the output we obtain a number of the events (signals, which are differed from background noise), which reach one of four categories in the limits of the time interval being investigated.

3.2 Extended Mode – "Mode B"

In the extended mode program of PSA the guarantee can conduct more detailed data processing. First according to registered data continuous spectrogram is created (without the spaces). After this, to each spectrum is advanced parametric estimation, similar to the process, described in the basic method, and subsequently is conducted the analysis of three-dimensional spectrum with plotting of the dependence of changes in intensity and signal frequency on the time.

This analysis can classify signal according to the following categories of events:

1. Pulse signal (short transient process).
2. Monotonic signal.
3. Scattered signal.
4. Signal of the wide range.

3.3 Programmed Mode of Data Collection

Using this operating mode it is possible depending on the entered telemetric commands to obtain data with the assigned degree of detailing wave form at the assigned point of the orbit.

4 Conclusion

The magnetic-wave complex of instruments, established on board the micro-satellite "Chibis-M" is directed first of all toward the solution of the problem of monitoring space weather for the purpose of finalizing the Procedure of prediction and analysis of the state of its changes, connected both with the actions "on top" – from the outer space – and by the disturbances, which enter "from below" – from the powerful natural catastrophes (earthquake, tsunami the rest) and anthropogenic actions (explosions, the starts of powerful rockets the rest). In connection with the available possibility of conducting the two-point synchronous observations from onboard of micro-satellite and s special hope is laid for a study of the methods of the separation of actions "from below" and "on top", which would give ex-nova stimulus for the development of the methods of predicting the natural catastrophes with the aid of the artificial earth satellites.

Work on the creation of the magnetic- wave complex of instruments is conducted in the international cooperation with the Hungarian and Ukrainian participants, which made it possible to obtain the level of the parameters of this complex, which is not inferior to the best world achievements. They are carried out with the support of the contract 1-05/08 s NKAU. Discussion on systematic questions of the study of atmospheric lightning discharges is carried out within the framework the working group of the international institute of space sciences, Bern (ISSI team: CARNES – Coupling of Atmosphere Regions with Near-Earth Space).

References

1. S. Maekawa, T. Horie, T. Yamauchi, T. Sawaya, M. Ishikawa, M. Hayakawa, H. Sasaki. Seismo-ionospheric phenomena: A statistical study on the effect of earthquakes on the ionosphere, based on the subionospheric LF propagation data in Japan. Annual Geophysics 24, 2219–2225, 2006.

2. J. Chum, O. Santolik, M. Parrot. Analysis of subprotonospheric whistlers observed by DEMETER: A case study. Journal of Geophysics Research 114, A02307, doi: 10.1029/2008JA013585, 2009.
3. V.D. Kuznetzov, Yu.Ja. Ruzhin, Yu.M. Mikhailov, G.A. Mikhailova, O.V. Kapustina, V.S. Dokukin, G.I. Druzhin, S.E. Smirnov, Sz. Ference, Ja. Lichtenberger, L. Bodnar, V.E. Korepanov. *VLF experiment on the Compass 2 satellite and ground-based measurements of the seismoelectromagnetic effects on Kamchatka.* 10th International Seminar "Low-frequency wave processes in space plasma", p. 6.1. Zvenigorod, November, 12–16, 2007.
4. В.Є. Корепанов, Є.І. Крючков, Г.В. Лізунов, А.К. Федоренко, Ф.Л. Дудкін, *Експеримент "Варіант" на супутнику "Січ-1М" – перші результати,* КНІТ.. Т. 13. №. 4. с. 10–17, 2007.
5. S.I. Klimov, V.G. Rodin, L.M. Zelenyi, V.N. Angarov. Development of the method of the creation of micro-satellite (~ 50 kg) platforms for the fundamental and applied research of the earth and near-earth outer space. In R. Sandau, H.-P. Röser, A. Valenzuela (Eds.) Small Satellites for Earth Observation, Dordrecht, Springer Science+Busines Media B.V. pp. 333–343, 2008.
6. V.N. Angarov, G.K. Garipov, V.M. Gotlib, A.V. Gurevich, S.I. Klimov, V.G. Rodin, S.I. Svertilov, L.M. Zelenyi. *Investigation of New Physical Phenomena in the Atmospheric Lightning Discharges. Micro-satellite "Chibis-M"* (in this issue).
7. S.I. Klimov, V.E. Korepanov, Y.V. Lissakov, O.V. Lapshinova, I.V. Sorokin, S.A. Belyaev, G.A. Stanev, K. Georgieva, B. Kirov, M.P. Gough, H.S.C.K. Alleyne, M. Balikhin, J. Lichtenberger, Cs. Ferencz, L. Bodnar, K. Szego, S. Szalai, J. Juchiewicz, H. Rothkaehl, K. Stasiewicz. *"Obstanovka" Experiment Onboard International Space Station for Space Weather Research, Space Weather Workshop: Space Weather Applications Pilot Project,* ESTEC, Noordwijk, The Netherlands, Proceeding, 16–18 December, 2002. www.estec.esa.int/wmwww/wma/spweather/workshops/spw_w4/proceedings/PaperFinal.pdf
8. S.I. Klimov. *Researches of Influence of Space Weather on a Condition of Middle Latitude and Near-Equator Ionosphere*, Tenth Jubilee International Scientific Conference "Contemporary problems of solar-terrestrial influences", Sofia, Bulgaria, pp. 12–13, 20–21 November, 2003.

SEPSAT – A Nanosatellite to Observe Parameters of Space Weather

Jens Rießelmann, Franziska Arlt, Klaus Brieß, Lars Dornburg, Kay Köhler, and Jana Weise

Abstract SEPSAT (Spherical EUV- and Plasma Spectrometer-Satellite) is a nanosatellite that will observe several parameters of space weather, especially solar Extreme Ultraviolet Radiation (EUV), the higher atmosphere and effects of solar fluctuation on parameters of the ionosphere. This helps to improve our knowledge of integrity and accuracy of GNSS signals, the origin and variation of EUV radiation and finally the composition of the solar and upper atmosphere. To detect and observe these parameters a concept for a nanosatellite was designed. SEPSAT is a boom stabilized, $35 \times 35 \times 35$ cm^3 satellite, weighing just 15 kg. Most of the subsystems are based on developments of Technical University of Berlin (TU Berlin), where the focus lies on the design of small satellites and their miniaturized components. The SEPSAT payload is a low-cost spherical EUV- and plasma spectrometer.

1 Introduction

Space weather has a direct effect on our daily routine. Interferences in communication and navigation as well as failures of satellites and many other things can be the result of space weather. It is defined analogue to the terrestrial weather and "describes the conditions in space that affect Earth and its technological systems" [1]. A multitude of physical phenomena like aurora, ionospheric disturbances and scintillations, geomagnetic storms etc. are associated with space weather. The main causes are structures and variability of the corona of the sun. Coronal mass ejections and flairs create shock waves, which are amplifying the solar wind abruptly. The interaction between these phenomena, the Earth and its atmosphere and magnetosphere is not thoroughly understood. Further measurements are necessary to improve existing models. For several parameters it is required to get in-situ data. Thus satellite missions are needed.

J. Rießelmann (✉)
Institute of Aeronautics and Astronautics, Technical University of Berlin, 13507, Berlin, Germany
e-mail: Jens.Riesselmann@ilr.tu-berlin.de

R. Sandau et al. (eds.), *Small Satellite Missions for Earth Observation*,
DOI 10.1007/978-3-642-03501-2_9, © Springer-Verlag Berlin Heidelberg 2010

To save costs, mass and volume of the satellite platform shall be reduced to a minimum. For this purpose university nanosatellites with a mass up to 20 kg are the preferred choice. The short development time, the use of inexpensive commercial off-the-shelf components and piggyback launch opportunities make SEPSAT a low-cost mission for space weather observation.

2 Mission Objectives

SEPSAT is used to observe solar EUV radiation and emissions of the atmosphere as well as parameters of ionospheric plasma. The following scientific objectives will be measured and observed during the mission.

2.1 *Variability of Solar Radiations*

Solar EUV radiation varies on timescales of minutes to years. New measurements can be used to describe the variability of EUV in spectral solutions. This helps to improve empirical models of EUV.

2.2 *Composition and Physics of the Upper Atmosphere in Models*

EUV radiation interacts with neutral higher atmosphere and ionosphere. Furthermore the variation of lower and middle atmosphere affects the higher atmosphere. A separation of these effects is necessary to explain the variability of the higher atmosphere. In addition the composition of the upper atmosphere will be measured in plasma modus. The gained measurement data will help to improve the knowledge of composition, variability and energy yield of upper atmosphere and their EUV parametrization.

2.3 *Composition of Solar Atmosphere and Generation of EUV-Radiation in Lower Chromosphere up to Corona*

Numerous EUV emissions arise in several regions of solar atmosphere, so that each atmospheric layer can be assigned a representative spectral line. By a continuous measurement of their chronological sequences it is possible to gains insight into the variation of solar atmosphere.

2.4 *Propagation of GNSS Signals in Ionosphere*

Empirical models of the ionosphere are necessary to forecast influences on the accuracy and reliability of GNSS signals. New measurements of EUV help to determine the interaction of ionosphere and the radiation more detailed. This yields in an improvement of the accuracy of these models.

3 SEPSAT Design

3.1 Mission

Various altitudes and inclinations has been analysed concerning their influence to the scientific objectives, the sun and umbra times and the contacts with the ground station. For global measurements of ionospheric plasma a polar orbit is required which is an essential constraint. Nadir pointing is also demanded, in order to measure the terrestrial irradiance. The main requirements of the mission are listed in Table 1.

Table 1 Mission requirements

Orbit	500–700 km
Inclination	85–98°
Mission duration	2 years
Pointing	nadir

In accordance with the space debris program of ESA it is necessary to deorbit the satellite not later then 25 years after operational phase. To avoid space debris the upper altitude of the SEPSAT orbit is limited to approximately 700 km.

3.2 Spacecraft Platform

The SEPSAT satellite bus has the shape of a 35 cm cube with a weight of 15 kg. During transport and launch no parts of the satellite will stick out. Many of the technologies being used for SEPSAT are based on proven heritage from the small satellites BeeSat and TUBSAT, developed and built by TU Berlin. Because the nanosatellite has a relatively large volume compared to its mass and a simple, non-specific architecture it can be adapted to other mission goals and a variety of different payloads easily. Figure 1 shows the in-orbit configuration with reaction wheel stabilization in the left and with the deployed gravity boom in the right picture.

3.3 Payload

The spherical EUV- and Plasma Spectrometer (SEPS) is the payload of SEPSAT and was designed by Fraunhofer Institute. As seen in Fig. 2 it consists of a metallic spherical shell and two metallic grids with a high optical transmission, which envelop the sphere concentrically.

All components are mounted isolated. By applying constant or variable voltage to the spheres it is possible to measure the inducted current of the EUV photons or the plasma. The spectral information of the photon flux is proportional to these current.

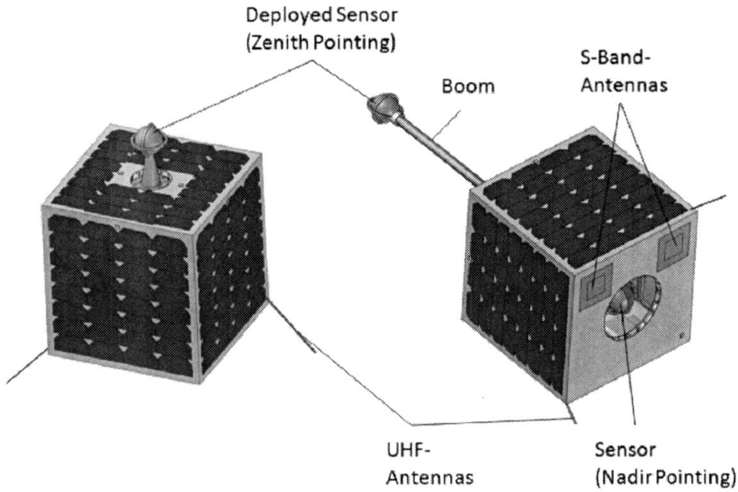

Fig. 1 SEPSAT

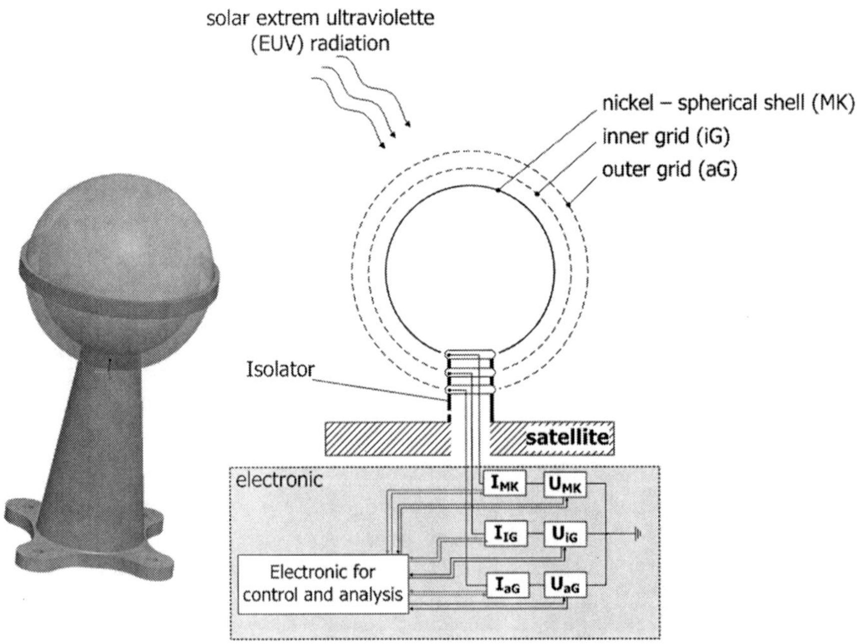

Fig. 2 SEPS-payload [2]

Thus it is possible to measure

- solar EUV-Radiation and its fluctuations,
- power distribution and fluctuation of density of electrons in plasma,
- energetic electron flux (e.g. solar wind) and
- charging of satellites

during the whole mission

To measure all parameters two sensors are foreseen to be mounted diametrically on the satellite. The nadir pointing sensor will be stored inside the satellite behind an opening for wide-field-of-view measurements of earth-emitted radiation. The other one is located outside SEPSAT, attached to the stabilization boom. The requirements of the SEPS-sensors are listed in Table 2.

Table 2 Requirements of SEPS-sensors

Power consumption	7.5 W(@ 28 V)
Pointing	nadir and zenit
Electrical interface	RS422
Data rate	up to 7.2 kbps
Total mass	3,000 g

3.4 Payload Control and On Board Data Handling

The onboard data handling subsystem (OBDH) and the payload control and data handling unit (PLC) will be designed at TU Berlin. To save costs they are built up similar. Both of them use an ARM-9 microcontroller as CPU and are laid-out redundant. The OBDH is intended to manage satellite housekeeping, while the PCL will manage the payload, data storage and transmission to Earth as shown in the block diagrams in Fig. 3.

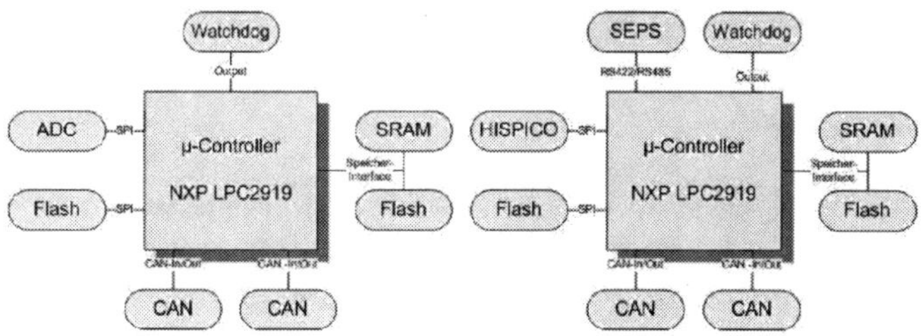

Fig. 3 SEPS – OBDH (*left*) and PLC (*right*)

A redundant CAN bus will connect all subsystems. The block diagram of SEPSAT is shown in Fig. 4.

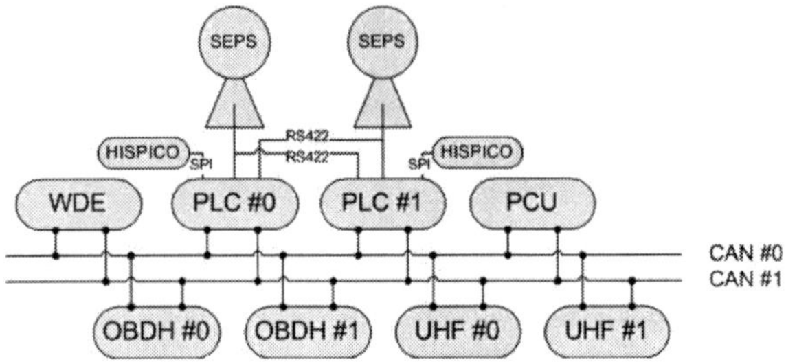

Fig. 4 SEPSAT – block diagram

3.5 Communication

SEPSAT will be capable of half duplex communication for Telemetry and Telecommand (TM/TC) during its mission. For uploading commands and software from Earth and downloading telemetry data and UHF-transceiver with a data rate up to 9.6 kbit/s and a transmission power of 27 dBm with a power consumption of 1.5 W in transmission mode will be used. The TM/TC system is already developed and space qualified for the cubesat BeeSat.

Additionally the S-band transmitter HISPICO (High Integrated S-Band Transmitter for PICO satellites) is intended for transmitting payload data. It is a development by TU Berlin and the German company IQ wireless GmbH and achieves 1 Mbps data rate with a transmission power of 27 dBm and power consumption of 5 W. HISPICO was already qualified before on ground by a dedicated space qualification program. Furthermore in October 2008 the system was successfully verified in space on the sounding rocket REXUS-4 conducted by Eurolaunch. The ground station at TU Berlin will be used for data reception as well as SEPSAT control. It supports both, the reception of a S-Band downlink and the communication to a satellite by an UHF up- and downlink.

3.6 Thermal Control

The operational temperature ranges of SEPSAT components determine the design of the thermal control subsystem. Table 3 shows ranges for the vital components.

Without any thermal control temperatures between –30°C and +30°C would occur inside the satellite which is not sufficient for some of the components. For SEPSAT passive thermal control, either of the whole satellite or of individual components is adequate to keep them within proper temperature ranges. Using only passive elements such as Multi-Layer-Insulation, heat pipes or black paint will minimize the cost and complexity of the satellite. Up to 25 thermal sensors will monitor the temperature of the satellite.

Table 3 Temperature ranges of SEPSAT components

Component	Operational temperature range [°C]
SEPS eject mechanism	–30 to +55
Battery	+10 to +30
HISPICO	–40 to +65
OBDH, PLDH, PCU	–40 to +85
Reaction wheels	–20 to +50

3.7 Attitude Determination and Control System (ADCS)

For the ADCS of SEPSAT three different types of control have been analyzed. This includes the use of three reaction wheels for 3-axis stabilization, the use of a deployable boom for gravity-gradient stabilization and a combination of both systems. The final choice for stabilization is the deployment of an extendable 3-meter-boom, designed by Institute of Aeronautics and Astronautics at TU Berlin. This will help to maintain a constant nadir-pointing toward Earth.

To determine the satellites attitude a set of sun sensors and a 3-axis magnetometer will be implemented.

A magnetorquer is used for a first stabilization of the satellite after deployment and to reorientate the satellite with extended boom if necessary. In the case of yaw-axis-stabilization a second magnetorquer is used for reaction wheel desaturation.

The design of the extendable boom is based on the over 30 years old STEM boom. In this concept a plain copper-beryllium sheet, coiled on a spool, forms a tube when unrolled from the spool. A tube has a higher bending stiffness than a plain sheet. Unlike the STEM boom the new boom contains a return shaft to reduce required space of the stowed package in the satellite. A power spring provides the driving torque for boom extension. According to that the boom can only be extended once. The concept is shown in Fig. 5 and its dimensions listed in Table 4. The SEPS-payload and a balance weight (if necessary) are mounted at the end of the boom. An own control unit including a microcontroller, a CAN-bus interface and voltage regulators allows using as a separate system.

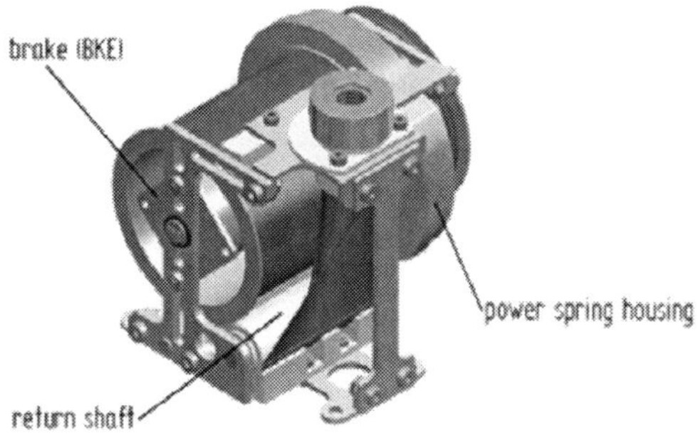

Fig. 5 Stowed boom system

Table 4 Dimensions of the boom system

Pack size	$110 \times 110 \times 120\ \text{mm}^3$
Mass	1.2 kg
Max. boom length	4 m

3.8 Power Subsystem

For a small satellite SEPSAT has a high demand on electrical power. Depending on the ADCS up to 11 W of continuous power is required, of which 7.5 W is needed by the payload. At the same time the spacecraft size is limited. Because of that the use of the latest technology of high-efficient triple junction solar cells will be indispensable. These cells have a high efficiency of 28% and a very low degradation rate over time. SEPSAT will be equipped with 156 solar cells, 28 of them mounted on the zenith pointing side and 32 on each adjacent side. The nadir pointing side carries no solar cells. The solar array output is regulated by a Peak Power Tracker.

In terms of energy storage a battery consisting of five Li-Ion cells, each with a capacity of 20 Wh, provides power during eclipses and supports peak power draws from equipment. The five cells are connected in series resulting in 100 Wh capacity at 18 V discharge voltage. To protect the battery against overcharging, which would cause serious damage, a charge regulator monitors battery charge current.

A redundant Power Control Unit (PCU) regulates the battery charge regulator, the Peak Power Tracker und provides three different bus voltages of 3.3, 5 and 28 V. Thereby a power adaptor for each component shall be avoided. The PCU also records data as current or voltage of different loads and transmits them to the OBDH. The PCU will be developed by the Institute of Aeronautics and Astronautics of the TU Berlin.

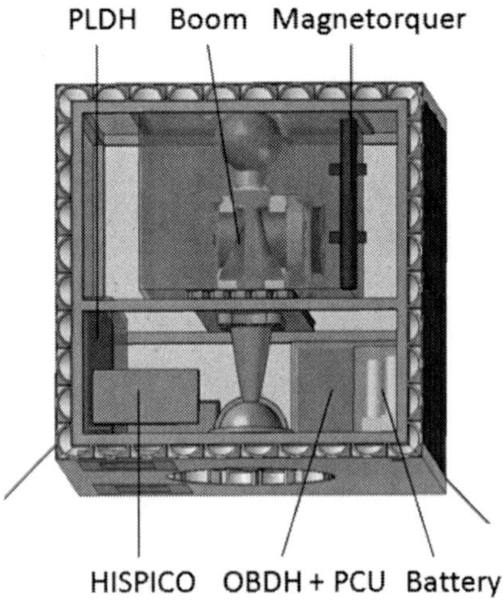

Fig. 6 Spacecraft assembly

3.9 Structure

The primary structure will be a skin frame structure made of aluminum and six sandwich panels, one on each side and with a thickness of 15 mm. The components of the other subsystems are inside attached to the sandwich panels (panel packing option), which makes additional decks expendable. Compared to other satellites this size the structural mass of approx. 6.3 kg is in the lower range. To keep disturbance torques low it is of great importance that the center of mass is within a small margin to the geometric center. The assembly of SEPSAT is shown in Fig. 6.

4 Conclusion

The SEPSAT mission will provide essential data on important space weather parameters and will therefore be of high value for the scientific community. In this study the design concept of a 3-axis stabilized nanosatellite and the necessary payload was presented. The low mass of 15 kg and the small size of the university satellite assure a low-cost mission for the observation of space weather.

References

1. National Academy of Sciences, Space Weather A Research Perspective, 2007.
2. R. Brunner, G. Schmidtke. Spärisches EUV- und Plasmaspektrometer, Final research report, 2007.

Small Satellite Constellations for Measurements of the Near-Earth Space Environment

Aaron Q. Rogers, Larry J. Paxton, and M. Ann Darrin

Abstract A constellation of small satellites is well suited to making measurements of the near-Earth space environment given the spatial and temporal variability that preclude solely independent, localized measurements. In this chapter we discuss some of the factors that must be considered when designing a system for space weather monitoring and the scientific investigation of the near-Earth environment. Some important parameters include: (1) the number flux and energy of energetic particles in the auroral region and/or the radiation belts, (2) the number density temperature and composition of neutrals and ions in the upper atmosphere, (3) the wind speed in the upper atmosphere, (4) the ion drift velocity and/or DC component of the electric field, and (5) small scale fluctuations in the ionosphere that lead to radio scintillation. An integrated solution will be addressed that spans the range of potential science objectives, enabling sensor technologies, triple-cubesats (<5 kg) and nanosatellite (<50 kg) system design, secondary launch options, and the associated data processing and distribution plan. In addition, details related to a JHU/APL effort with a planned 2010 mission schedule to serve the requirements of user communities for near-Earth space environment and space weather data will also be discussed.

1 Introduction

"Space weather" refers to conditions in the space environment that can influence the performance and reliability of space-borne, Earth-based, and planetary technological systems and endanger human life or health. Adverse conditions in the space environment can cause disruption to satellite operations, communication, aviation, navigation, electric power distribution grids, and space travel leading to a variety of socioeconomic losses.

Lack of essential observations from space is currently a major limiting factor in space weather research – particularly the measurement of ion and neutral density

---

A.Q. Rogers (✉)
The Johns Hopkins University Applied Physics Laboratory, Laurel, MD, 20723-6099, USA
e-mail: Aaron.Rogers@jhuapl.edu

R. Sandau et al. (eds.), *Small Satellite Missions for Earth Observation*,
DOI 10.1007/978-3-642-03501-2_10, © Springer-Verlag Berlin Heidelberg 2010

distributions, and neutral winds. Under geomagnetically disturbed conditions, there is currently a very limited capability for accurately predicting the nature and extent of the response of the upper atmosphere. This is because the near-Earth space environment is undersampled. Specifically, the ion and electron properties must be known in order to be able to forecast or "nowcast" the global state of the ionosphere. Radio-frequency (RF) communications and surveillance activities are particularly sensitive to ionospheric conditions, where RF propagation is complicated by ionospheric structure, since large and mesoscale irregularities affect our ability to predict the SNR and receivability of transmissions [1].

Recent advances in sensor and spacecraft technologies now make it feasible to obtain key measurements from low-cost, small satellite missions. A particularly promising aspect of this development is the prospect of obtaining multi-point observations in space that are critical for addressing many outstanding problems in space science and, in particular, space weather.

In this chapter we will address the fundamental physical processes and interactions affecting the Earth's ionosphere and upper atmosphere, as well as those specific parameters and measurements critical to developing a spatially and temporally relevant space weather models. Building upon this scientific foundation, a Discussion is included related to a novel mico-electro-mechanical system (MEMS) based instrument suite suitable for an emerging class of small spacecraft platforms, that is currently under development by JHU/APL. Lastly, details relating to the mission design and current program status are provided.

2 Investigations to Date

One of the principal problems encountered by radio frequency communications users is the occurrence of radio scintillation. Scintillation arises due to turbulence in the ionosphere. The temporal and spatial variability of scintillation varies with local time, season and location. Predicting the occurrence, location, and extent of scintillating regions is a challenge, The main ground-based system now operational is the network of SCINDA (Scintillation Network Decision Aid) ground stations, which has been operated by the United States Air Force Research Laboratory (AFRL) for over a decade. It began as a research system, but soon became quasi-operational. SCINDA has 12 stations now operating, with two more planned as of 2004 [2]. The stations are located between –15 and +15 degrees magnetic latitude; greater coverage is desirable, but placement of stations is limited by geography and political considerations. Data from the stations is used both to describe the immediate state of the ionosphere and to make short-range forecasts.

The Defense Meteorological Satellite Program (DMSP) satellites have provided a large database of plasma density latitudinal profiles at the altitude of the DMSP orbit [3]. This data was used to characterize the long-term behavior of the ionosphere, and ionospheric variability can be seen to correlate with season and solar cycle epoch.

AFRL added a space-based component to SCINDA to form C/NOFS (Communication/Navigation Outage Forecasting System) [4, 5]. The system, which launched in April 2008, is comprised of a single satellite orbiting at an inclination of 13° at an altitude of 375–710 km. Following commissioning, the space-based observations will be combined with ground station data to produce both real-time ionospheric descriptions and forecasts up to 12 h in the future. However, the system will be limited by the orbital gap between observations that is imposed by having just one satellite. An attractive, viable alternative, is a constellation of small satellites known as "cubesats."

3 Cubesats as a Mission Enabler

Earth orbiting small satellites that are capable of conducting valuable experiments have become increasingly useful and efficient. According to the current CubeSat Standard [6], a single unit (U) cubesat is a picosatellite with 10 × 10 × 10 cm dimensions and mass equal or less than 1 kg. To facilitate launch of the satellites as secondary or piggyback payloads on existing missions, accommodation of the satellites utilize the standardized cubesat deployment system: the Poly Picosatellite Orbital Deployer (P-POD) [7]. The P-POD packages cubesats in a linear configuration of either three individual satellites or as a single triple cubesat termed a "3U." By utilizing the entire dispenser provision, a space vehicle can be developed to a slightly larger configuration in terms of size, with a maximum length of approximately 34 cm, and a mass in the 4–5 kg range.

With advances in technology, the potential of these small satellites has dramatically increased enabling a large variety of useful missions. The first Cubesat missions were launched in 2,000, with nearly 40 more since then on ten different manifests (see Fig. 1). Of these, several 3U designs have been developed and successfully operated to date [8], including GeneSat [9] and QuakeSat [10].

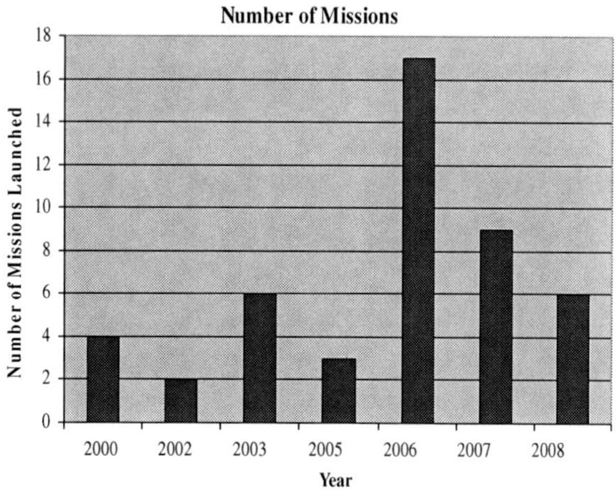

Fig. 1 Number of cubesat missions launched by year

Table 1 Summary of desired measurement parameters and ranges of interest

Measurement parameter	Requirement
Ion number density	10^3–10^7 cm^{-3}
Electron number density	10^3–10^7 cm^{-3}
Ion composition	1–16 amu
Ion temperature	800–4,000 K
Electron temperature	800–4,000 K
Ion drift rate	100 mV/m
Neutral density	10^{11}–10^7 cm^{-3}
Neutral composition	1–32 amu
Neutral temperature	800–2,000 K
Neutral velocity (wind)	10–2,000 m/s
Magnetic field changes	5–30 nT

Many government organizations are now actively engaged in cubesat programs in support of science and technology initiatives. Among them, the National Space Weather Program (NSWP) is a multi-agency Federal program that promotes basic research focused on space weather. A high-level panel assessing the NSWP recommended the creation of a small satellite program to address space weather goals in 2008, with additional opportunities made available recently [11, 12].

We intend to specifically address the need to develop small, low-power, low-cost, robust, sensors that make in situ measurements of the key signatures of space weather in the upper atmosphere (about 110–600 km). The choice of parameters with their associated measurement goals are listed in Table 1, and reflect a technology development plan that builds, in successive steps, on advances in MEMS technology. The use of this technology enables, for the first time, an affordable constellation of nanosats that are tailored to measure just a few key parameters.

In order to adequately specify the upper atmosphere, we have to make a number of measurements that are sufficiently large to adequately sample the spatial and temporal variability of the system. It is more valuable to have more measurements with slightly higher measurement errors than a few, widely dispersed, but precise measurements: certainly MEMS sensors on cubesats offers a solution to this problem.

3.1 Instrument Description

Among the envisioned instrument suite needed to fully capture the dynamics of these effects, JHU/APL is currently developing two of the critical elements: The Wafer Integrated Spectrometer (WISPER) and a Carbon nanotube field emitter.

WISPER, shown in Fig. 2, is designed using advanced micro-fabrication techniques, and was developed as an instrument demonstration capable of making fine

Fig. 2 The Wafer Integrated SPectromtER (WISPER) currently manifested on FalconSat-5

resolution measurements of the kinetic energy spectra and angular distributions of ions in a space environment in the range of 0–2,000 eV with a fixed FOV spanning a region of $\pm 15°$ in azimuth and elevation [13]. The total volume occupied by the WISPER instrument including sensor head array, control electronics and high voltage power supply is 10 cm^3, and packaged to weigh a total of 0.5 kg while consuming 0.7 W of power.

A prototype version of WISPER will be demonstrated during the FalconSat-5 mission, manifested for the STP-S26 launch in February of 2010, where it will assist in providing RF and plasma measurement diagnostics by characterizing the perturbed plasma environment created by an on-board ion source.

The needed measurements require a sensitive instrument capable of obtaining several types of information. While WISPER is capable of obtaining ion densities, composition and energies, it does not provide the same information for neutrals. This can be accomplished via the addition of an ionizing stage in front of the instrument; the neutrals are then ionized and their densities and mass can then be obtained by the instrument.

In order to keep power requirements low, an efficient ionization source based on carbon nanotube (CNT) field emitters [14] will be employed. CNTs are ideal field-emitter materials. They are naturally high-aspect-ratio objects with small-radius ends, which Results in a large electric field enhancement at the Tip, reducing required extraction voltages. Furthermore, in the array geometry under development, they are extremely redundant, as only a small fraction of the CNTs are emitting electrons at any time, affording a very long lifetime for the devices. As shown in Fig. 3, when packaged, the CNT field emitter payload, is 1 cm^3, has a mass less than 100 g, and draws approximately 3 mW of power while operating.

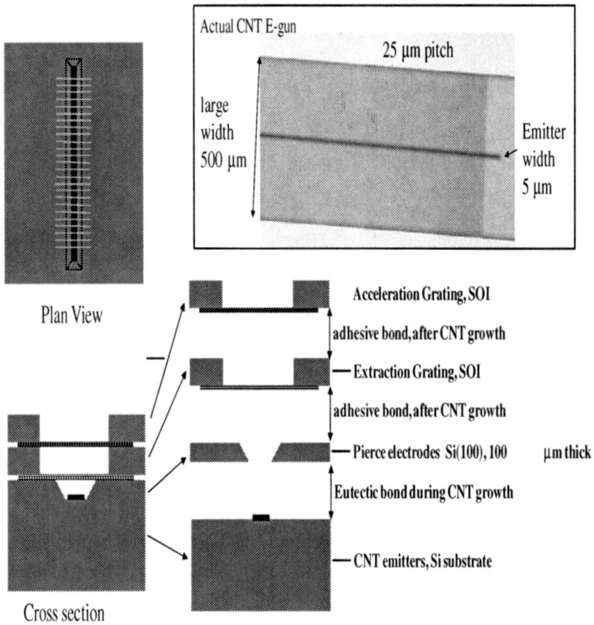

Fig. 3 Carbon Nanotube (CNT) field emitter to be used for local ionization of neutrals

3.2 Spacecraft Design

As shown in Fig. 4, the 3U spacecraft design features a simple, low-power guidance, navigation, and control (GN&C) design that combines gravity-gradient stabilization with an active magnetic control design. Attitude, orbit ephemeris, and precise time knowledge are provided by a sensor suite consisting of a dual-head star tracker, three-axis magnetometer, six sun sensors, and GPS receiver. This sophisticated implementation is required in order to provide the accurate pointing knowledge required for both measurement correlation and control of the sensor aperture relative to the ram-face (oriented about the +Y-axis). For the command and data handling (C&DH) subsystem, a rad-hard 32-bit LEON3FT processor suitable for long missions is employed, which provides scalable processing capability with 1 MB EDAC SRAM. The electrical power subsystem (EPS) utilizes four deployed, canted, two-sided solar arrays. Operated in a peak-power tracking mode with an ample battery, the system can deliver up to 60 W peak power for a portion of the orbit, with provision for 10 W orbit average power available throughout many orbital configurations. Communications with the ground are conducted in the UHF band.

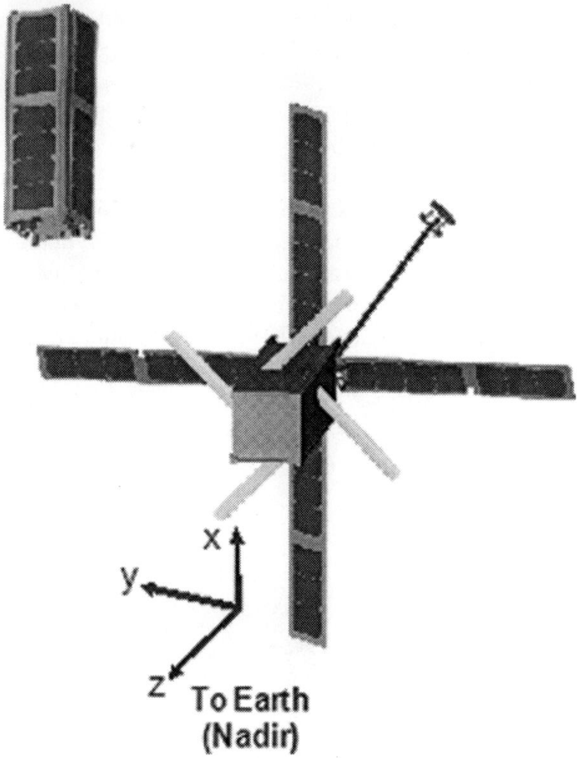

Fig. 4 Stowed and deployed configurations of triple-cubesat space vehicle concept

3.3 Mission Design

The Earth's equatorial ionosphere shows tremendous variability. The peak of the ionosphere can vary in location from 300 km to well over 600 km in height, The spatially averaged number density near the equatorial arcs (two regions of increased electron number density on either side of the magnetic equator that are created by the action of the eastward-directed electric field and the local magnetic field) can vary by more than a factor of ten throughout the day. In addition, just after sunset, the disappearance of the E-region ionosphere can lead to the creation and non-linear growth of plasma instabilities. These regions of plasma instability are evident as radio frequency scintillation and have spatial scales that range from centimeters to kilometers. In order to sample this dynamic and highly variable phenomena, the ideal mission will involve one or more small satellites that orbit near the equator through the F layer region of the atmosphere. This region is the most critical to current missions for other spacecraft, and is also largely uncharacterized.

There are many ways that the discussed mission measurement goals can be achieved. Utilizing the 3U cubesat space vehicle concept, a multi-point system or constellation, could be established to provide the critical spatial and temporal measurement distribution. While ordered constellations have some scientific and logistical advantages, a measure of heterogeneity is preferable for information diversity reasons. By varying the separation velocity or time, altering drag characteristics, or carrying propulsion these small satellites can be dispersed throughout the equatorial region. Similarly, by covering a wide range of local solar times and even altitudes, these new cubesat missions can deliver unique, key observations of the ionosphere that are vital to improving our ability to predict space weather.

Ultimately, the goal of this work is to demonstrate a simple, robust, low-cost architecture that will provide the required measuremens. We find that these cubesats can fly these MEMS-based sensors as part of an inexpensive constellation of satellites, which could be populated by a number of means, including deployment from a multi-P-POD manifest on an EELV Secondary Launch Adapter (ESPA) as shown in Fig. 5, or other emerging secondary opportunities, including those on the Minotaur I and IV, and the Space-X Falcon 1E and 9. The constellation is required because it reduces the revisit time over any one location, enabling the operator to separate out local solar time effects, thereby allowing predictions by extrapolation or incorporation in models.

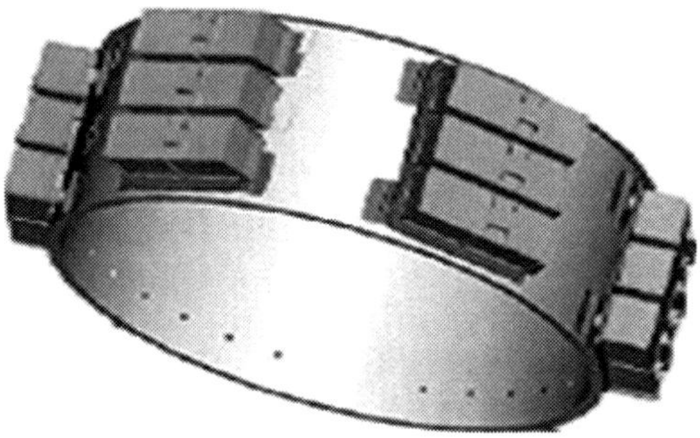

Fig. 5 A conceptual design of multiple P-POD's manifested on an EELV Secondary Launch Adapter (ESPA) that would enable deployment of large-scale cubesat constellations (courtesy of United Launch Alliance)

New approaches that allow the deployment of the elements in the same plane or in orbits that slowly process in local solar time to provide measurements widely separated in longitude. One can readily envision a single launch that could, in principle, lead to a population of space weather sensors that would be as dense as the Iridium constellation. This distribution would enable, for the first time, continuous observations of the global distribution of density, temperature, and/or composition within this region which is literally the gateway to space.

4 Conclusions and Future Work

There is a critical need to improve the temporal and spatial measurement fidelity of our ionosphere to enable our improved knowledge and reaction to space weather effects. Specifically, cubesats offer an affordable mechanism for conducting the necessary multi-point, in situ science necessary to address the needs of the National Space Weather Program. To support the NSWP, JHU/APL is currently aggressively proceeding towards a baseline Preliminary Design Review (PDR) in the late-2009 timeframe, focusing on system design, WISPER and CNT field emitter sensor development, and other associated risk-reduction and technology maturation activities.

References

1. A.Q. Rogers, L.J. Paxton. "Small Satellite Constellations for Space Weather and Space Environment Measurements", International Astronautical Congress, Glasgow, Scotland, September 29th–October 4th, 2008.
2. 1000 Beaujardiere, de La 2004; 998 Rishbeth, Henry 1969.
3. 1009 Huang, C.Y. 2002.
4. C/NOFS details: http://en.wikipedia.org/wiki/C/NOFS
5. O. de La Beaujardière et al. "C/NOFS: A mission to forecast scintillations". Journal of Atmospheric and Solar-Terrestrial Physics, 66, 17 November 2004, 1573–1591.
6. CubeSat Design Specification: http://cubesat.atl.calpoly.edu/media/CDS_rev10A.pdf
7. P-POD Mk. III Interface Control Document: http://cubesat.atl.calpoly.edu/media/P-POD%20Mk%20III%20ICD.pdf
8. Michael's List of Cubesat Satellite Missions: http://mtech.dk/thomsen/space/cubesat.php
9. GeneSat mission website: http://www.crestnrp.org/genesat1/
10. Stanford University QuakeSat website: http://www.quakefinder.com/services/quakesat-ssite/
11. Space Weather Program Strategic Plan and Implementation Plan program website: http://www.nswp.gov
12. CubeSat-based Science Missions for Space Weather and Atmospheric Research solicitation: http://www.nsf.gov/pubs/2009/nsf09523/nsf09523.htm
13. D.M. Wesolek, M.A.G. Darrin, R. Osiander. Wafer Scale Integration Enabling Space Science, IEEE Aerospace Conference, 1–8 March 2008.
14. S. J. Papadakis et al. "Carbon-nanotube field-emitter driven compact frequency-scanning THz source", Proceedings of SPIE, vol. 6549, 65490 N, 2007.

Part III
Distributed Space Systems

Satellite Formation for a Next Generation Gravimetry Mission

Stefano Cesare, Sergio Mottini, Fabio Musso, Manlio Parisch, Gianfranco Sechi, Enrico Canuto, Miguel Aguirre, Bruno Leone, Luca Massotti, and Pierluigi Silvestrin

Abstract The technique called "Low-Low Satellite-to-Satellite Tracking" makes use of two-satellite loose formations for detecting the Earth's gravity field and its space/time variations: the effect of the geopotential shows up as a variation of the inter-satellite distance, which is measured by a suitable metrology. Accelerometers are utilized on each satellite for measuring and separating the effect of the non-gravitational forces. Thales Alenia Space Italia (TAS-I) has studied for the European Space Agency a gravimetric mission of this kind, in which the inter-satellite distance variation is measured by a laser interferometer. The reference mission scenario that has been defined and studied consists of two satellites flying along the same circular orbit at 10 km relative distance and 325 km altitude. The formation control for this mission shall be designed to work in synergy with the drag-free control (necessary for providing quiet operational environment to the accelerometers), to not interfere with the scientific measurement and to minimize the use of the thrusters. Another control system is in charge of maintaining the fine pointing of the interferometer laser beam from one satellite to the other. This chapter summarizes the main results of the studies performed by TAS-I and its team on these subjects 1 2 3.

1 Introduction

GOCE, the first "Earth Explorer Core Mission" of the European Space Agency (ESA) launched on March 17th 2009, during its 20-month nominal mission lifetime will provide the first global *static* gravity field map with homogeneous quality of unprecedented accuracy (1 mGal on gravity anomaly, 1–2 cm on geoid height) and resolution ($\sim$100 km), flying at an altitude between 278 $\div$ 268 km.

Another mission, GRACE (launched on March 17th 2002), during the last 7 years has monitored the *variations in time* of the Earth's gravity field at medium spatial

S. Cesare (✉)
Thales Alenia Space Italia, Strada Antica di Collegno 253, 10146 Turin, Italy
e-mail: stefano.cesare@thalesaleniaspace.com

R. Sandau et al. (eds.), *Small Satellite Missions for Earth Observation*, 125
DOI 10.1007/978-3-642-03501-2_11, © Springer-Verlag Berlin Heidelberg 2010

and temporal resolution (∼500 km, 1 month), flying at an altitude between 500 ÷ 450 km.

Since 2003 ESA has promoted studies to establish the scientific requirements, to identify the most appropriate measurement techniques and start the associated technology developments, and to define the system scenarios for a "Next-Generation Gravimetry Mission" (NGGM). To satisfy the requirements of the geophysical applications (solid Earth science, glaciology, hydrology, oceanography, atmosphere circulation, etc.), this NGGM shall provide the *temporal variations* of the Earth's gravity field over a long time span (>6 years) with *high spatial resolution* (as good as GOCE), with *high time resolution* (weekly or better, to reduce the level of aliasing of the high frequency phenomena found in the time series of the Earth's gravity field variation provided by GRACE), and it shall improve the separability of the observed geophysical signals [4]. A resolution equal to 0.1 mm/year of the geoid height variation rate at spherical harmonic degree $\ell = 200$ (corresponding to a 100 km spatial resolution) was preliminarily identified as the mission performance needed for detecting the finest geophysical phenomena of interest [1].

2 Earth Gravity Measurement Techniques

For "observing" the Earth's gravity field from space, GOCE makes use of the "gradiometry technique": the components of the gravity acceleration gradient (second spatial derivative of the gravitational potential) along the orbit velocity, orbit normal and radial direction are measured using three orthogonal pairs of ultra-sensitive accelerometers (forming the "gradiometer"), spaced by 0.5 m. Since the anomalous gravitation potential of degree ℓ rapidly decreases with the orbit radius r ($U_\ell \propto r^{-(\ell+1)}$), it is necessary to fly at very low altitudes in order to achieve a high spatial resolution: consequently, the mission duration cannot be very long because of the atmosphere drag.

GRACE instead makes use of the so called "Low-Low Satellite-to-Satellite Tracking" (LL-SST) technique, which exploits the satellites themselves as proof masses immersed in the Earth gravity field. In a loose formation of two satellites, the distance variation between their centers of mass (Δd, produced by gravitational and non-gravitational forces together) is measured by a length metrology. The drag accelerations (D_1, D_2) experienced by the two satellites are separately measured by means of accelerometers. From $\Delta \ddot{d}_D = D_1 - D_2$, the non-gravitational component of the distance variation is obtained ($\Delta\ d_D = \iint \Delta \ddot{d}_D$) and subtracted from $\Delta\ d$ to isolate the component ($\Delta\ d_G$) produced by Earth's gravity field only (Fig. 1). Thanks to the separation between the satellites, this "measurement instrument" (which can be regarded as a kind of one-dimensional, very long baseline gradiometer) has a higher sensitivity then a gradiometer embarked on a single satellite. Then a good resolution on the gravity field mapping can be achieved at relatively higher altitudes (>300 km) where, with an affordable amount of propellant, it is possible to achieve a mission lifetime ≥6 years.

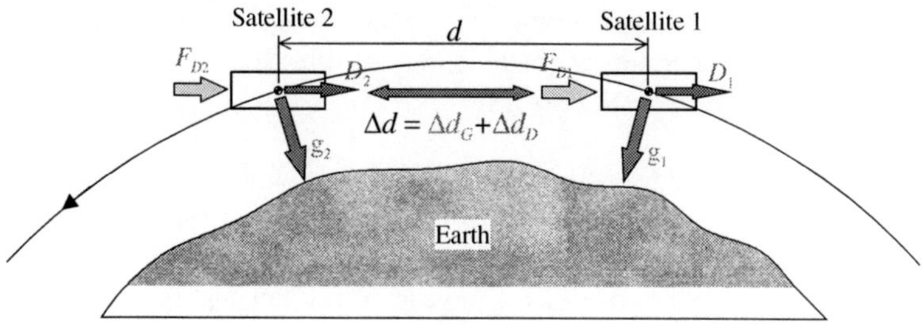

Fig. 1 Principle of the LL-SST technique: the inter-satellite distance variation Δd is measured by a metrology system. The distance variation (Δd_D) produced by the drag forces F_{D1}, F_{D2} is measured by means of accelerometers and removed from Δd so to separate the contribution (Δd_G) produced by the gravity accelerations (g_1, g_2) only

3 Mission Scenario Outline

Given the objectives established for the NGGM (Earth's gravity temporal variation monitoring over a long time span), the LL-SST has been identified as the most appropriate measurement technique. The simplest mission scenario implementing this technique consists of a loose formation of two co-orbiting satellites, as in GRACE. In particular, the main parameters of the mission scenario defined in [1] are (Fig. 2 left):

- Circular orbit with mean spherical altitude $h = 325$ km (actual value to be adapted to the drag force variation experienced along the mission).
- Orbit inclination/node $i = 96.78°$, $\Omega = \text{RA}\odot \pm 90°$ (sun-synchronous, dusk-dawn or dawn-dusk: significant simplification in the satellite design, with still wide latitude range coverage).

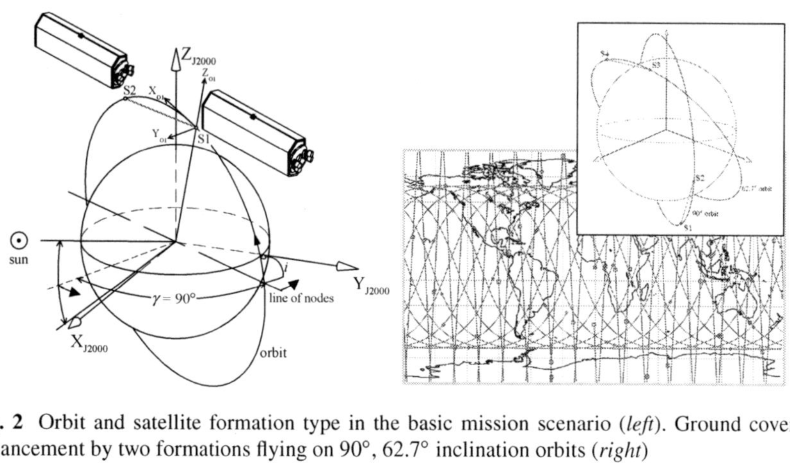

Fig. 2 Orbit and satellite formation type in the basic mission scenario (*left*). Ground coverage enhancement by two formations flying on 90°, 62.7° inclination orbits (*right*)

– In-flight measurement phase duration = 6 years (achievable with ion propulsion and an amount of propellant comparable to that embarked on GOCE).
– Inter-satellite distance $d = 10$ km.

These parameters have been chosen in conjunction with the requirements on the measurement errors of the fundamental observable utilized for the determination of the Earth's gravity field ($\Delta\ d$ and $Delta\ddot{d}_D$, Fig. 3), so to fulfill the requirements on the geoid variation rate (0.1 mm/year) at 100 km spatial resolution. These requirements have been derived using analytical models and numerical simulations of the gravity field determination through the LL-SST technique, with the support of geodesy institutes [1].

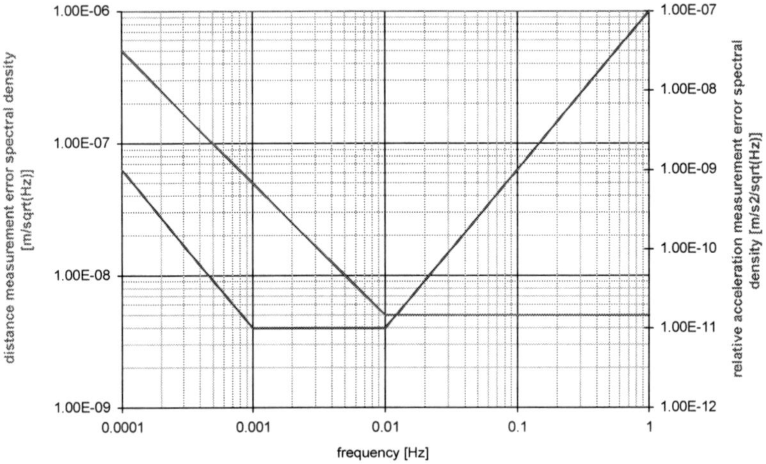

Fig. 3 Requirement on the measurement error spectral density of Δd (*left axis* of the chart) and of $\Delta\ddot{d}_D$ (*right axis* of the chart)

This basic mission scenario still does not meet the objectives of improving the time resolution (i.e. reducing the time interval between two determinations of the Earth gravity model) and the separability of the geophysical signals. But this two-satellite formation can provide the "building block" for the realization of more complex scenarios which can potentially fulfill these objectives. For instance, a dense and uniform coverage of the Earth surface which allows generating gravity field solutions in less than one week can be obtained by flying two pairs of satellites in two circular orbits with altitude 312 km and 90°, 62.7° inclination respectively [5] (Fig. 2 right).

4 The Metrology System

The distance variation measurement requirement provided in Fig. 3 is by far beyond the performance of the K-band ranging system utilized on GRACE, having a noise floor $\cong 10\ \mu\text{m}\Big/\sqrt{\text{Hz}}$. The NGGM shall therefore use a satellite-to-satellite laser tracking.

The metrology system designed for the NGGM [2] is based on a Michelson-type heterodyne laser interferometer. A "measurement laser beam" produced by a frequency-stabilized Nd:YAG source (wavelength = 1,064 nm, output optical power = 750 mW) is emitted by the satellite S1 (*follower*) towards S2 (*leader*). Here it is retro-reflected by a corner-cube optical system towards S1, where it is superimposed to with a "reference laser beam" generated by the same source but frequency shifted relatively to the measurement beam. The measurement beam is amplitude modulated in on-off mode with a period corresponding to the round-trip time to avoid the occurrence of spurious signals and non-linearity caused by the unbalance between the optical power of the strong reference beam and of the weak measurement beam when it comes back from S2.

The distance variation is obtained from the phase variation of the interference signal formed by the measurement and the reference laser beams. A breadboard of such interferometer has been realized and subject to functional and performance tests up to a ~90 m distance. The test results (Fig. 4) prove that the requirement of Fig. 3 can be met.

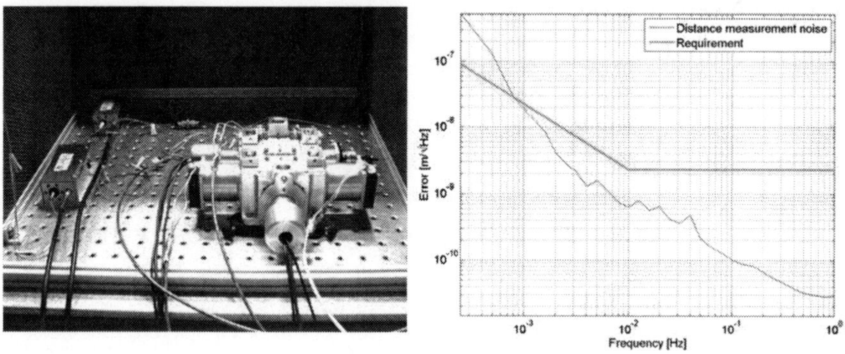

Fig. 4 Breadboard of the laser interferometer for the satellite-to-satellite distance variation measurement (*left*). Result of the breadboard performance test (*right*)

The specified measurement performance of the drag accelerations is within the reach of the accelerometers utilized on GOCE, having a floor noise $<3 \cdot 10^{-12}\text{m}\Big/\left(\text{s}^2\sqrt{\text{Hz}}\right)$ between 1 and 100 mHz. Two accelerometers are used on each satellite, arranged symmetrically about the center of mass (occupied by the laser retro-reflector) on the same optical bench of the laser interferometer (Fig. 5). The bench includes also two sets of three small telescopes utilized for the measurement of the satellites orientation relatively to the laser beam, an ancillary data utilized for determining the relative distance variation. The same optical bench is replicated on both satellites even if the laser is emitted only by satellite S1, because in case of failure of the interferometer on S1, the satellite S2 can take the role of S1 after the position exchange between S1, S2.

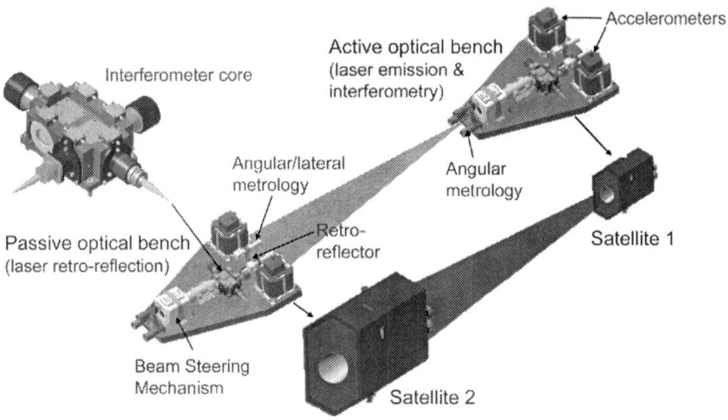

Fig. 5 Layout of the optical bench with the laser metrology and the accelerometers and its accommodation on the satellite

5 The Control System

5.1 Requirements

In order to operate the accelerometers at their maximum performance level, each satellite shall be endowed with a drag-free control system to reduce the level of the linear non-gravitational accelerations below a maximum value of 10^{-6} m/s^2 and a spectral density level of $10^{-8}\mathrm{m}\Big/\left(\mathrm{s}^2\sqrt{\mathrm{Hz}}\right)$ between 1 and 10 mHz, along each axis.

A smooth satellite attitude control (in terms of orientation angles stability, angular rates, and angular accelerations) is also necessary to achieve the specified performance on the distance variation and non-gravitational relative acceleration measurement (Fig. 3). In particular, the variation of the orientation angles of each satellite shall have a spectral density $\leq 10^{-5}\mathrm{rad}\Big/\sqrt{\mathrm{Hz}}$ between 1 and 100 mHz, and the angular accelerations shall be $\leq 10^{-8}\mathrm{rad}\Big/\left(\mathrm{s}^2\sqrt{\mathrm{Hz}}\right)$ between 1 and 10 mHz. The absolute pointing requirements are instead not very stringent (each satellite must be kept Earth pointing, aligned to its Local Orbital Reference Frame – LORF – within 0.5°) thanks to the utilization of a specific mechanism for controlling the orientation of the laser interferometer beam (the Beam Steering Mechanism shown in Fig. 5). The formation control requirements are also not very stringent (*loose* formation flying). In fact, the "control box" inside which the motion of the Satellite 2 relatively to Satellite 1 must be bounded is established by the working range for which the optical metrology system has been sized: $\Delta d_X = \pm 500$ m (with respect to the nominal distance, $d = 10$ km); Δd_Y, $\Delta d_Z = \pm 50$ m along the Y, Z axes of the Satellite 1 LORF (Fig. 6). The challenge of the formation control design for this mission consists in keeping the relative motion within these boundaries without interfering with the scientific measurements (the satellites must be "free" to move under the effect of the gravity field over time scales of 1,000 s; see Fig. 6), without spoiling the drag-free environment (i.e. the formation control accelerations must fulfil the drag-free requirement too) and minimizing the thrusters use (in terms of dynamic range, propellant consumption). For these reasons, the satellite formation cannot be tight.

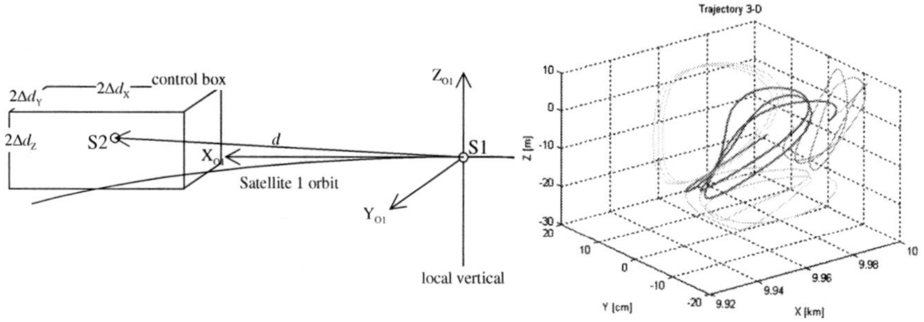

Fig. 6 Definition of the formation control box (*left*). Free, gravitational motion of Satellite 2 in the in the LORF (X_{O1}, Y_{O1}, Z_{O1}) of Satellite 1 (*right*)

5.2 Formation and Drag-Free Control

The formation and drag-free control architecture is shown in Fig. 7. The drag-free is an inner, larger bandwidth (BW) loop of the formation control, with a BW upper frequency of ~0.8 Hz. There is an independent drag-free control system on each satellite, fed by the accelerometer measurements. The formation control is an outer,

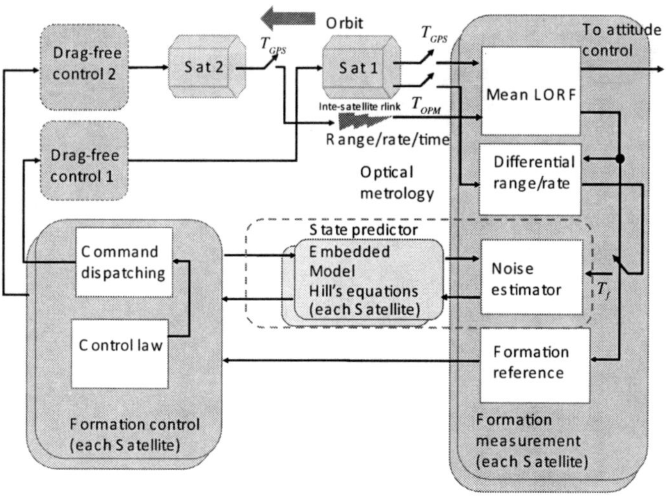

Fig. 7 Formation and drag-free control architecture

narrower BW loop (BW upper frequency <1 mHz), providing also the reference acceleration profile which is tracked by the drag-free control. In fact, to avoid that the drag-free control cancels out the formation control accelerations, the latter ones must be used as "references" in the drag-free loops. The formation control acceleration must be very smooth in order to fulfil the drag-free requirements, and this imposes limits to the satellite relative position and velocity measurement noise (<2 mm/Hz$^{1/2}$, <0.005 mm/s/Hz$^{1/2}$) and to the noise of the thrusters in charge of the formation control (<0.5 μN/Hz$^{1/2}$, including quantization). The relative position and velocity can be in principle measured with the required accuracy and noise by the differential GPS technique. The satellite position measurements shall be used also for estimating the bias of the accelerometers installed on the satellites and feeding the respective drag-free control loops. In fact, the effect of these biases (unless they are identical for all the accelerometers) is to drive apart the two satellites (through the drag-free control loop). Therefore, to avoid wasting propellant for their compensation (through the formation control), they must be estimated and removed from the drag-free commands sent to the thrusters.

A specific simulation tool has been implemented to assess the performance of the formation, drag-free and attitude controls in a realistic mission environment. The time histories of the relative motion between the satellites in open-loop condition (i.e. without formation control) and in closed-loop condition (i.e. under the formation control) are shown in Fig. 8. The residual non gravitational accelerations of the satellite, consequently to the formation control (also shown in Fig. 8), satisfy the drag-free requirements. The maximum force utilized for the formation control is <1 mN.

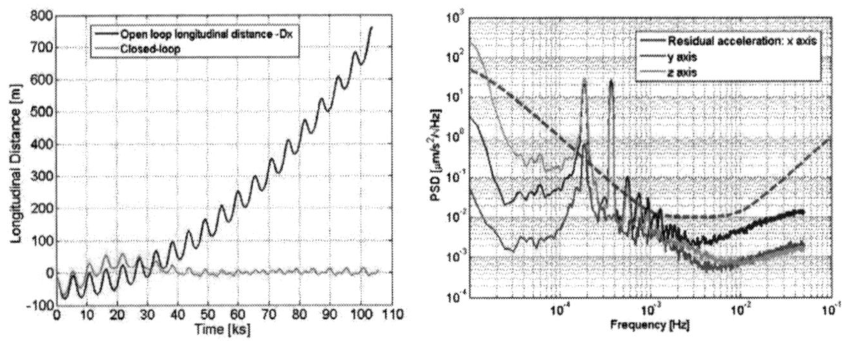

Fig. 8 Open and closed-loop longitudinal relative motion (*left*) and residual non gravitational acceleration of the satellite compared to the drag-free requirement (*right*)

5.3 Attitude Control and Laser Beam Pointing

In the control scheme that has been devised, the pointing of each satellite in its LORF and the pointing of the laser beam emitted by Satellite 1 towards the retro-reflector installed on Satellite 2 are nearly decoupled functions. The satellite pointing is performed using the same set of proportional thrusters in charge of the drag-free and formation control, driven by the measurement of star trackers, hybridized with the angular accelerations provided by the accelerometer pair. Since the Satellite 1 attitude control does not take care of the laser pointing, it can track a "synthetic", smooth LORF estimated from the GPS measurements, rather than the instantaneous position of the Satellite 2. Such a smooth attitude reference is essential to fulfil the very stringent stability requirements on the satellite angles, angular rates and angular accelerations.

The satellite-to-satellite laser pointing is kept by the Beam Steering Mechanism (BSM, an opto-mechanical device constituted by two wedge prisms arranged back-to-back and rotating about a common axis), driven by the measurements of a metrology that, on Satellite 2, senses the lateral displacements of the incoming laser beam by measuring the optical power in three different points of the beam cross section. A feed-forward action is also implemented using the Satellite 1 attitude measurements and the relative position between the two satellites. Breadboards of the BSM and of the lateral displacement metrology have been realized and are under test at TAS-I premises (Fig. 9).

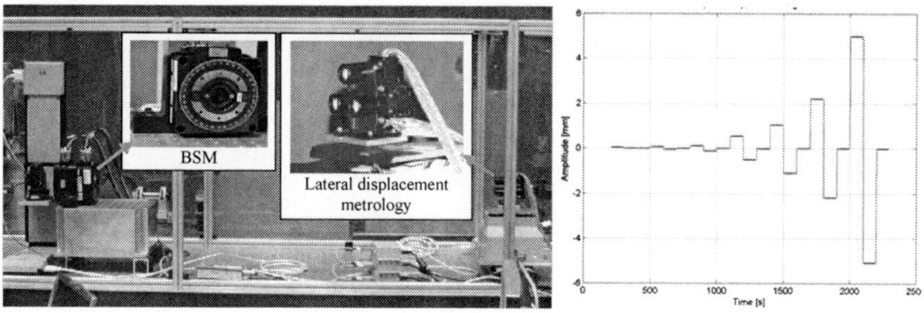

Fig. 9 Breadboards of the BSM and of the lateral displacement metrology (*left*). Lateral displacement steps measured by the lateral displacement metrology (*right*)

References

1. Laser Doppler Interferometry Mission for determination of the Earth's Gravity Field, ESA Contract 18456/04/NL/CP, Final Report, Issue 1, 19 December 2005.
2. Laser Interferometry High Precision Tracking for LEO, ESA Contract No. 0512/06/NL/IA, Final Report, July 2008.
3. System Support to Laser Interferometry Tracking Technology Development for Gravity Field Monitoring, ESA Contract No. 20846/07/NL/FF, Final report, September 2008.
4. R. Koop, R. Rummel. The Future of Satellite Gravimetry, Final Report of the Future Gravity Mission Workshop, ESA/ESTEC, Noordwiik, Netherlands, 12–13 April 2007.
5. P.L. Bender, D.N. Wiese, R.S. Nerem. "A Possible Dual-GRACE Mission with 90 Degree and 63 Degree Inclination Orbits", Proceedings of the 3rd International Symposium on Formation Flying, Missions and Technologies, Noordwijk (NL), April 2008.

EO Small Satellite Missions and Formation Flying

Tony Sephton, Alex Wishart, Helmut Rott, Thomas Nagler, Bernhard Grafmueller, David Hall, Alice Robert, Marline Claessens, Cristina de Negueruela Alemán, Karsten Strauch, and Kristof Gantois

Abstract The aim of the ESA study on concepts for demonstration of advanced techniques and technologies on an EO small mission is to assess ideas for EO missions compatible with implementation on a small satellite such as PROBA, and which may benefit from Formation Flying. "Small Mission" here implies small satellite platforms in the 150–200 kg spacecraft mass range. The output of the study is a definition of various small satellite missions and their required developments, including costs. The study has been led by Astrium Ltd, with support from Astrium SAS, Astrium GmbH, ENVEO, GMV and Verhaert Space. Following initial selection in the first part of the study, 3 candidates have been analysed in more detail as discussed below.

1 Introduction

In Phase 1 of this Definition Study of Concepts for Demonstration of Advanced Techniques and Technologies on an EO Small Mission about 40 ideas for EO missions compatible with implementation on a small satellite and which may benefit from formation flying were identified, assessed and traded off. This has led to the selection of 3 candidates to be studied further in Phase 2: the Topographic Levelling mission (TOPOLEV), a C-band PAssive RAdar Satellite(s) (C-PARAS) mission, and the Aerosol Land Mission (AERL).

The study comprises a definition of the missions and payloads, accommodation requirements, performance and exploitation, along with associated development plan.

2 TOPOLEV Mission

The Topographic Levelling Mission (TOPOLEV) proposed by ENVEO is aimed at mapping surface topography and its temporal change at very high accuracy

T. Sephton (✉)
Astrium Ltd., Gunnels Wood Road, Stevenage, SG1 2AS, England

R. Sandau et al. (eds.), *Small Satellite Missions for Earth Observation*,
DOI 10.1007/978-3-642-03501-2_12, © Springer-Verlag Berlin Heidelberg 2010

(below 1 m in height) and high spatial resolution. The focus of the mission is on low and moderate relief terrain, including water surfaces. The mission concept applies single-pass SAR interferometry with 2 satellites flying in close formation in a cartwheel configuration. The active SAR (Master) operates in Strip-map mode, and the receive-only SAR (Slave) on the second satellite (note that for redundancy purposes, and to maintain the same ballistic coefficient, the second satellite will also have a transmit capability). Ku-band (13.6 GHz) is proposed as the SAR frequency because surface penetration (e.g. in snow, ice, dry soil, etc.) is reduced, and the backscatter signal of water and other surfaces is higher compared to lower radar frequencies. At this frequency a relatively small antenna size can be used.

Higher backscattering at water surfaces and swamps is also the driver for the selection of VV polarisation. Atmospheric effects at Ku-band are higher than at X-band but are still sufficiently small. Signal degradation is expected in areas of heavy rain, which will be excluded in the processing of the DEM's (Fig. 1).

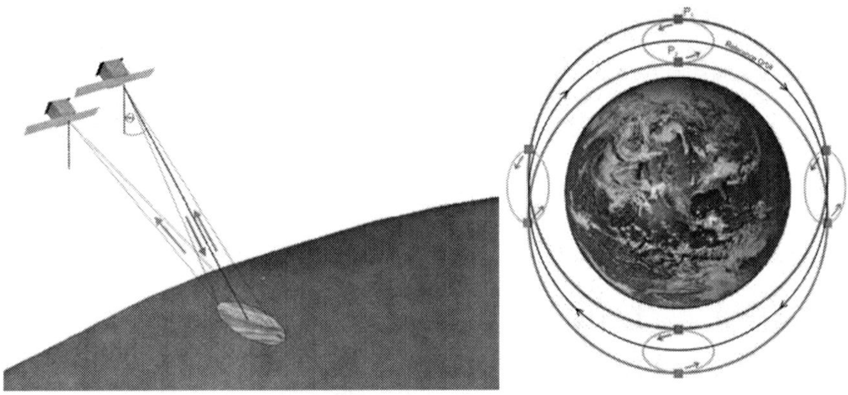

Fig. 1 TOPOLEV mission configuration (*LHS*) and sketch of TOPOLEV Cartwheel orbit configuration with two satellites (*RHS*)

For phase referencing, the two satellites will be equipped with 6 horn antennas each, allowing the transmission and reception of reference signals. The horn antennas will provide near omni-directional coverage, but only the horn that is looking towards the tandem satellite is activated through a switch matrix. In regular intervals, pulses (5–10 pulses per second) dedicated to phase referencing are exchanged, and it is possible to synchronise both phase and PRI timing for offline on-ground processing.

The instruments are based on long and narrow slotted waveguide antennas (5.4 × 0.54 m). To avoid highly complex folding mechanisms a fixed antenna configuration is proposed. This allows for dual launch of the 2 satellites in a single VEGA

Table 1 TOPOLEV payload characteristics

Frequency, polarisation	Ku-band, VV polarisation
Measurement geometry	31° incidence, stripmap, 25 km swath width
Key performance parameters	NESZ ≤ –20 dB
Mass	81 kg (incl. 20% margin)
Power (payload)	922 W peak/128 W mean over orbit (incl. 20% margin)
Antenna dimension	5,400 × 540 × 50 mm (fixed, non-deployable)
Data rate	254 Mbps (compressed)
Data/orbit	55 Gbits (compressed, incl. 20% margin)
Telemetry	X-band

launcher. For the required high-accuracy baseline determination, a dual-frequency GPS receiver will be used. The GPS receiver consists of a receiver box, a GPS antenna and a small external LNA. TerraSAR-X/TANDEM-X experience has shown that the required baseline accuracy of 1 mm can be achieved on-ground by fitting the GPS measurements to a highly complex orbit model (Table 1).

Investigations performed on some typical scenarios indicate the feasibility of achieving the main TOPOLEV mission goal in generating digital elevation models with high accuracy. Based on the SAR design and its sensitivity, the interferometric performance has been studied assuming backscatter properties of bare soil and rock, and taking various error sources (including baseline uncertainty) into account. The 90% point-to-point error was estimated for two proposed DEM products, the DEM-GEODETIC and the DEM-GEOSCIENCE product. It was found that baseline uncertainty is a major contributor to the DEM relative height error, a baseline error of 1 mm increasing the height error by about 0.5 m.

The proposed DEM GEODETIC products have a horizontal resolution of 12 m at 9.6 independent looks. The 90% point-to-point height error is ≤1.55 m, assuming a baseline uncertainty of 1 mm. This height error is well within the DTED-3 specifications. In stable areas, the vertical height error can be further reduced to below 1 m by combining 2 or more interferograms from different over-flights. The DEM GEOSCIENCE product is dedicated to various applications in geosciences including ecology, hydrology, erosion mapping, glaciology, etc. The product has a resolution of 25 m, with 41.4 independent looks. The 90% point-to-point height error is <1 m (in case of a baseline uncertainty of 1 mm). By combining 2 interferograms from different over-flights the vertical relative height error can be reduced to 0.6 m, and with 5 interferograms to 0.4 m (Fig. 2).

Almost all instrument equipment can be derived from systems with flight heritage (e.g. ERS, TerraSAR-X/TanDEM-X), or Sentinel-1 which is currently in Phase C. The only part which needs a new development is the slotted waveguide radar antenna. Here it should be investigated if manufacturing in aluminium would be sufficient (from the thermal stability point of view) or if CFRP technology would be needed.

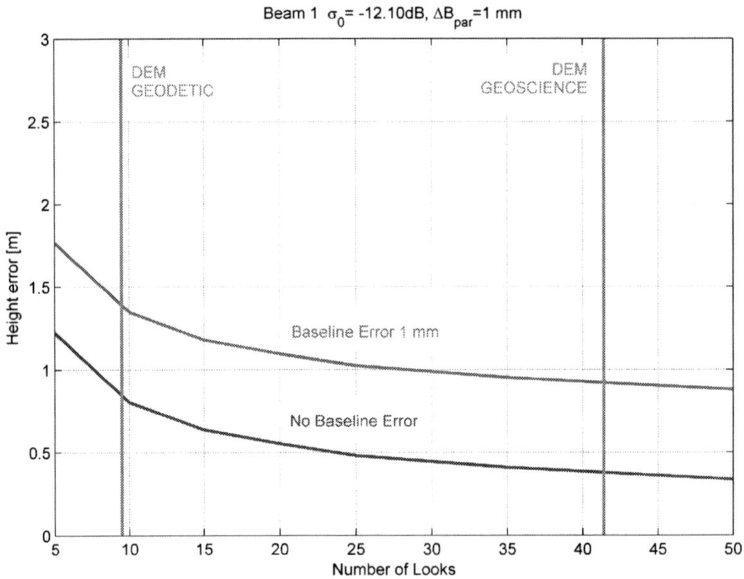

Fig. 2 Height error in dependence on number of looks for bare soil and rock (–95% confidence backscatter signature, at 30° incidence angle, flat terrain)

3 C-PARAS Mission

C-PARAS is another mission for precise topographic mapping, being a free-flying formation of 3 small SAR spacecraft (with radar receiver only) flying in a cartwheel configuration with each other and providing a bistatic SAR observation capability at C-band in formation with Sentinel 1 which provides the scene illumination. Potential benefits of C-PARAS include acquisition of multiple interferograms in a single pass (overcoming problems of phase unwrapping in steep terrains), along-track and multi-baseline interferometry for ocean current and forestry applications, and spatial resolution enhancement (Fig. 3).

The hub SC1 follows the orbit of Sentinel-1 at an along-track separation of ~150 km and the two 'wheel' spacecraft SC2 and SC3 orbit the hub at radii of 120 and 480 m, respectively. The three spacecraft perform periodic propulsive manoeuvres to maintain both the formation and its along-track position with respect to Sentinel-1.

Sentinel-1 has no provision for bistatic operations so, unlike TOPOLEV, there is no direct means of establishing synchronisation between C-PARAS and Sentinel-1. C-PARAS receivers record continuously and ground processing recovers timing from Sentinel-1 datation. Inter-satellite links between the C-PARAS spacecraft are used for the symmetric exchange and recording of their respective oscillator signals to provide phase compensation information for use in ground interferometric product generation.

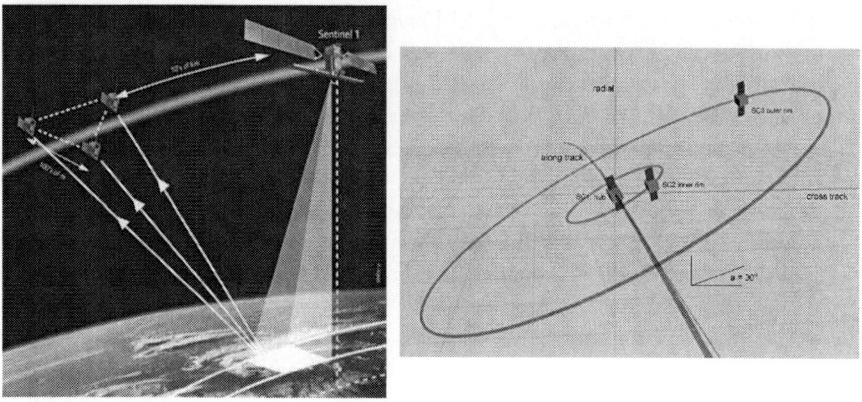

Fig. 3 C-PARAS concept of operation and cartwheel orbit configuration

It is likely that Sentinel-1 will operate primarily in the Interferometric Wide Swath (IWS) mode. This is a ScanSAR TOPS mode in which Sentinel-1 implements electronic steering of both azimuth and elevation beams. In order to provide simplification, mechanical beam steering in *elevation* is proposed in the C-PARAS satellites enabling imaging of the selected sub-swath during any given imaging sequence. In *azimuth*, mechanical steering in yaw is proposed, together with electronic scanning to track the Sentinel 1 beam in its ScanSAR mode. Antenna timing control is derived from the platform GPS.

Satellite configurations suitable for triple launch in VEGA have been identified, and lead to the recommendation of an antenna with dimensions 3.6 × 1.4 m. The performance achievable with such an antenna has been shown to provide a sensitivity ($NE\sigma_0$) of about –23 dB with the addition of Scan-in-Receive (a technique for electronically implementing very fast elevation scanning, in which the receive beam is scanned across the image swath in synch. with the transmit pulse). For high accuracy baseline determination (within 3 mm), a dual-frequency GPS receiver will be used (Table 2).

Table 2 C-PARAS payload characteristics

Frequency, polarisation	C-band, VV polarisation
Measurement geometry	IWS mode, incidence angle 30° for IWS1
Key performance parameters	NESZ ≤ –23 dB (including Scan-in-Receive)
Mass	111.8 kg (incl. 20% margin)
Power (payload)	112 W peak/6.2 W mean over orbit (incl. 20% margin)
Antenna dimension	3,600 × 1,400 mm (fixed, non-deployable)
Data rate	<200 Mbps
Data/orbit	17 Gbits
Telemetry	X-band

Interferometric performance for DEM generation has been studied based on various options of the SAR antenna design and its sensitivity. The 90% point-to-point height error was estimated for 2 DEM products (DEM-BASIC and DEM-PREC) for mean backscatter and 90% probability, assuming backscatter properties of short vegetation and taking various error sources (e.g. baseline uncertainty) into account.

The DEM-BASIC product (30 m resolution and 9 independent looks) has a 90% height error of about 3.5 m for mean backscatter conditions. At 95% probability backscatter the 90% point-to-point height error is below 5 m for most parts of the swath using the 3.6 × 1.4 m antenna with Scan-in-Receive. By assuming a baseline error of about 1 mm, the height error increases by 0.5 m at near range of IWS1 to about 1.5 m at far range of IWS3. The DEM-PREC product (resolution of 90 m and 81 independent looks) provides improved height accuracy at reduced spatial resolution. By neglecting the baseline inaccuracy, the 90% point-to-point error is below 2.1 m for the two antennas with Scan-in-Receive. Increasing the baseline error by 1 mm, the height error increases by about 0.5 m at near range of IWS1, and by about 1 m at far range of IWS3 (Table 3).

Table 3 Relative height error of DEM–BASIC (for 600 m baseline)

	Sensitivity (NEα0)		
Error in baseline	–23dB	–20dB	–17dB
0 mm	2.8 m	3.4 m	4.3 m
1 mm	3.7 m	4.2 m	5.2 m
2 mm	4.5 m	5.1 m	6.1 m

Summary of height errors predicted versus sensitivity baseline error (α^0-mean).

The major development activities required for C-PARAS are associated with the radar antenna. About half of the instrument equipment can be derived from systems with flight heritage (e.g. ERS, TerraSAR-X/TanDEM-X), or Sentinel-1. The parts which need new development are the lightweight patch array antenna, the Scan-in-Receive module and the intra-CP satellite sync link antenna.

4 AERL Mission

The Aerosol Land Mission (AERL) has been proposed by ENVEO to provide improved information on optical depth, type and size of aerosol needed for climate and environmental research and for air pollution monitoring. The AERL mission would enable the study of aerosol distribution and properties at regional scales with high spatial detail by measuring directional reflectance, polarization and aerosol layer height near-simultaneously. For retrieval of aerosol properties over land, it is necessary (by multiple viewing angle measurements of both intensity and polarizations) to separate the atmosphere signal from the surface signal of unknown or poorly known reflection properties. For this mission, it is proposed to have a triple-view imaging spectrometer (LAS-1, for the retrieval of columnar aerosol properties) in combination with a dual-view, narrow-band imaging spectrometer (LAS-2, for retrieving aerosol layer height and stratification), both mounted on a single spacecraft (Fig. 4) (Table 4).

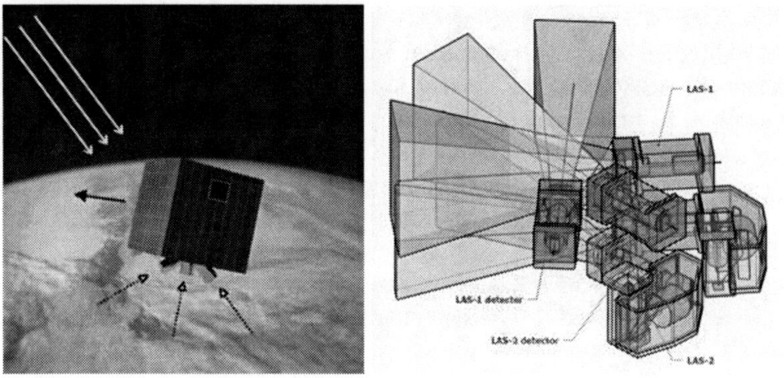

Fig. 4 AERL mission – single spacecraft accommodating LAS-1 and LAS-2

Table 4 AERL payload characteristics

Spectral range	Spectral radiances in 420–910 nm band
Measurement geometry	Push broom, viewing angles nadir and ±50°
Key performance parameters	Spectral resolution, SNR, radiometric stability
Mass	39 kg (incl. 20% margin)
Power	55 W (incl. 20% margin)
Instrument dimension (mm)	400 × 460 × 210 mm (optics), 320 × 480 × 210 mm (electronics)
Data rate	21.8 Mbps (incl. 15% overhead for calibration and packetisation)
Data/orbit	13 Gbits (uncompressed, 10% orbit duty cycle)
Telemetry	X-band

LAS-1 is an imaging spectrometer with 7 channels in the 420–910 nm range. A spectrometer with narrow band-pass filters provides the most efficient concept. The multi-angle viewing is performed by implementing 3 LAS-1 payloads, looking in different directions. LAS-2 is a spectrometer operating in the 757–775 nm Oxygen A2 band. A dispersive spectrometer with grating is the most appropriate concept. The measurements in the Oxygen A2 band require comparatively high spectral resolution (about 1 nm). For the northern hemisphere, the preferred view direction of the tilted sensor is –50° (towards south) in combination with the nadir view because it is more favourable in terms of scattering angles. A calibration unit is also implemented, common to all the payloads, and placed in their common entrance pupil plane. This calibration unit includes a diffuser for sun calibration, a shutter to close the instrument and a free space to perform imaging measurements.

In order to study the feasibility of AERL to retrieve aerosol properties, simulations of the TOA radiance and reflectance have been performed using the

6S radiative transfer model. The simulations for LAS-1 suggest that the sensor will offer good sensitivity for retrieving the Atmospheric column aerosol Optical Thickness (AOT) over land surfaces. The multiple-viewing-angle measurements of intensity and polarization are important for decomposing the aerosol and surface signals and discriminating different aerosol types. This is mainly due to the angular behaviour of scattering, with increased scattering intensities in the forward and backward direction due to Mie scattering and longer path length. The simulations show that radiometric errors $\leq 1\%$ have only a small impact on the retrieval accuracy.

Simulations of radiative transfer in the O_2 band were performed with a generic radiative transfer model that does not fully account for the coupling between scattering and absorption. Nevertheless, the model is able to provide important insights on the measurement technique, suggesting the feasibility to discriminate different aerosol layers by means of LAS-2 measurements. The simulations show significant sensitivity of the forward beam to aerosol layering, whereas the sensitivity of the nadir view is reduced. The sensitivity of the LAS-2 measurements in respect to aerosol layering will depend on the strength of the Earth surface reflectance and the observation geometry vs. the solar beam, and therefore will vary with both season and geographic latitude. In order to fully assess the achievable quality for the retrieval of aerosol layers, simulations need to be performed with an improved RT model that more completely accounts for the interaction between scattering and absorption (Fig. 5).

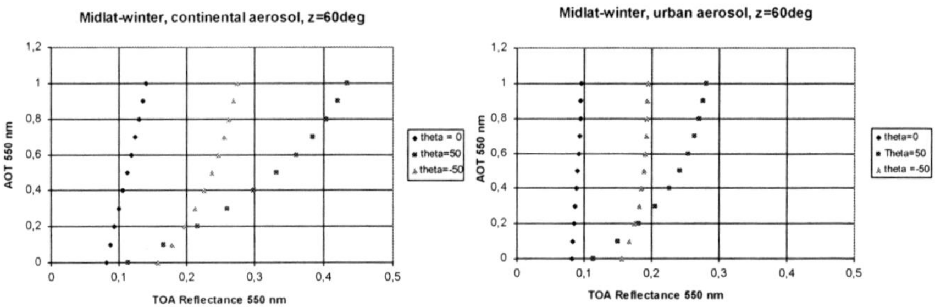

Fig. 5 TOA reflectance at 550 nm at the 3 LAS-1 view directions, versus AOT(550 nm), for two different aerosol types, winter conditions ($\theta s = 60°$)

The main items of hardware requiring further development for AERL are the LAS-1 spectral filters, LAS-1 polarization device, CMOS detectors, and calibration diffuser technology.

5 Spacecraft Design

The three missions are designed for PROBA-class small spacecraft, with conventional platform equipments and structure as far as possible to minimise development costs and risk. AERL is entirely conventional, accommodated in a standard 1 m^3 bus. The driver in the case of TOPOLEV and C-PARAS is accommodation of the relatively large, non-deployable SAR antenna which is fixed to the body of the spacecraft. The large antenna requires spacecraft structural stiffness qualification for a VEGA launch.

All three missions require propulsion capability for initial orbit acquisition (launcher dispersion), orbit maintenance, formation control (TOPOLEV and C-PARAS) and end of life de-orbit. An HPGP chemical propulsion system using 1 N thrusters is proposed as the baseline. Each mission requires ~10 kg of propellant. For the two formation flying missions, an additional cold gas propulsion system with 10 mN thrusters is an option to provide finer control capability.

The ACS (Attitude Control System) uses off-the-shelf equipment wherever possible to minimise cost and development schedule. The control law is nadir pointing, 3-axis stabilised. The Relative Pointing Error (RPE) requirement is 20 arsec (~5 m°), and the Absolute Pointing Error (APE) requirement is 100 arsec (~0.028°). The APE and RPE are important factors in the accurate calculation of the baseline between the SAR antenna phase centres, as the actual measurements are made between the phase centres of the GPS antennas.

All three mission baselines have a high data rate X-band TM subsystem for the payload telemetry. This is mandated for the SAR missions because of the large volume of data (tens of Gbits) generated in a radar data take, and the limited viewing opportunity with the ground station (baseline ground stations are Redu and Svalbard).

Finally, all three missions use body-mounted solar array panels. The power requirements of AERL and C-PARAS are relatively modest (C-PARAS has a purely passive receiver). TOPOLEV has relatively high power but this has been minimised by trading off against the antenna length. Li-Ion batteries are used. The radar orbits are dawn-dusk so eclipse periods are short.

6 Launcher and Orbit

The missions are designed to use the VEGA launch vehicle. AERL is a small, single spacecraft which will fit on the VESPA adapter as a secondary payload for a "piggyback" launch. TOPOLEV and C-PARAS are dual and triple spacecraft launches, respectively, and in both cases the large size of their fixed (non-deployable) SAR antenna means that the complete VEGA fairing volume is required to accommodate the spacecraft. The three missions all employ sun synchronous orbits at altitudes between 650 and 700 km, and inclinations of approximately 98°.

7 Summary

This study has identified 3 EO candidates which can be accommodated on small satellite platforms and therefore benefit from heritage such as PROBA. The TOPOLEV mission presents an excellent example of an EO mission requiring very closely coordinated formation flying. The C-PARAS mission requires no active instrument as it takes advantage of formation flying with Sentinel 1. The AERL mission addresses the urgent need for improved information on optical depth (AOD) and properties of aerosols over land surfaces at a regional scale.

Acknowledgements Grateful acknowledgement is given to ESA for funding this study under Contract number 20395/06/NL/JA.

Relative Trajectory Design for Bistatic SAR Missions

Marco D'Errico and Giancarmine Fasano

Abstract Trajectory design for formation-based bistatic SAR missions is presented for both close (cross-track interferometry) and large (radargrammetry) formations. Trajectory design process is simplified by using analytical relative motion models that include Earth oblateness perturbations and optimize mission performance. In addition, proposed design reduces trajectory establishment and maintenance efforts. examples are briefly summarized where best relative trajectory is derived starting from application models.

1 Introduction

Spaceborne bistatic synthetic aperture radar (BSAR) has a positive impact on Earth observation applications thanks to both potential development of novel techniques and improvements of already assessed procedures (contemporaneous data acquisition [1]). Processing algorithms strongly depend on satellite distance (baseline). Close formations enable cross-track Interferometric SAR (INSAR), i.e. coherent processing accounting for phase information to derive digital elevation models (DEMs) [2]. Large formations allow for applications based on non coherent processing of received radar echoes [3].

In particular, INSAR allows to build DEMs provided that effective baseline (i.e. projection of baseline vector onto the normal to viewing direction) is selected within a minimum (adequate sensitivity) and a maximum (to limit decorrelation) values. INSAR error budgets are deeply discussed in [4–6] for both extended and point targets. Large baseline BSAR also allows for DEM generation with standard [7] or innovative radargrammetric techniques [8] which do not allow for phase processing.

M. D'Errico (✉)
Department of Aerospace and Mechanical Engineering, Second University of Naples, Via Roma 29, 81131 Aversa (CE) , Italy
e-mail: marco.derrico@unina2.it

R. Sandau et al. (eds.), *Small Satellite Missions for Earth Observation*,
DOI 10.1007/978-3-642-03501-2_13, © Springer-Verlag Berlin Heidelberg 2010

Despite great potential applications, Shuttle-based SRTM mission is the only one which has implemented BSAR, with Tandem-X [9] being the next. Nevertheless, bistatic SAR missions can be studied in conjunction with any of the high-performance SAR satellites currently flying/planned, with the advantage of the potential integration of receiving-only (Rx) radars to enable small satellite utilization. To study such missions the authors have developed analytical relative motion models based on orbital parameter differences and including J_2 secular effects [10]. This chapter presents application of the model to design Rx orbit for both INSAR and radargrammetric applications.

2 Cross-Track INSAR Formation Design

The main geometric parameter to be optimized in cross-track INSAR formation design is the effective baseline (Fig. 1). Although larger baselines allow better height measurement sensitivity, spatial decorrelation determines larger phase measurement noise, and moreover height of ambiguity decreases, which is a drawback especially in areas with steep slopes [4–6]. Thus, an "optimal" value of the effective baseline can be usually identified, and orbit design is aimed at achieving this baseline with small changes at different latitudes along the orbit.

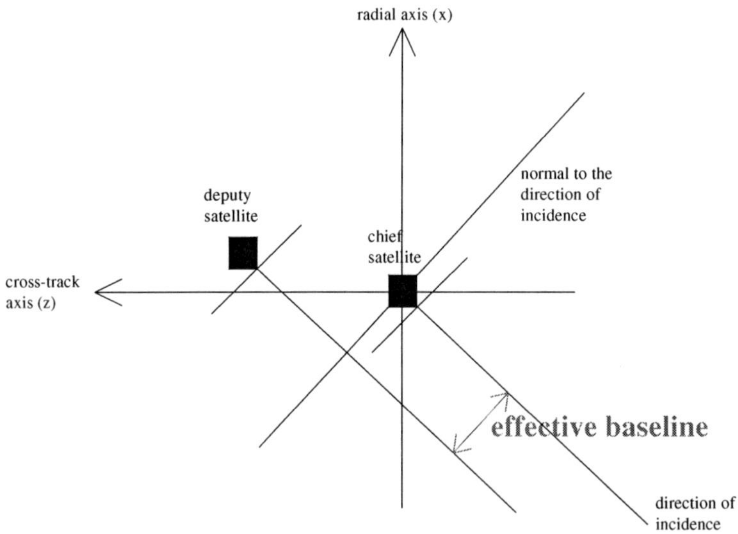

Fig. 1 Effective baseline

In particular, a performance quantitative analysis can be performed using height accuracy models in literature [6] and setting some typical radar parameters (range resolution 5 m, off nadir angle 33.5°, SNR 15 dB, 4 looks). Results are reported in Fig. 2 considering different altitudes and frequency bands. It is worth noting that only phase noise is considered in height accuracy and height of ambiguity is evaluated over a 2π phase shift. Uncertainties in slant range, baseline components, and attitude also affect height accuracy but are not considered here because they depend on the navigation system. Nevertheless, it must be noted that phase noise effect is the main source of error if a mm-level error on baseline determination is assumed, which is considered to be feasible in post processing analyses.

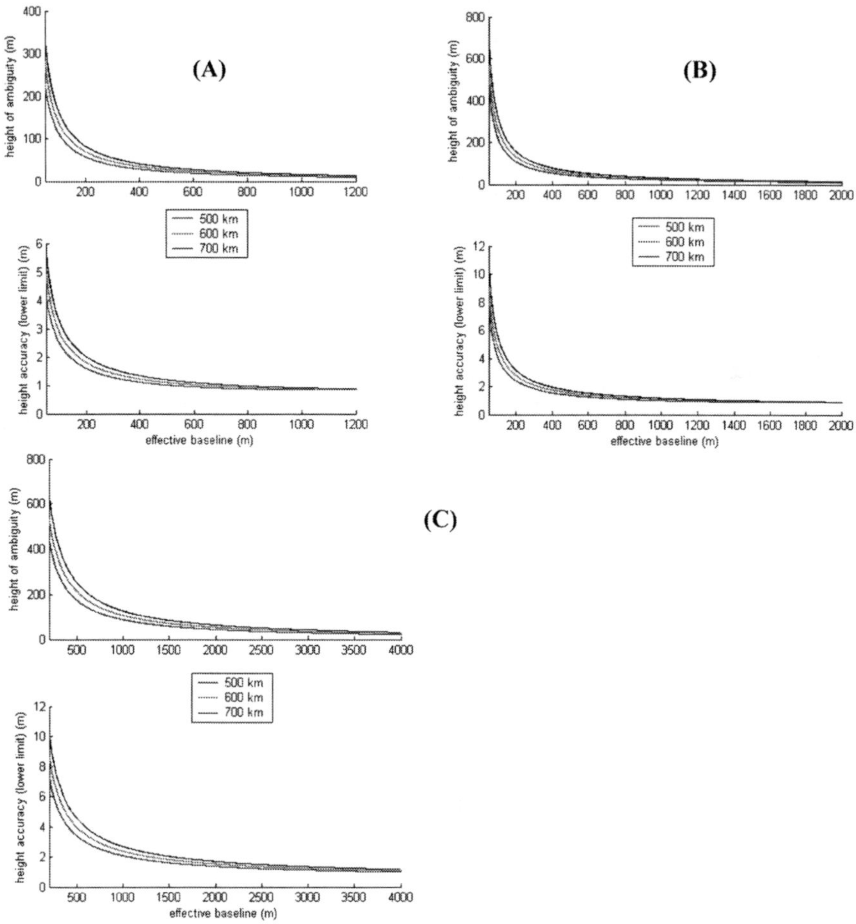

Fig. 2 Interferometric capabilities as a function of effective baseline, X- (**A**), C- (**B**), and L- (**C**) bands

For the considered radar parameters, the best achievable height accuracy is less than 1 m, however this condition leads to an unacceptably small height of ambiguity. Thus, it is supposed that the latter must not be less than 20 m. At the same time, it is supposed that height accuracy must not be degraded substantially with respect to 1 m. On the basis of these considerations short ranges for effective baseline are identified to meet required height accuracy and height of ambiguity ranges (Table 1). Thus, if such interferometric performance are to be achieved, formation tuning must

Table 1 Interferometric performance and effective baseline ranges

Band	Effective baseline (m)	Height accuracy (m)	Height of ambiguity (m)
X	400÷700	0.97÷1.24	21.5÷36
C	700÷1,300	0.99÷1.3	21.5÷39.5
L	2,800÷5,000	1÷1.3	22÷39.5

design a relative trajectory which guarantees that the effective baseline varies within the required range for most part of the orbit and on the desired latitudes.

To perform formation design, the 1st order model for relative trajectories presented in [10] is used and herein reported. In particular, subscript D identifies the deputy satellite parameters, xyz represents the Deputy coordinates in the chief Hill's reference frame (origin in the chief position, x-axis in the radial direction, z-axis along the angular momentum vector, and y-axis lying in the orbit plane), and δ identifies parameter variation between deputy and chief. M is the mean anomaly, u the argument of latitude, and [a, e, i, Ω, ω] the classical orbital parameters. Mean anomaly, argument of perigee, and argument of latitude rates are related to Keplerian motion and J_2 secular perturbation.

$$\begin{bmatrix} x \\ y \\ z \end{bmatrix} \cong a \begin{bmatrix} \frac{\delta a}{a} - \delta e \cos(M_0 + \delta M_0 + \dot{M}_D t) + 2e \sin\left(\frac{\delta M_0}{2}\right) \sin\left(M_0 + \dot{M}_D t + \frac{\delta M_0}{2}\right) \\ 2\delta e \sin(M_0 + \delta M_0 + \dot{M}_D t) + 4e \sin\frac{\delta M_0}{2} \cos\left(M_0 + \frac{\delta M_0}{2} + \dot{M}_D t\right) \\ +\delta(\omega_0 + M_0) + \delta\Omega_0 \cos i + t\left(\delta\dot{u} + \delta\dot{\Omega}\cos i\right) \\ -(\delta\Omega_0 + \delta\dot{\Omega}\, t)\sin i \cos(\omega_{D0} + M_{D0} + \dot{u}_D t) + \delta\, i \sin(\omega_{D0} + M_{D0} + \dot{u}_D t) \end{bmatrix} \quad (1)$$

It is worth noting that to achieve the desired effective baseline range, in-plane radial motion and out-of-plane oscillation must be combined. To this end, since a difference in inclination is to be avoided for stability reasons (as well as δa), it derives that in plane oscillation must be coupled with a difference in right ascension of the ascending node. Two alternatives are possible: the first one consists in using differences in mean anomaly and in argument of perigee, the second one builds on a difference in eccentricity. Considering frozen orbits in order to minimize perigee precession, it comes out that in plane oscillation is to be achieved by a slight difference in eccentricity. The resulting cross-track plane trajectory is roughly an ellipse whose principal directions coincide with the coordinate axes and whose semi-major axes depend on δΩ and δe (Fig. 3). Since the effective baseline affects height accuracy and height of ambiguity, the latter can be selected as the drivers to determine δΩ and δe.

Baseline range amplitude is proportional to wavelength. In order to select, for each band, values for δΩ and δe which maximize the orbit fraction where baseline range is satisfied, it suffices to note that effective baseline expression, given by $|x \sin\vartheta - z\cos\vartheta|$($\vartheta$ is the off nadir angle) seen as a function of time, in the considered case is a sinusoid with amplitude given by $a\sqrt{\delta e^2 \sin^2\vartheta + \left(\delta\Omega \sin i_{ref} \cos\vartheta\right)^2}$

Fig. 3 Radial/Cross-track ellipse

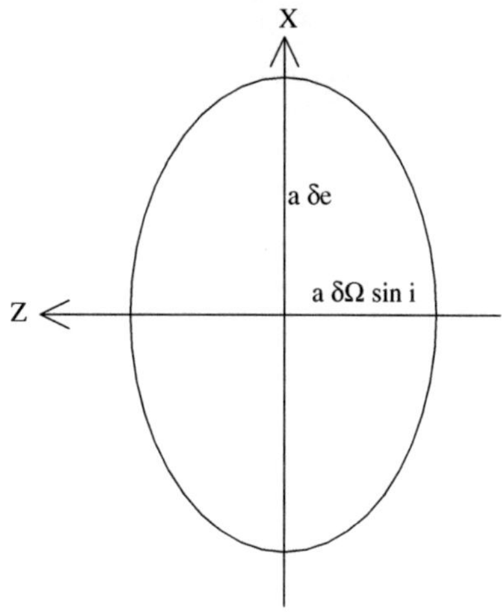

while the phase is connected to the latitude where maximum effective baseline is achieved. Thus, imposing that amplitude equals the maximum baseline limit, and as a second condition that semi-axes ratio equals off nadir angle tangent (maximum effective baseline achieved at intermediate latitudes), one obtains values in the three cases shown in Table 2. As in the case of effective baselines, it can be noted that required differences in Ω and eccentricity are roughly proportional to wavelength.

Table 2 "Optimal" formations for the three considered frequency bands

Band	$\delta\Omega$ (°)	δe
X	±0.0049	$\pm 1.28 \cdot 10^{-4}$
C	±0.0091	$\pm 2.38 \cdot 10^{-4}$
L	±0.035	$\pm 9.15 \cdot 10^{-4}$

In order to show the potential of these orbital configurations, the X-band study case is analyzed, assuming COSMO/SkyMed as chief satellite and computing the interferometric performance as function of latitude (Fig. 4) to verify that almost the same interferometric performance are achieved. Half orbit is characterized by small percentage changes of effective baseline, whereas smaller baseline value and fast baseline variation occur in the other half orbit. From the system point of view, it is important to analyze stability and collision risk since the considered formation is not J_2 invariant, although δe weakly affects J_2 effects (after 100 nodal periods $\delta\Omega = -3.8 \cdot 10^{-6\circ}$, for $\delta\omega = 1.26 \cdot 10^{-5\circ}$, and $\delta M = 9.8 \cdot 10^{-6\circ}$). Line of apsides precession poses a major problem to collision avoidance. In fact, if on the one hand a $\delta\Omega$–δe formation has not intersecting orbits, on the other hand perigees rotation deforms the xz trajectory reducing minimum distance between the two platforms. Time evolution of xz trajectory is shown in Fig. 5.

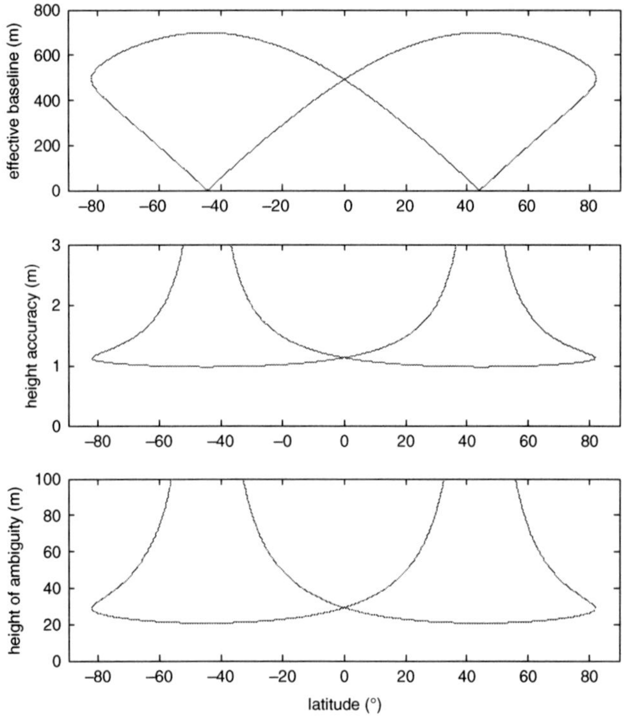

Fig. 4 Interferometric performance as a function of time, X-band, COSMO/SkyMed as chief satellite

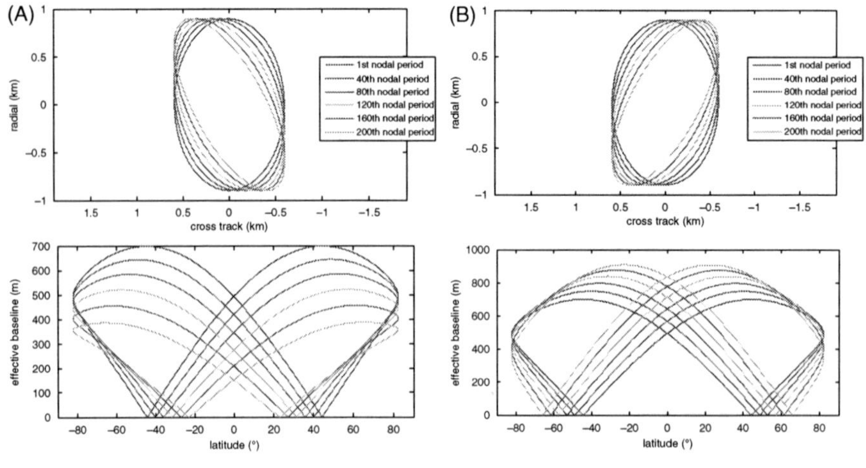

Fig. 5 Time evolution of $\delta\Omega$–δe formation with $\delta\Omega = 0.0049°$ and $\delta e = 1.28 \cdot 10^{-4}$ (**A**) or $-1.28 \cdot 10^{-4}$ (**B**)

3 Radargrammetric BSAR Formation Design

DEM generation has been also explored by means of radargrammetric techniques, and recently novel ideas and new error budgets have been proposed. In the following radargrammetric performance are evaluated as proposed in [8], where a large distance in the cross-track direction is required. In addition, the authors proposed new bistatic configurations in [11] and proposed a 2nd order relative motion model for large formations [10].

If a relatively large bistatic angle is needed among platforms and the initial configuration is a close formation devoted to INSAR, a feasible solution from the orbital mechanics point of view is to generate a separation in the orbital planes by properly exploiting differential J_2 nodal precession. Differential nodal precession can be activated by means of a slight difference in inclination which does not have a significant impact on observation geometry. Accurate control of semi-major axes and eccentricity allows a fine tuning of along-track drift so as to minimize propellant expense and attitude/pointing requirements. The potential of this drifting large baseline formation is showed in Fig. 6, where best height accuracy achievable along the orbit is plotted as a function of time considering $\delta i = 0.1°$ as out-of-plane drift generator. Figure 7 shows the relative distance vector for 6 orbits at different times during drifting.

The basic principle in such trajectory is the establishment of $\delta\Omega$ drift by means of δi, and the control of the desired along-track shift at nodal crossing by means of δa. At maneuver start (close satellites, 1st order model) we get:

$$\begin{bmatrix} x \\ y \\ z \end{bmatrix} \cong a \begin{bmatrix} \frac{\delta a}{a} - \delta e\, cos(\mathrm{M}_0 + \dot{\mathrm{M}}_D t) + 2e\, sin\left(\frac{\delta M_0}{2}\right)\, sin\left(\mathrm{M}_0 + \dot{\mathrm{M}}_D t\right) \\ 2\delta e\, sin(\mathrm{M}_0 + \dot{\mathrm{M}}_D t) + 4e\, sin\frac{\delta M_0}{2}\, cos\left(\mathrm{M}_0 + \dot{\mathrm{M}}_D t\right) \\ +\delta\Omega_0\, cosi + t\left(\delta\dot{u} + \delta\dot{\Omega}\, cosi\right) \\ -(\delta\Omega_0 + \delta\dot{\Omega}\, t)\, sini\, cos(\omega_{D0} + \mathrm{M}_{D0} + \dot{\mathrm{u}}_D t) + \delta\, i\, sin(\omega_{D0} + \mathrm{M}_{D0} + \dot{\mathrm{u}}_D t) \end{bmatrix} \tag{2}$$

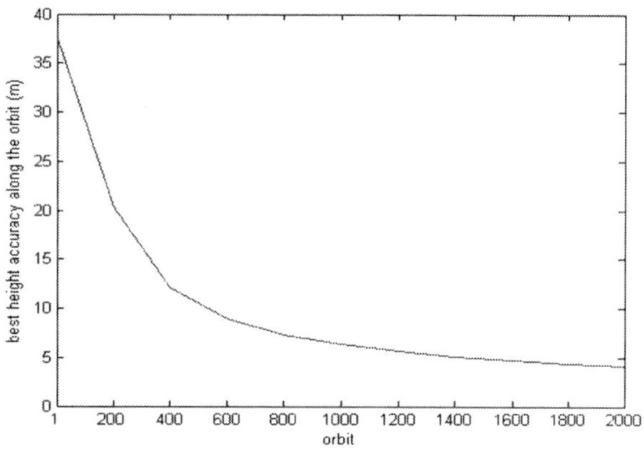

Fig. 6 Radargrammetric accuracy as a function of time for slowly drifting bistatic formation

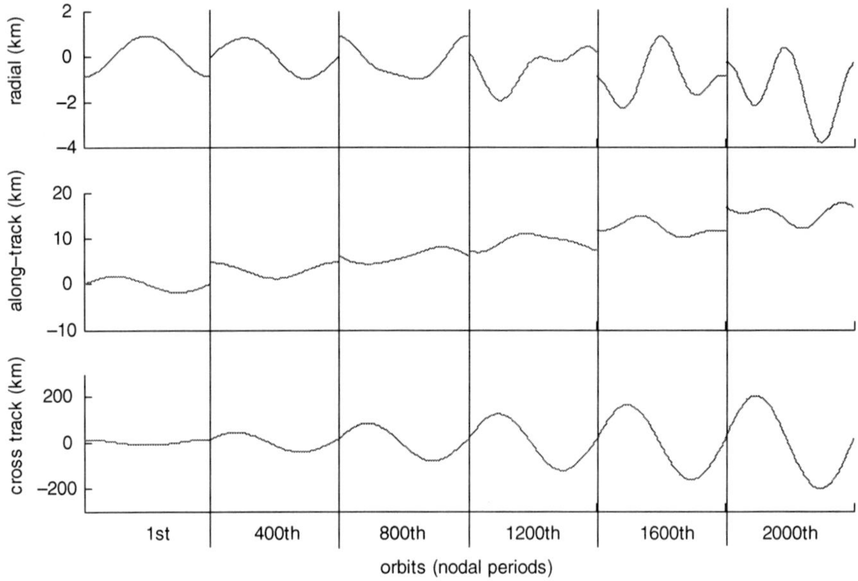

Fig. 7 Relative satellite distance for slowly drifting bistatic formation

Therefore, it must be noted that with establishment of δi the cross-track plane baseline can go to zero potentially creating a collision risk since the two orbits intersect and satellite separation is only based on along-track baseline control (Fig. 8). Such risk can be mitigated by applying a very small δi to let δΩ increase for some time before going to the operational δi. For example we can select:

δi of 0.005° (δe $= \pm 1.28 \cdot 10^{-4}$ still stands): min distance in cross-track plane is slightly less than 500 m (Fig. 9). In 50 orbits δΩ increases from 0.0049° to 0.007°.

δi increased to 0.01: minimum distance in cross-track plane is about 600 m (Fig. 9).

δi to 0.1°, δa = 9.7m: drifting formation. After 1500 orbits δΩ ≈ 1.27° (Fig. 10).

Drifting bistatic formation stability in along-track is evident in Fig. 10b, which is a very good results because at each nodal passing the deputy remains in the vicinity of the chief radar elevation angle allowing for a feasible pointing strategy and SNR. It is worth noting that a certain along-track drift is needed to ensure that for increasing nodal separation the passive satellite lies near the active radar elevation plane at equator passing.

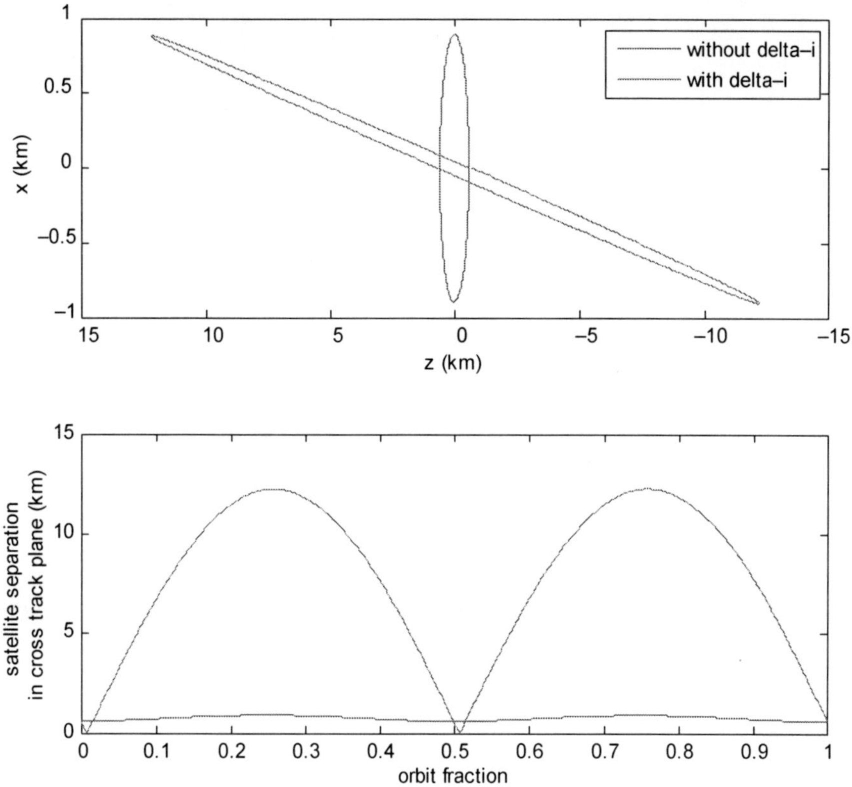

Fig. 8 Effect on relative motion of $\delta i = 0.1°$

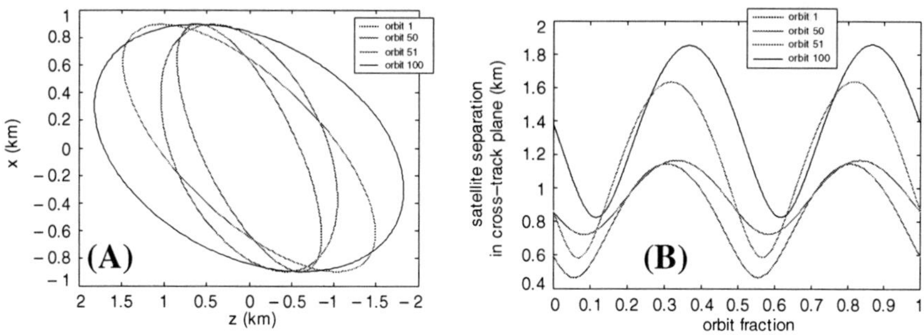

Fig. 9 Cross-track trajectory (**A**) and satellite cross-track plane separation (**B**) during phases 1 and 2

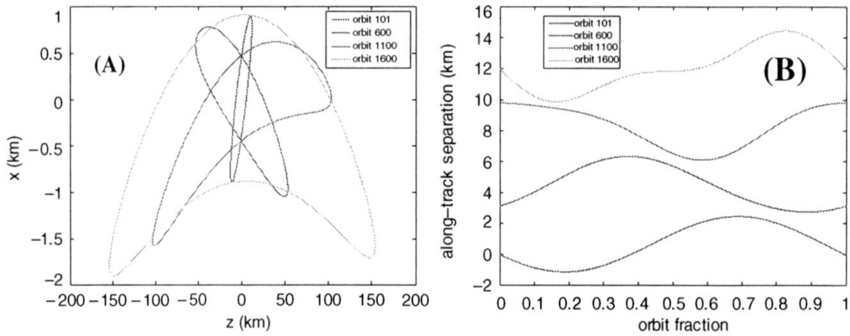

Fig. 10 Cross-track trajectory (**A**) and along track separation (**B**) during phase 3

4 Conclusions

Formation design based on analytical models allows to properly exploit the potential of bistatic SAR. It is possible to design stable formations with very low collision risk and satisfying performance at all the achievable latitudes for INSAR. For large baselines, a bistatic angle range can be explored by properly exploiting differential J_2 effects.

References

1. M. Cherniakov (Ed.) Bistatic Radars: Emerging Technology, New York, Wiley & Sons, 2008.
2. P.A. Rosen, S. Hensley, I.R. Joughin, F.K. Li, S.N. Madsen, E. Rodríguez, R.M. Goldstein. Synthetic Aperture Radar Interferometry. Proceedings of the IEEE, 88(3), 333–382, 2000.
3. A. Moccia, N. Chiacchio, A. Capone. Spaceborne bistatic synthetic aperture radar for remote sensing applications. International Journal of Remote Sensing, 21(18), 3395–3414, 2000.
4. F. Li, R.M. Goldstein. Studies of multibaseline spaceborne interferometric synthetic aperture radars. Transactions on Geoscience and Remote Sensing, 28(1), 88–97, 1990.
5. E.R. Rodriguez, J.M. Martin. Theory and design of interferometric synthetic aperture radars. Proc.-F, 139(2), 147–159, 1992.
6. A. Moccia, S. Esposito, M. D'Errico. Height measurements accuracy of ERS-1 SAR interferometry. EARSeL Advances in Remote Sensing, 3(1), 94–108, 1994.
7. F. Leberl. Radargrammetric Image Processing, Norwood, Artech House, 1990.
8. A. Renga, A. Moccia. Performance of stereoradargrammetric methods applied to spaceborne monostatic–bistatic synthetic aperture radar. Transactions on Geoscience and Remote Sensing, 47(2), 544–560, 2009.
9. G. Krieger, A. Moreira, H. Fiedler, I. Hajnsek, M. Werner, M. Younis, M. Zink. TanDEM-X: A satellite formation for high-resolution SAR interferometry. Transactions on Geoscience and Remote Sensing, 45(11), 3317–3341, 2007.
10. G. Fasano, M. D'Errico. Modeling orbital relative motion to enable formation design from application requirements. Celestial Mechanics and Dynamical Astronomy, 105(1–3), 113–139, 2009.
11. M. D'Errico, G. Fasano. Design of interferometric and bistatic mission phases of COSMO/SkyMed constellation. Acta Astronautica, 62(2–3), 97–111, 2008.

Conceptual Design of the FAST-D Formation Flying Spacecraft

D. Maessen, J. Guo, E. Gill, B. Gunter, Q.P. Chu, G. Bakker, E. Laan, S. Moon, M. Kruijff, and G.T. Zheng

Abstract The chapter presents the latest results in the design of FAST-D, the Dutch micro-satellite for the Dutch–Chinese FAST (Formation for Atmospheric Science and Technology demonstration) formation flying mission. Over the course of the 2.5 year mission, the two satellites, FAST-D and FAST-T, will demonstrate various new technologies and perform observations of atmospheric aerosols and seasonal variations of height profiles in the cryosphere using spectropolarimeter and altimeter payloads on both spacecraft. A conceptual design for the Dutch spacecraft, FAST-D, is presented. Special focus is laid on the design of the attitude determination and control subsystem and on the space-based computing experiments to be performed on this spacecraft. Furthermore, new results in the development of the science payloads on FAST-D, the aerosol characterisation instrument SPEX (Spectropolarimeter for Planetary Exploration) and the altimeter SILAT (Stereo Imaging Laser Altimeter), are described. For SPEX, several design changes have been made to make the instrument more compatible with the FAST mission. For SILAT, an instrument re-design for Earth missions is presented, which results in considerable mass savings as compared to the earlier design.

1 Introduction

The FAST (Formation for Atmospheric Science and Technology demonstration) mission is a collaboration between Delft University of Technology in the Netherlands and Tsinghua University in Beijing, China. It is a formation flying mission consisting out of two micro-satellites: FAST-D (Delft) and FAST-T (Tsinghua). Mission objectives exist in the fields of technology, science, and education. Detailed information on the mission can be found in [1–4]. The following sections focus on the conceptual design of FAST-D that was developed during phase A of the

S. Moon (✉)
Cosine Research B.V., Niels Bohrweg 11, 2333 CA, Leiden, The Netherlands
e-mail: s.moon@cosine.nl

R. Sandau et al. (eds.), *Small Satellite Missions for Earth Observation*,
DOI 10.1007/978-3-642-03501-2_14, © Springer-Verlag Berlin Heidelberg 2010

project and on the progress made in the development of the two science payloads of FAST-D.

2 FAST-D Conceptual Design

The conceptual design of FAST-D is explored in the following sections. First, the overall system design is briefly discussed, followed by a more in-depth treatment of the attitude determination and control subsystem (ADCS) and the planned space-based computing experiments to be performed on the spacecraft.

2.1 Overall System Design

The current design for the 50 kg FAST-D, cf. Fig. 1, foresees in a $0.5 \times 0.5 \times 0.7\ m^3$ box-shaped structure with three body mounted solar panels and a secondary Li-ion battery for power supply during eclipse. The internal structure consists of multiple load carrying trays for easy spacecraft integration. A GNSS (Global Navigation Satellite System) receiver will function as absolute navigation sensor for onboard orbit determination, which will be enhanced through uploading of enhanced GNSS ephemerides from ground. Relative navigation with respect to FAST-T, required

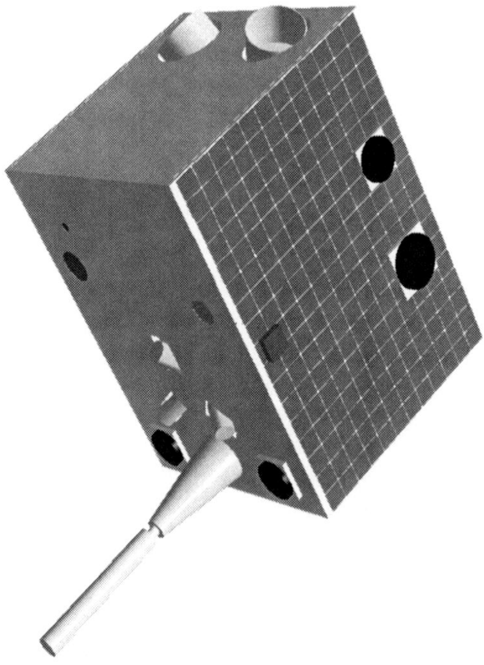

Fig. 1 FAST-D conceptual design

for autonomous formation flying, is achieved by exchanging GNSS data products between the two satellites using an S-band inter-satellite link. Communications with the ground stations will also be performed in S-band. FAST-D will be primarily nadir pointing and will utilize three axis attitude control. Formation and orbit control is achieved by means of a cold gas propulsion system capable of providing a total ΔV of 12 m/s.

2.2 Attitude Determination and Control Subsystem

Conventional attitude control of satellites has been based on classical techniques which, together with a conservative trend of thought, form bottlenecks for ADCS to achieve a higher performance rating, flexibility in mode switching, and fault tolerance. FAST, being an educational and technology demonstration mission, allows the application of innovative ADCS technologies aimed towards solving these bottlenecks. The FAST-D ACDS currently utilizes the following sensor types for attitude determination: Earth horizon sensor, fine sun sensor, autonomous star tracker, 3-axis magnetometer, and 3-axis rate sensor. Attitude control is achieved by means of four reaction wheels in a pyramid configuration maximizing control effectiveness, cf. Fig. 2, together with three magnetic coils for de-saturating the reaction wheels when necessary.

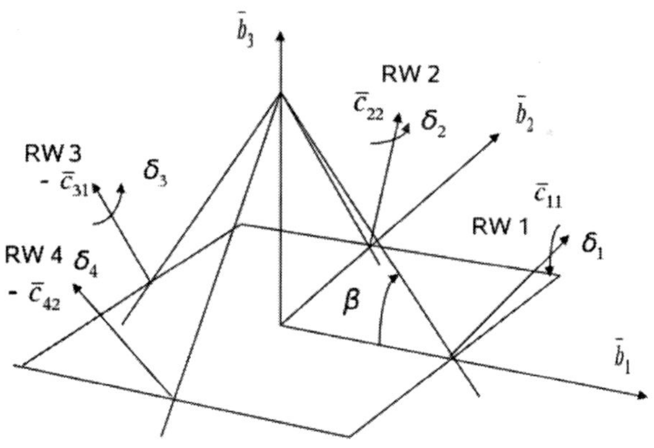

Fig. 2 Four reaction wheels in a pyramid configuration [5]

An integrated attitude determination system combines the rate and vector sensor data in an Extended Kalman Filter (EKF) to obtain highly accurate attitude and rate estimates for the entire mission duration, cf. Fig. 3. Furthermore, the ADCS system utilizes a double-loop Nonlinear Dynamic Inversion (NDI) technique, cf. Fig. 4. This allows flexible control of a nonlinear system in different flying modes,

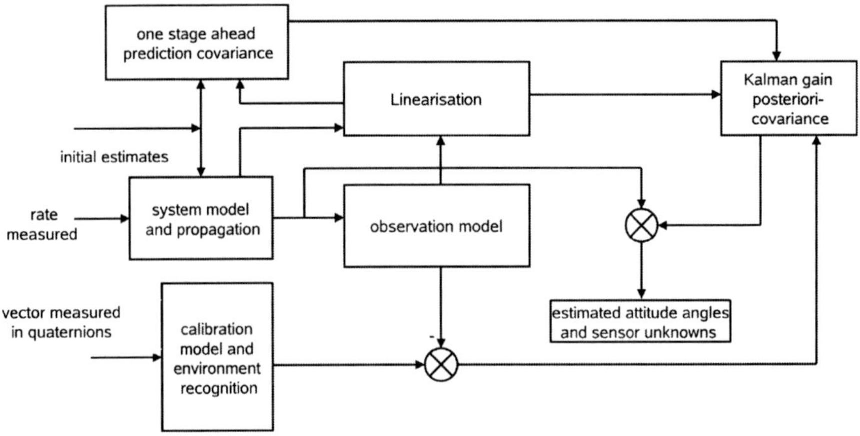

Fig. 3 Sensor integration structure with EKF

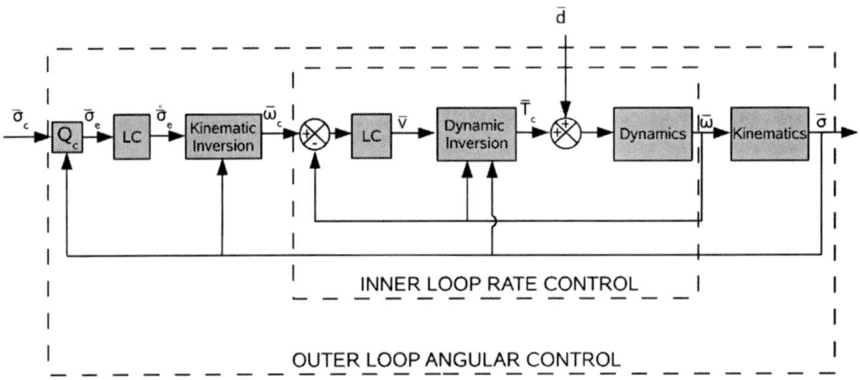

Fig. 4 Double loop NDI nonlinear attitude control

separating spacecraft dynamics from its kinematics, and providing high pointing accuracy in a dynamic environment. A reconfigurable control allocation system is utilized to optimally assign angular momentum profiles to each reaction wheel, accounting for over- and under-actuated situations maintaining control effectiveness during potential actuator failure modes.

The combination of these techniques provides various advantages to the ADCS design: The double-loop NDI technique avoids strong nonlinear feedback which causes unnecessary actuator saturations. Furthermore, it provides flexibility for multiple operational modes by separating rate and angular control. The reconfigurable control allocation contributes to the overall ADCS robustness.

2.3 Space-Based Computing

The space-based computing experiments for the FAST mission are split into two components: on-board and distributed computing. On-board computing includes those computations that are done on a single satellite; however, this can also include parallel processing methods if the on-board processor has multiple cores. The primary goal of the on-board experiments is to reduce the amount of data that needs to be downloaded by conducting some degree of data processing in flight. For example, this could include the pre-processing of data generated from one of the on-board instruments, such as the SPEX instrument, cf. Sect. 3.1, or a GNSS receiver. This pre-processing has the potential to dramatically reduce the amount of data that needs to be transferred to the ground stations, without sacrificing the quality of the final science products. The primary concerns with such an experiment are the processor speed, and consequent power usage. Fortunately, advances in computer hardware for commercial mobile devices have produced low-power processors that may be well suited for small satellite systems. For example, though not space-certified, products such as the Intel Atom[1] processor are small (13×14 mm) and powerful (2 GHz) with relatively low power requirements (<2.5 W).

One example that highlights the potential benefit of conducting on-board computing involves the SPEX payload. This instrument samples a spectrum from approximately 400–800 nm at 1 nm intervals, resulting in roughly 400 data points that are then convoluted by the instrument electronics into a sinusoidal function. The instrument has nine fields of view (FOVs), with each FOV measuring a spectrum across a $7^\circ \times 1^\circ$ swath, sampled into 8 separate pixels. Data is collected in two sets for each swath, at a rate of 0.5 Hz, resulting in 72 total data sets (400 points each) per second. Using the standard processing approach, the data points collected are transmitted to the ground for post-processing, which primarily consists of fitting a sinusoidal or high-order polynomial (i.e., 5th order) curve through the data using standard least squares. The raw SPEX data has 10-bit precision, meaning that 49 minutes of SPEX data collection (assuming a 50% duty factor for SPEX) generates approximately 106 megabytes (MB) of data. If the least squares operation is performed in space, then the transmitted data will only consist of the estimated curve parameters, which for a 5th order polynomial (i.e., 6 parameters) will be on the order of 5.1 MB (assuming conversion to single precision numbers). This translates into a 95% reduction in the total transmitted data. Furthermore, while a standard space-certified processor (e.g., the LEON3FT,[2] operating at 4 MFLOPS (floating-point operations per second)) does not have the speed to keep pace with the processing demands, a processor such as the Intel Atom or similar, operating at $\sim$3 GFLOPS, is able to manage well. The fact that the processor is not fault-tolerant can potentially be addressed through the application of software (i.e., algorithmic)

[1] www.intel.com

[2] www.gaisler.com

fault-tolerance [6]. This approach requires ~10% overhead, but verifies that the calculations are correct, while significantly reducing the cost and performance loss incurred by utilizing space-ready hardware components.

The distributed computing experiments planned will be a first among small satellite formations. As such, the inter-satellite communication will be an integral part of the process. Multiple communication lines will potentially be utilized throughout the mission lifetime, including ground-based relays, direct inter-satellite links, and possibly through commercial satellite phone networks [2]. Different links will be utilized at different stages of the mission, since the separation distance of the two satellites will change [4]. Ideally, the use of the various communication links will be selected automatically depending on which link offers the highest bandwidth at a given time. All these features require an uncommon and highly flexible communications subsystem design. The primary goal of the distributed computing experiment on FAST will be to exercise the process and identify performance boundaries. Once established, future experiments with this approach could involve topics which make use of large, dense linear systems that are well suited for distributed computing, such as real-time high accuracy orbit determination or real-time gravity field determination.

3 Science Payloads

For FAST-D, the opportunity exists to fly two advanced Dutch science payloads: SPEX (Spectropolarimeter for Planetary Exploration) and SILAT (Stereo Imaging Laser Altimeter). Although these instruments are ultimately intended for planetary science missions, some minor adjustments allow them to be used as Earth observation payloads on a micro-satellite. Since the FAST mission is not totally geared towards scientific observations and since the science payloads are not tailored explicitly for Earth use, the FAST mission cannot compete with 'big' science missions in terms of science output, but it can certainly provide supplementary science data. The next two sections will discuss the science payloads of FAST-D.

3.1 SPEX

The shoebox-sized SPEX instrument, cf. Fig. 5, is under development by a consortium of Dutch companies and knowledge institutes. SPEX is capable of accurate full linear spectropolarimetry without moving parts or liquid crystals, making it an ideal instrument to measure and characterize aerosols from space. Being developed using a stepwise approach, its use on the FAST mission presents an intermediate step towards its ultimate goal: to act as payload on a future Mars observation satellite to study Martian dust and clouds.

The SPEX version to be used for the FAST mission is known as the SPEX breadboard (BB). SPEX BB is a modified version of SPEX, which is tailored for Martian

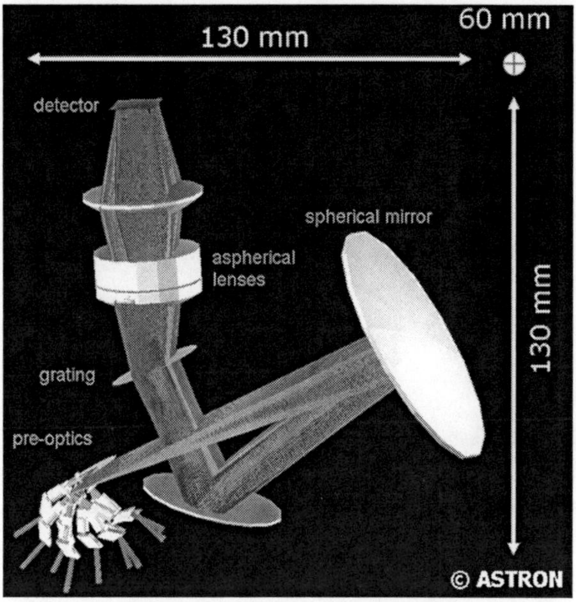

Fig. 5 SPEX optics

use and discussed extensively in [7]. An important modification of SPEX BB compared to SPEX is the change from seven planet-looking FOVs and two limb-looking FOVs to nine planet-looking FOVs. Reason for this change is that SPEX BB will focus on characterizing aerosols, leading to a preference for planet-looking FOVs over limb-looking FOVs. The SPEX BB FOVs measure $7° \times 1°$, resulting in a spatial resolution of 11.4 km from a 650 km high orbit, and have angles of $\pm 56°$, $\pm 42°$, $\pm 28°$, $\pm 14°$, and 0° with respect to nadir.

While SPEX will be a fully space-qualified instrument, SPEX BB will be as much space qualified as possible to reduce development time and costs. Pragmatically, this means that it is foreseen to use space qualified optics, a space qualified detector and a housing that can withstand the launch environment of the Ariane 5 launcher. This specifically does not include space qualification of the polarisation optics and space qualified read-out electronics.

3.2 SILAT

SILAT, cf. Fig. 6, is a miniaturized, integrated instruments suite consisting of a high resolution camera (HRC), a stereo imaging camera (SCAM) and a photon counting laser altimeter (LAT). The instrument is treated in detail in [8]. A structural re-design of the instrument has been made for the FAST mission, which results in considerable mass savings as compared to the earlier design for a mission to Jupiter's moon Europa. The drivers for the mass changes are changes in the thermal

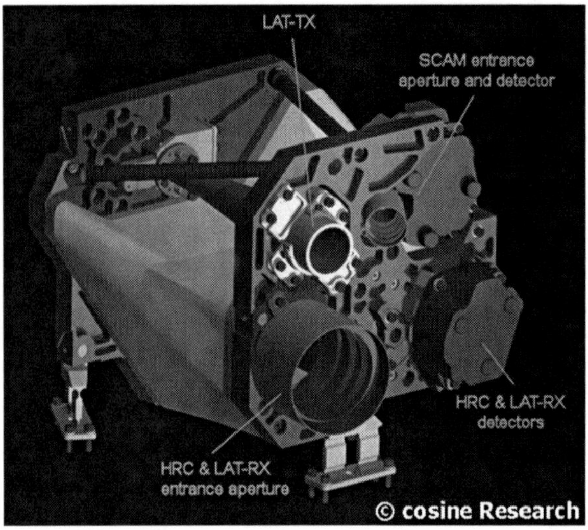

Fig. 6 SILAT

and harmful radiation constraints due to the shift in operational orbit from Europa to Earth. The main changes to the structure are the replacement of Densimet, a high density tungsten alloy used for radiation shielding of the SILAT electronics in the high radiation environment of Europa, by aluminium and the reduction in thickness of other structural components made out of aluminium. This leads to a reduction in mass of ~1.5 kg. Due to this mass reduction, the thickness of the SiC baseplates of SILAT can also be reduced, resulting in an additional mass saving of ~1 kg, leading to a total mass reduction from the initial 7.8 kg for the Europa design to 5.2 kg for FAST. Other reductions will be investigated as the FAST project moves forward into the detailed design phase.

Performance evaluations conducted for the FAST mission indicate that the optical elements of SILAT as designed for Europa will perform well in an Earth environment. Therefore, no modifications to the optical elements of SILAT are required for the FAST mission.

A prototype of the complete LAT is currently being developed at cosine Research. As of writing, all of the components in the prototype have been completed; these are the micro-chip laser, the detector array, readout electronics and the controlling firmware. These components are being integrated at tested at the moment, with prototype performance results expected by the end of May, 2009.

4 Conclusions

The conceptual design of FAST-D has been presented, with special emphasis paid to the novel attitude control system and the impact of the space based computing experiment on the FAST-D system design. Furthermore, the two science payloads on FAST-D have been treated in detail, highlighting changes in their design for the FAST mission.

The 50 kg FAST-D spacecraft differs from most other micro-satellites in that it is capable of formation flying with another spacecraft, namely FAST-T, by means of exchanging GNSS data products for relative navigation and through a cold gas propulsion system for formation and orbit control. The ADCS of FAST-D will make use of a double loop nonlinear dynamic inversion technique for the attitude control, which has never before been implemented on a spacecraft. FAST-D will also serve as test platform for several space-based computing demonstrations such as onboard processing of scientific data and distributed computing between two satellites. Lastly, the scientific payloads of FAST-D, both designed for inter-planetary missions, are specially modified for this mission to increase science output and limit their impact on the overall spacecraft design.

References

1. D.C. Maessen et al. Preliminary Design of the Dutch-Chinese FAST Micro-Satellite Mission. In Proc. of the 4S Symposium, Rhodes, Greece, 26–30 May, 2008.
2. D.C. Maessen et al. Increasing System Performance and Flexibility: Distributed Computing and Routing of Data within the FAST Formation Flying Mission, IAC-08-D1.4.6, In Proc. of the 59th IAC, Glasgow, Scotland, September 29–October 3, 2008.
3. E. Gill et al. Atmospheric Aerosol Characterisation with the Dutch-Chinese FAST Formation Flying Mission, IAC-08-B1.I.1. In Proc. of the 59th IAC, Glasgow, Scotland, September 29–October 3, 2008.
4. D.C. Maessen et al. Mission Design of the Dutch-Chinese FAST Micro-Satellite Mission. To be published in Proc. of the 7th IAA SSEO Symposium, Berlin, Germany, May 4–8, 2009.
5. Y. Choi et al. Dynamic Control Allocation for Shaping Spacecraft Attitude Control Command, AIAA 2006-6040, AIAA GNC Conference and Exhibit, Keystone, Colorado, USA, August 21–24, 2006.
6. J.A. Gunnels, D.S. Katz, E.S. Quintana-Orti, R.A. van de Geijn. *Fault-Tolerant High-Performance Matrix-Matrix Multiplication*, FLAME Working Note #2. The University of Texas at Austin, Department of Computer Sciences, Technical Report TR-2000–34, December, 2000.
7. E. Laan et al. The Spectropolarimeter for Planetary Exploration – SPEX. In Proc. of the 7th ICSO, Toulouse, France, October 14–17, 2008.
8. S. Moon et al. Towards a Miniaturized Photon Counting Laser Altimeter and Stereoscopic Camera Instrument Suite for Microsatellites, To be published in Proc. of the 7th IAA SSEO Symposium, Berlin, Germany, May 4–8, 2009.

Earth Observation Using Japanese/Canadian Formation Flying Nanosatellites

Marleen van Mierlo, Keisuke Yoshihara, Alfred Ng, Linh Ngo Phong, and François Châteauneuf

Abstract Japan Canada Joint Collaboration Satellites (JC2Sat) is a joint project between the Canadian Space Agency (CSA) and the Japan Aerospace Exploration Agency (JAXA). The main objective of the project is to design, build, launch and operate two 18 kg nearly identical nanosatellites that will be launched as a stack and separated in space to demonstrate the feasibility of Autonomous Formation Flight (AFF) based on aerodynamic differential drag only. The specific configuration of the JC2Sat nanosatellites serves as an ideal technology demonstration platform for the Miniature far Infra-Red Radiometer (Mirad) instrument developed jointly by CSA and Institut National d'Optique (INO). Each nanosatellite carries a Mirad instrument for the purpose of Earth's limb sounding. Based on un-cooled far infrared micro-bolometers, these low mass and low power payloads will co-register the limb profiles in the emission bands of greenhouse gases CO_2 and H_2O, respectively centered at wavelengths of 15 and 25 μm. The development of JC2Sat is carried out by a united small team consisting of engineers and researchers from both CSA and JAXA. JC2Sat is planned to be ready for launch in 2011. Mission duration is defined to be one year. At the time of writing, the project is in phase C.

1 History

From April 2006 to March 2007 JAXA engineer Mr. Yoshihara worked at the Canadian Space Agency as part of an Exchange/Training program. During his stay he and his CSA supervisor Dr. A. Ng initiated the conceptual study of JC2Sat as a nanosatellite project within the CSA in-house Space Technology Research Program (STRP). The study quickly identified areas of common interest to both JAXA [1] and CSA and a nanosat formation flying mission based on differential drag technique was finally proposed by CSA visiting Post-Doc Fellow Dr. B. Shankar Kumar [2].

M. Mierlo (✉)
Canadian Space Agency, 6767, Route de l'Aéroport, St-Hubert, QC, J3Y 8Y9, Canada
e-mail: marleen.vanmierlo@asc-csa.gc.ca

R. Sandau et al. (eds.), *Small Satellite Missions for Earth Observation*,
DOI 10.1007/978-3-642-03501-2_15, © Springer-Verlag Berlin Heidelberg 2010

An H-2A launch opportunity presented itself in May 2006. In October 2006, CSA submitted a proposal to JAXA asking for consideration of launching this joint Japan Canada development on a future H2A rocket. In July 2007 an MOU was signed by JAXA and CSA on the development of JC2Sat.

2 Introduction

JC2Sat is a formation flight nanosatellite mission consisting of two nanosatellites. This project is an international collaboration project between JAXA and CSA. The three programmatic objectives of the JC2Sat project are:

(a) To demonstrate on-orbit specific novel technologies in order to increase their Technology Readiness Level (TRL) [3].
(b) To research, to implement and to establish a practical streamlined process required to perform in-house development of small-scale spacecraft with low cost and short delivery time.
(c) To provide hands-on training to CSA and JAXA engineers and researchers, while strengthening relationship between the two agencies.

This chapter begins with a description of the three novel technologies, in particular the earth observation instrument Mirad, that will be demonstrated on-orbit as part of the first programmatic objective of JC2Sat. This is followed by an overview of the JC2Sat on-orbit operations. Finally several aspects of the JAXA-CSA collaboration are discussed.

3 On-Orbit Demonstration of Novel Technologies

The first programmatic objective of the JC2Sat mission is to demonstrate three novel technologies on-orbit in order to increase their TRL. These three novel technologies are described in detail below.

3.1 Autonomous Formation Flight Using Differential Drag

The first technology to be demonstrated is the use of differential drag to autonomously control the two formation flying satellites over distances varying from 100 to 5 km. To generate a differential aerodynamic drag force the cross-sectional area of the rigid spacecraft is varied by changing the satellite's attitude around the pitch axis [4]. Performing this technology successfully on-orbit will raise the TRL from 3 to 5.

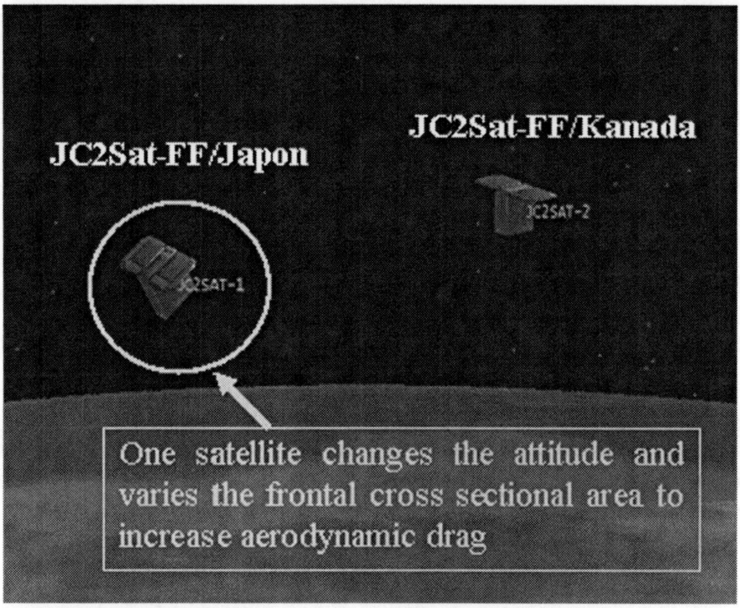

Fig. 1 Autonomous Formation Flight (AFF) using atmospheric drag control

3.2 Relative Navigation Using Commercial GPS Receiver

Fig. 2 NovAtel OEM4-G2L GPS receiver

GPS based relative positioning using dual-frequency carrier phase is one of the essential technologies for future spacecraft formation flying missions. Compared with the system with single-frequency GPS sensors, GPS dual-frequency carrier-phase observation can provide higher accuracy, adapt to a larger baseline application up to few hundreds of kilometres, and also improve the identification of systematic measurement errors. Despite significant advantages of the GPS dual-frequency sensor system, only few space missions have adopted this system due to practical limitation on the available selection of space-capable dual-frequency receivers. In this project, a commercial GPS receiver with several modifications on its firmware will be used [5]. Current TRL of this component is 5. After successful operation in orbit TRL will be increased to 6.

3.3 Miniature Far InfraRed RADiometer (Mirad)

The earth radiation budget is a balance between incoming radiation from the sun and outgoing radiation emitted by the Earth and atmosphere. The budget governs climate behaviour and provides insight into greenhouse effects. It is estimated that over 50% of Earth's outgoing radiation and 75% of atmosphere's outgoing radiation are emitted at far InfraRed (FIR) wavelengths (>15 μm). Despite this fundamental fact space based far infrared measurement has not been performed routinely due to the lack of sensors with adequate capability beyond 15 μm. CSA and Institut National d'Optique (INO) have jointly developed a new generation of un-cooled micro-bolometer detectors that is capable of FIR sensing. The TRL of the micro-bolometer is 5. Testing it on-orbit will bring up its TRL to 6.

The Miniature far InfradRed RADiometer (Mirad) instrument is a FIR radiometer miniaturized for operation with the limited resources of nanosatellite. In the JC2Sat mission, each satellite will be equipped with one Mirad instrument. The optics of each Mirad will be different such that each instrument measures radiance in different spectral bands.

The purpose of the two Mirad instruments in the JC2Sat mission is to co-register the Earth's limb profiles with global and seasonal variations in the emission bands of greenhouse gases CO_2 and H_2O, centered respectively at wavelengths of 15 and 25 μm (Fig. 3). The two Mirad units have been designed to be compatible with the nanosatellite allocated budgets: an average power consumption of 500 mW, a mass of 500 g, and an envelope of 100 $\times$ 105 $\times$ 105 mm [6].

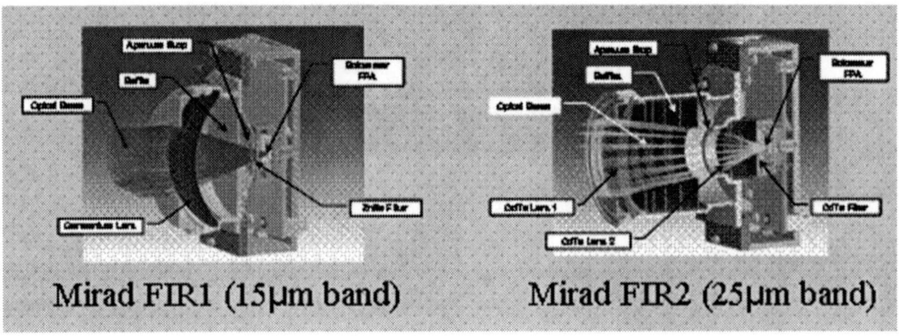

Fig. 3 The JC2Sat formation carries two versions of the miniature far InfraRed RADiometer (Mirad)

The ultimate goal of the measurements by the distributed far infrared radiometers is twofold. First, these instruments are designed to produce a new database of navigational horizon characteristics for the developers of horizon sensors. Second, they provide an assessment of the atmospheric outgoing radiation in support of planned missions exploring the far infrared spectrum.

Mirad makes use of a custom designed linear array of 256 × 2 un-cooled VO_x micro-bolometers for FIR sensing. The first row of this array is made of pixels sensitive to the thermal target and the second row contains pixels sensitive only to thermal drifts of the die. This arrangement alleviates the need for temperature control of the die so that power consumptions resulting from the use of an otherwise required thermoelectric cooler can be avoided.

The limited power budget of the spacecraft has prevented as well the use of an on-board blackbody for in-flight calibration. Therefore, Mirad will rely extensively on frequent cold space measurements and extensive pre-flight characterization to minimize radiometric errors. The limited power has driven as well the electronics design of Mirad towards the use of commercial electronics.

In operation, each Mirad unit will be mounted at the rear of a nanosatellite and backward staring at the Earth's limb behind the satellite. One advantage of the proposed configuration is that the unit is allowed to view the cold space to perform calibration each time the satellite pitches down. Another advantage is that Mirad is shielded against drag and particles by the sides of the satellite. Two nanosatellites, the leader with the CO_2 band radiometer and the follower with the H_2O band radiometer, allow co-registration of the limb radiance profiles in these two emission bands.

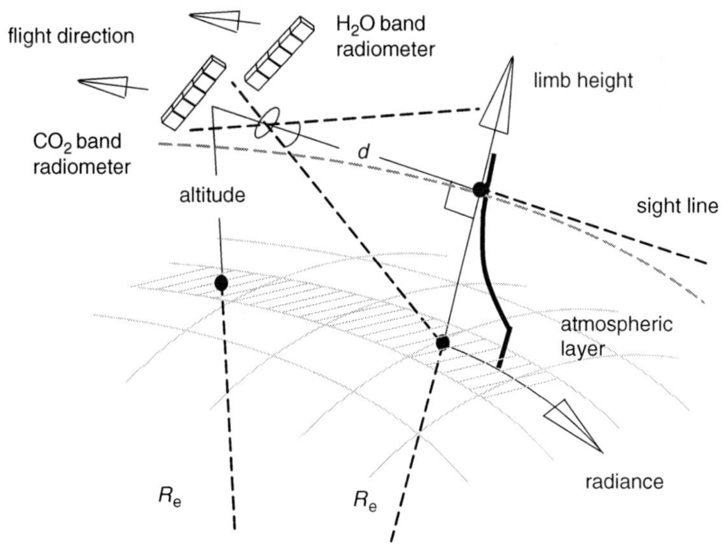

Fig. 4 Operating configuration of the radiometers for the Co registration of limb profiles in the emission bands of CO_2 and H_2O

To perform co-registration, the limb radiance is measured successively by each unit when the latter is at a given distance from the tangent point. Each unit starts acquiring data when turned on and stops when powered down by the spacecraft computer. The focal plane array operates in the staring mode at a frame rate of 10 Hz, and the time required to make one measurement is the time required to acquire and average out a predetermined number of frames. During a sequence of measurements a latitude portion is covered and this sequence is repeated periodically until full latitude coverage is achieved.

4 JC2SAT Operations

The two JC2Sat satellites are planned to be launched, stacked together, into a sun-synchronous orbit at an altitude between 600 and 700 km in 2011. Spacecraft operations will be performed mainly from two ground stations located in Canada and Japan. The JC2Sat mission duration is defined to be one year and split up in four phases:

Phase I: Stacked Phase – 1/2 Month

Once released in orbit the satellite stack will be de-tumbled into a so-called Power Positive Parking (P3) mode, using the attitude control system of only one of the two satellites. While in P3 attitude, power generation is sufficient to allow for a proper check-out of the bus and Mirad payload of each spacecraft.

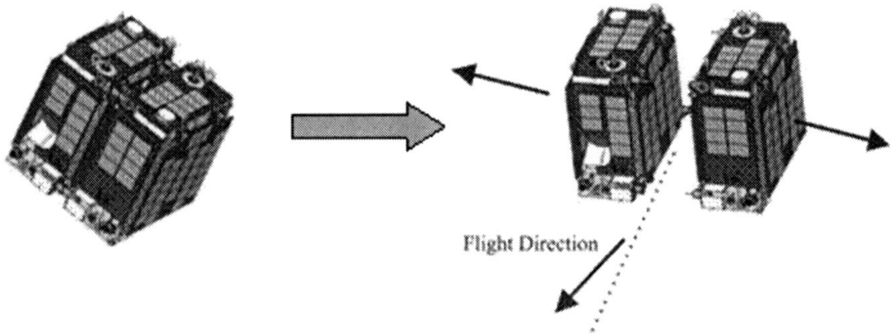

Fig. 5 JC2Sat Inter-satellite separation

Finally the stack will be given the Inter-satellite Separation Pointing (ISP) attitude and after activation of the individual navigation system (GPS) on each satellite the satellites will be separated by activating the Inter-satellite Separation Mechanisms (ISMs) at a specific point in the orbit when the separation direction is at a 10° angle with the orbit normal [4].

Phase II: Preparation Phase – 1/2 Month

During preparation phase, drag panels will be deployed. To ensure that the satellites will not drift apart too far before Autonomous Formation Control (AFC) will be activated; the Follower spacecraft is given a pitch command directly from ground which will start drift reduction. Afterward the Inter-Satellite Link (ISL) and Relative Navigation System (RNS) [5] will be activated and checked out. AFC will then be activated to bring the drift autonomously to a stop.

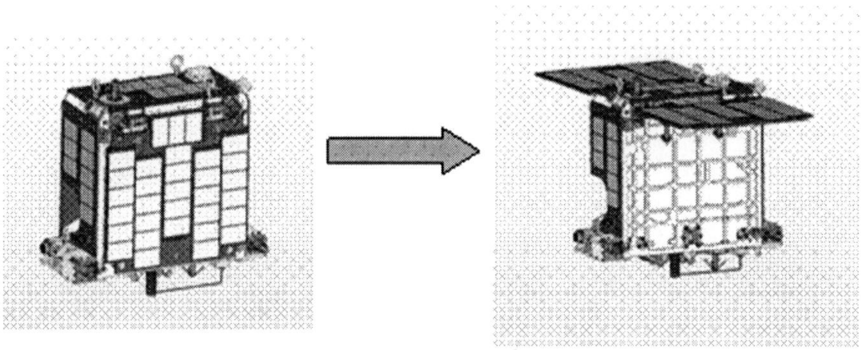

Fig. 6 JC2Sat drag panel deployment

Phase III: Autonomous Formation Flight Phase – 5 Months

During this phase AFC is exercised by performing Formation Keeping at a distance between 1 and 5 km (Wide Formation), 1 km (Close Formation), and 100 m (Tight Formation) and performing Formation Manoeuvring between these distances. A separation accuracy of 1% (with a minimum of 10 m) of the separation distance is targeted. During this AFF phase random Mirad data collection takes place.

Phase IV: Mirad Phase – 6 Months

During the last phase of the mission, priority is given to the collection of Mirad data. To acquire a complete set of scientific Mirad data within 30 days, both Mirad instruments need to make one sequence of measurements (~14 s total) per orbital period (~100 min) at a pre-defined location per orbit. As such, both satellites need to be kept at an un-pitched attitude, i.e. AFC needs to be interrupted, for the very short duration of this measurement. As this interrupted period is extremely short compared to the control cycle of the formation control, the effect on the accuracy of the constant distance between the two satellites will be negligible.

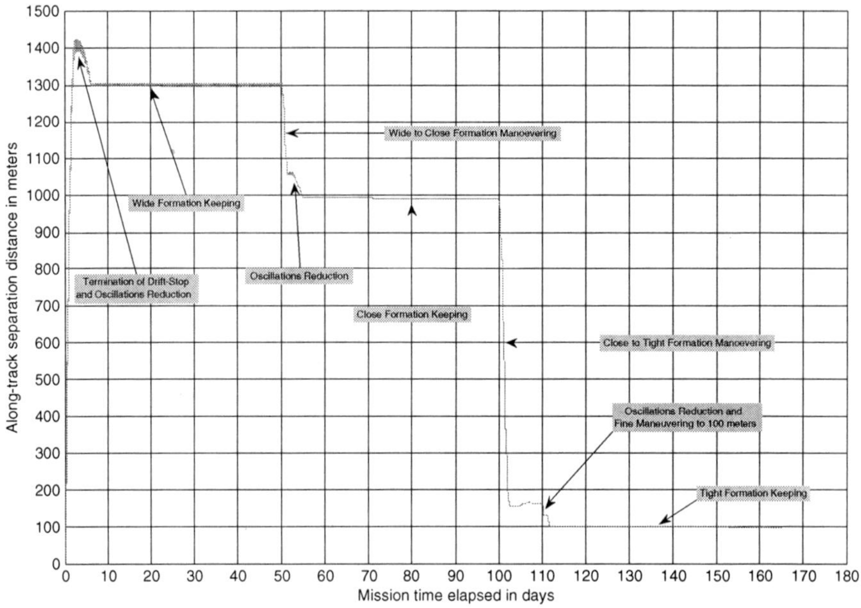

Fig. 7 Autonomous Formation Flight (AFF) phase

5 Collaboration Between JAXA and CSA

5.1 Set-Up Between JAXA and CSA

About 80% of the manpower in JC2Sat comes from CSA while the rest comes from JAXA. Regular face-to-face meetings are held at CSA and as such, most personnel travel cost is covered by JAXA. CSA is responsible for the procurement of manufacturing material and satellite components which are mostly Commercial-of-the-Shelf (COTS). Both satellites are being manufactured (S/C structure) and assembled at CSA. Integration and test activities take place in Canada as well, partly at CSA, partly at the David Florida Laboratory (DFL). The satellites are designed to be compatible with JAXA's H2A launch vehicle.

5.2 Development Philosophy

The efforts in JC2Sat are concentrated on specific areas which are mission design (Formation Flight), systems engineering, ACS simulator development, relative navigation technique, software development, structural manufacturing and general Assembly Integration & Test (AIT). Development and manufacturing activities at component level are avoided as much as possible by purchasing COTS components where possible. Activities in areas for which expertise is unavailable in CSA or

JAXA (such as thermal analysis) or dedicated small activities that cannot be performed in-house efficiently (such as test harness manufacturing) are outsourced. The project tries to involve Canadian and Japanese universities as much as possible, for example, for spacecraft real-time operations.

An overview of the JC2Sat system design can be found in [6].

5.3 One Single United Team

The JC2Sat team consists of a mix of CSA and JAXA engineers and researchers who are together responsible for the complete end result. Each subsystem or mission/payload is represented by one person in the team. This person receives support from other CSA/JAXA personnel with specific expertise, where necessary.

The advantages of having one single united team in the collaboration are:

1. Direct technical exchange in this international team gives stimulation to the members and generates a synergistic effect for the project.
2. The division of work between the team members can be flexible throughout the project, i.e. there is no need to officially re-negotiate between agencies every time a change in the division of work needs to take place.
3. Because of the interesting international mix, team members enjoy the project which generates motivation, the most essential key to success.
4. By working closely together, team members are building up personal relationships and knowledge of and respect for the differences between each other's work culture. As a consequence competition or rivalry is non-existing and there is a general satisfaction about the positive effect one can have on the other agency.
5. Personal contact between all team members facilitates agile decision making and reduces the amount of necessary documentation.

5.4 CSA – JAXA Communications

JAXA and CSA team-members communicate daily via email and video conferences (VC's) addressing specific technical topics are held on a weekly basis. To allow for frequent face-to-face exchange, JAXA team-members travel to Canada once every two months. During these 3- or 4-day visits work progress is reviewed and several sub-team meetings are held to discuss the various technical topics and problems that are being worked. Any necessary meetings with external Canadian partners are typically planned during these visits as well.

6 Conclusion

JC2Sat is a unique international collaboration on formation flying nanosatellites. Information is shared openly between JAXA and CSA and each side contributes to the project its own unique expertise. The JC2Sat platform is ideal for technology demonstration of the Mirad Earth observation instrument which is low power, mass and volume. The nanosatellite formation allows for distribution of multiple Mirad instruments measuring at different wavelength which would traditionally require a large single spacecraft. Once demonstrated in space Mirad will open up opportunities for future low cost formation flight Earth observation missions.

Acknowledgements The authors wish to acknowledge the dedication of three JC2Sat team-members who have been instrumental in the JC2Sat project but nevertheless had to move on to the next phase in their career. Mr. Yutaka Komatsu played a significant role in the design of the spacecraft structure and mechanisms until he was promoted to a new position within JAXA. The two other team-members worked on JC2Sat during their Post-Doc Fellowship at CSA; Dr. Balaji Shankar Kumar has been responsible for the autonomous formation flight part of JC2Sat and now works as a mission analyst at ComDev Ltd., Dr. Anton de Ruiter designed the JC2Sat attitude control and relative navigation subsystems and became an assistant professor at the Carleton University in Ottawa. We congratulate both organizations with the new additions to their team.

The authors further want to acknowledge the contribution of all JC2Sat partners that have made this collaborative project to a success so far, in particular SpaceQuest, UTIAS Space Flight Laboratory, Sinclair Interplanetary, MAYA, NovAtel, NGC Aerospace Ltd. and Ryerson University.

References

1. H. Horiguchi, K. Hirako, H. Hashimoto. Small satellites development program in JAXA. Proceedings of the CASI ASTRO 2008, Montreal Quebec, 2008.
2. C.L. Leonard. Orbital formation-keeping with differential drag. Journal of Guidance, Control, and Dynamics, 12(1), 108–113, Jan.–Feb. 1989.
3. A. Ng, M. van Mierlo, K. Yoshihara, H. Horiguchi, H. Hashimoto. JC2Sat: A unique international collaboration. Proceedings of the CASI ASTRO 2008, Montreal Quebec, 2008.
4. B.S. Kumar, A. Ng, K. Yoshihara. Flight dynamics and control of the JC2Sat mission. Proceedings of the AAS/AIAA Astrodynamics Specialist Conference 2007, Mackinac Island, MI, Paper no. AAS 07–410.
5. K. Yoshihara, T. Yamamoto, Y. Kondoh, H. Hashimoto, A. Ng, A. de Ruiter, J. Lee. Performance Assessment of Single and Dual Frequency, Commercial-Based GPS Receivers for LEO Orbit. 21st Annual Conference on Small Satellites, Utah State University, Logan, Utah, August 13–16, 2007.
6. K. Yoshihara, M. van Mierlo, A. Ng, B. Shankar Kumar, A. de Ruiter, Y. Komatsu, H. Horiguchi, H. Hashimoto. JC2Sat: An international collaboration Nano-Sat project – overview of the system analyses and design. Proceedings of the 4S Conference, Greece, 2008.
7. L.N. Phong, F. Châteauneuf. Nanosatellite distributed far infrared radiometers. Proceedings of the SPIE, vol. 7208, 2009.

A Bi/Multi-Static Microsatellite SAR Constellation

Tippawan Wanwiwake and Craig Underwood

Abstract Microsatellites are already playing a major role in operational optical-imaging remote sensing applications, *viz.* the Disaster Monitoring Constellation (DMC) and RapidEye. However, such satellites are currently limited to providing useful images over cloud free regions during daylight. Whilst the use of constellations does increase the probability of achieving useful images within short timescales over most regions, there are still areas, such as the tropics, that have significant cloud cover, or extreme latitudes, which can have low light levels, where the all weather, day/night imaging capability of Synthetic Aperture Radar (SAR) would be beneficial.

Current SAR satellites typically have complex, large volume, large mass and high power payloads. Usually they operate singly with relatively long repeat cycles. Recent missions have shown that SAR payloads can be flown on smaller satellites, but there are still significant size/cost barriers to being able to set up affordable SAR satellite constellations. To this end, a feasibility study has been undertaken to examine incorporating a low cost L-band SAR payload into a 100 kg class microsatellite – primarily for disaster monitoring and forestry/flood monitoring. The radar makes use of continuous wave (CW) technology and bi-static pairs, to provide modest resolution (30–100 m) images over 30 km swaths.

1 Introduction

Microsatellites have already demonstrated their capabilities in optical-imaging applications, *viz.* the Disaster Monitoring Constellation (DMC) and RapidEye [1]. However, there are many occasions when useful imagery cannot be achieved due to cloud or low light conditions. For an all-weather, day/night imaging capability, an active radar payload – e.g. a synthetic aperture radar (SAR) – becomes necessary.

T. Wanwiwake (✉)
Geo-Informatics and Space Technology Development Agency, Bangkok, Thailand; Surrey Space Centre, University of Surrey, Guildford, UK
e-mail: T.Wanwiwake@surrey.ac.uk

R. Sandau et al. (eds.), *Small Satellite Missions for Earth Observation*,
DOI 10.1007/978-3-642-03501-2_16, © Springer-Verlag Berlin Heidelberg 2010

The next challenge then is to be able to fit such payloads onto a microsatellite at a cost which is commensurate with the low-cost approach typical of microsatellite missions.

In Earth Observation (EO) missions, wide coverage, high spatial resolution and frequent re-visits are all desirable qualities. Optical microsatellite constellations have shown how, with some compromise in resolution, wide coverage and rapid re-visits can be achieved. To date, SAR satellite missions have tended to be based on single, large, and expensive satellite platforms, carrying complex, expensive and high power demand payloads (e.g. ERS-1/2 and ENVISAT with 35-day repeat cycles; RADARSAT-1/2 with 24-day repeat cycles and ALOS with a 46-day repeat cycle. Recent missions, such as SAR-Lupe (2008 ~770 kg) have demonstrated the use of a constellation to improve the revisit time and coverage, and TECSAR (2008 ~ 300 kg) has shown that a SAR payload can be flown on smaller satellites. However, there is a significant size/cost barrier to be able to set up affordable SAR satellite constellations.

If we are to fit a SAR payload onto a 100 kg class microsatellite, the payload itself must be small, and necessarily be limited in its power consumption. This poses a great challenge, however, once achieved, a constellation of such microsatellites could provide the necessary coverage and repeat visit capability necessary for an affordable all-weather day/night global disaster monitoring capability [2].

We believe a possible design solution to this problem is to split the SAR payload across two such platforms, and to operate the radar in a bi-static mode. The physical isolation between the transmitter and receiver means that a continuous wave (CW) radar solution becomes possible, eliminating the need to generate significant high peak power radar pulses, and potentially simplifying some of the radio-frequency elements of the payload. We recognise that this introduces the problem of synchronisation and control across the two platforms – and so this is the trade space which needs to be examined.

A bi-static configuration can also be considered the first step towards a multi-static SAR system (e.g. achieved by flying a cluster of such microsatellites), which offers great system flexibility, robustness and potential for offering more complex image products.

2 Micro SAR Satellite Mission Definition

Multiple purpose SAR satellites are often designed with high spatial and radiometric resolution, large swath, with multi-polarization and/or multi-frequency payloads to suit many applications. We chose to simplify the payload by aiming at a more narrow range of applications to give a specific focus for the complex trades among parameters such as power, size of solar panel, battery selection, antenna, processing capability, cost and detection capabilities, etc. We have selected disaster monitoring (to compliment the optical DMC), forestry and flood monitoring to be key applications. These applications have similar requirements (see Table 1) with relatively

Table 1 Minimum requirements for forestry and flood monitoring

Application	Required frequency band	Polarisation	Swath (km)	Resolution (m)
Forestry	*P, L*		*medium*	*medium*
Deforestry and clear cut	P	HV or VH	30	50–100
Tropical/Temporate forest monitoring	P, L	HV or VH	30	100
Soil moisture	P, L	HV or VH	–	25–50
Forest biomass estimation	VHF, P, L	HV	–	25–50
Forest classifica-tion/canopies	VHF, L	HH, VV and HV or VH	–	25–50
Forest wetland hydrology	L, C	HH and VV	–	–
Forest fire	L	–	–	–
Flood monitoring	*P, L*	–	*medium*	*medium*
Post flood inventory	P, L, C	HH, HV or VH	30	25–50
Inundation and extent flood	P, L, C	HH and HV	30	25–50

Source: Summarized from [3–9].

modest spatial and radiometric resolution, which makes them particularly suited to microsatellite implementation.

3 Microsatellites SAR Payload Design Issue and Trade Offs

3.1 Frequency Selection

In forestry and flood monitoring, a longer wavelength is preferable due to the greater canopy penetration giving the ability to image the forest floor [3, 4]. However, there are number of challenges such as the effect of Faraday rotation and scintillation in the ionosphere, and the requirement for a large physical antenna size. In Table 2 the ionosphere propagation effects for different SAR bands are calculated under extreme conditions according to the ITU models [10–12].

This shows that the higher frequency bands (S-, C-, X-) are not suitable for our application as they do not penetrate to the forest floor, even though they have advantages with regard to propagation and antenna size. P-band penetrates very well, but is very problematical with regard to propagation and antenna size. Therefore, for our application, L-band (approx. 1.28 GHz) is considered to be the most suitable, as it provides the best penetration with manageable ionosphere propagation effects [13] and a feasible antenna size requirement – particularly when this antenna is split over two platforms.

Table 2 Frequency band comparison

Band	Nominal frequency range (GHz)	Mid frequency range (GHz)	Wavelength (m)	Penetration	Faraday rotation (deg)	Scintillations (dB)	Antenna diameter (m) to achieve the same gain (36 dB)
UHF/P	0.3–1	0.65	0.46	High (to ground level)	255.62	>20	16.92
L	1–2	1.50	0.20	Medium (to trunk level)	48.00	<20	7.33
S	2–4	3.00	0.10	Low (to top of tree level)	12.00	10	3.67
C	4–8	6.00	0.05	Low (to top of tree level)	3.00	>4	1.83
X	8–12	10.00	0.03	Low (to top of tree level)	1.08	4	1.10

Typically, low frequency (L-band) spaceborne radars use long, narrow rectangular phased array antennas with high gain (i.e. large aperture) to minimise power constraints and to provide suitable resolution and received signal strength. The antenna must be mounted so as to look to one side of the satellite's track so as to resolve left-right ambiguity. A typical antenna might be 3 × 10 m.

The antenna width determines the swath (the wider antenna the smaller the swath), whilst the antenna length determines the azimuth ambiguity, which impacts the pulse repetition frequency (PRF) in a traditional pulsed radar. Provided the PRF is set correctly, the best azimuth resolution, r_a (parallel to the satellite's track) that can be achieved is equivalent to half the antenna length (*L*/2) – independent of the distance to the target (provided the return signal is strong enough). The range resolution, r_r (orthogonal to the satellite's track) is dependent on the speed of light (c), the pulse period (τ), and the incidence angle (θ) measured from the vertical, $r_r = c\tau/2\ \sin\theta$.

A phased array gives an electrical steer capability that can maintain power on the target during imaging in scarSAR mode, and this is more favoured than a complex (and heavy) mechanically steered antenna.

Nevertheless, due to the mass, volume, complexity and cost of the antenna compared to microsatellite itself, we propose a simple fixed-mount, deployable parabolic mesh antenna. Analysis shows that a light weight, low cost design is possible, with suitable side-lobe performance. A 3.66 m diameter antenna would be mounted on each spacecraft.

3.2 Polarization

The correction for Faraday rotation can be performed by using multi-polarizations or a full polarization technique, given an accurate model for Faraday rotation [10, 13, 14]. Different polarizations (horizontal – H; vertical – V) can be combined to provide additional information about the imaging area [15, 16]. For instance, the cross-polarized HV or VH data are better for detecting clear cuts, tropical forest, and temperate forest as well as post flood inventory. For inundation and extended flood, HH is better [5, 6]. Circular polarization has been proposed in order to avoid the impact of the Faraday rotation [3], as well as to reduce the complexity of the antenna design. However there is currently little support for this from the user community. Therefore we propose a linearly polarised radar, preferably with a co- (HH) and cross- (HV) polar capability.

3.3 Imaging Mode, Swath and Coverage

For global forestry and flood monitoring, the ability to image large areas is important – but high spatial resolution is not so critical, therefore we propose a medium

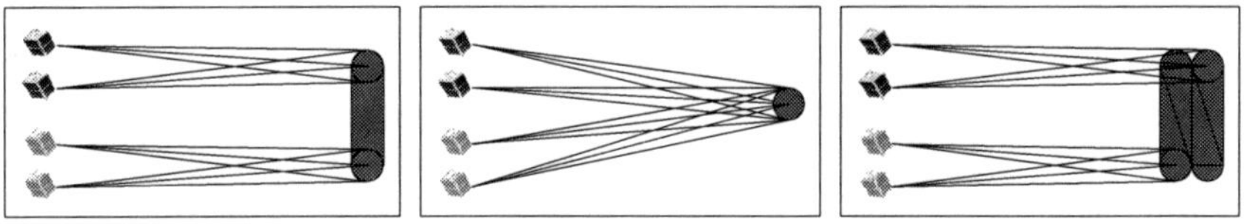

Fig. 1 Strip-map, spotlight and ScanSAR modes.
Source: Adapted from [18, 20]

swath of at least 30 × 30 km with a resolution of 30 m. There are at least three distinct bi-static imaging modes we could choose, including strip-map, spotlight and scanSAR (see Fig. 1).

Strip-map SAR is considered as the best option as the passive antenna does not need to change its orientation during imaging. This also simplifies attitude control [17]. The spotlight mode is most often used to map small specific areas. The beam direction moves as the satellites fly by to keep pointing at the target. In this way, higher (azimuth) resolution can be achieved by increasing the illumination time [18, 19]. This mode would be possible with a fixed antenna, if the satellite bodies can be rotated in a coordinated way. ScanSAR increases the swath width at the expense of azimuth resolution. However this mode requires a more complex antenna design with beam steering capability, which would not be possible with our simple antenna configuration.

3.4 Platform

The initial design assumptions including orbit, power, mass and structure were made based on the 100 kg DMC platform (SSTL-100) developed by Surrey Satellite Technology Limited (SSTL). Whilst the platform can house the payload and antenna, the four body-mounted solar panels could not generate sufficient power, so we propose a small modification involving a deployed (but otherwise fixed) solar panel. 200–300 W of raw electrical power should be sufficient to operate the platform and SAR payload (~60 W transmit power – 50 dBW effective isotropic radiated power (EIRP)).

3.5 Orbit Design (Altitude and Inclination)

A circular low Earth orbit (LEO) is preferred to keep consistent radar path lengths. An altitude between 650 and 700 km would keep the radar path lengths to a minimum, whilst also ensuring that the effects of atmospheric drag would not unduly limit the mission lifetime. If we want to maintain a Sun-synchronous orbit (SSO) – as we already do with the DMC satellites – we will need to carry sufficient propellant to maintain the orbit altitude against the effects of atmospheric drag. At this altitude, the DMC platform can give a useful mission lifetime of at least 7 years.

For forestry and flood observations, a SSO inclination around 98° does allow the micro SAR satellites to spend time over the tropics and other areas of interest [15].

Whilst optical satellites prefer a SSO with a local time of ascending node (LTAN) around 10:30–11 a.m. to give good viewing conditions on the ground, for SAR satellites a "dusk-dawn" orbit (LTAN = 6 a.m.) is best, as this minimises any eclipsing of the satellite – allowing it to image with sunlight falling on its solar panels whilst also minimising the deleterious ionospheric effects [13]. However, even in such an orbit, eclipses of up to 17 min (at 700 km altitude) are experienced, and so it would not be too detrimental to have the micro SAR satellites co-orbital with optical satellites.

4 Bi-static Configuration Design Issue and Trade Offs

Since power is one of the major limiting factors in microsatellites, having a physically separated transmitter and receiver on different platforms provides good isolation and allows the use of continuous wave (CW) radar rather than the traditional pulsed radar techniques.

In traditional pulsed radars, the range resolution depends on the pulse having a short duration, which implies very large peak radio-frequency (RF) powers have to be generated – typically of the order of a kW or so. For example, to obtain a 30 m range resolution at a 30° angle of incidence requires a pulse duration of 10 μs. Even using chirped signals to extend the pulse length (whilst maintaining adequate bandwith to keep the resolution), the peak power required does not reduce to the levels commensurate with that available from a typical microsatellite.

However, by separating the transmitter and receiver, the transmitter can transmit power continuously over the PRF interval (typically a few kHz implying an interval of a few hundred microseconds) reducing the peak RF power demand to a continuous RF power demand of a few 10's of watts. There is also no need for a rapid-acting transmit-receive (T/R) switch or duplexer to provide the high isolation crucial to protect receiver from damage when transmitting high power. Therefore, high power efficiency can be maintained, which is a critical because it has direct implications on the power dissipation, system power generation requirements, thermal design and reliability of the payload [21].

Figure 2 shows the bi-static configuration we have adopted. For simplicity, a flat Earth surface is assumed. According to Willis, the baseline (D_L) can be defined as "long" or "short" depending upon the angle subtended at the target [22].

A short baseline will provide essentially mono-static characteristics and seems to provide more advantages in term of minimising range, maximising signal-to-noise ratio (SNR), and maintaining similar orbit dynamics and disturbances. Synchronization and ensuring antenna footprints overlap are also easier. In our case the baseline is "short" when (D_L) < 80 km (i.e. when $\beta < 5^\circ$). Under these conditions the bi-static look angle is approximately the same as the mono-static look angle which we propose to be 30–40°.

Fig. 2 Single orbit geometry

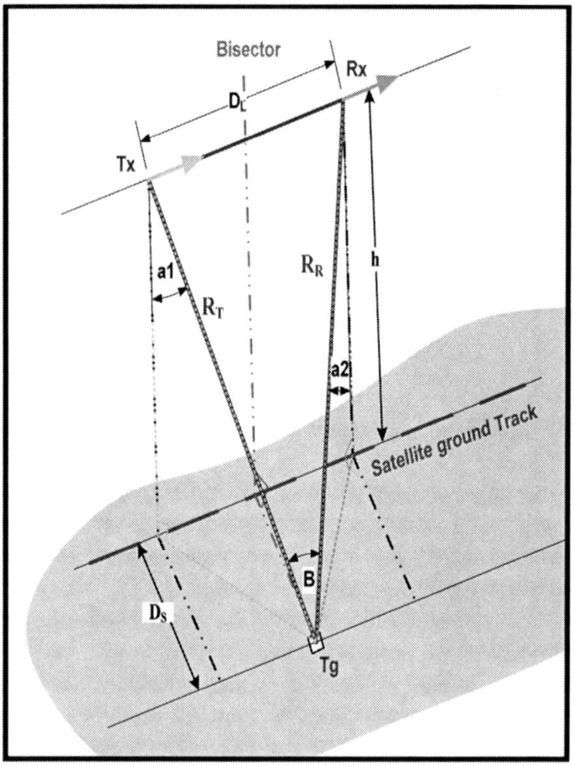

From orbital dynamics theory, the distance between the two satellites could be as close as few tens of metres (i.e. the satellites could be placed in tight formation), however, this would be too close from an RF perspective – the transmitter would interfere with the receiver. We need to separate the satellites sufficiently so that we can use the "far field" radar equations, and also minimise side-lobe interference. In order to ensure we are in the far field with a 3.66 m parabolic dish antenna, we need a separation of ~5–10 km (to be safe) – which sets a minimum separation. As the maximum separation of 5° approximates to ~70–80 km, we propose a mean separation of ~30 km.

Even at this separation, there is potentially a problem of interference due to the transmit power leaking into the receiver through side-lobes, particularly if uniform illumination is assumed for the antennas. Therefore a more precise antenna design and potentially a side-lobe suppression technique is required in order to avoid this interference.

4.1 Synchronisation

Good synchronization between the transmitter and receiver is required, and high accuracy attitude control (and stability) is required to point the antennas on the two spacecraft to the same instantaneous foot print. This synchronization could be achieved by using extremely accurate independent time and frequency standards onboard each satellite, or by the continuous transfer of time and frequency standards between satellites.

For L-band micro SAR satellites, time and phase stability requirements are not as severe as they are in higher frequency systems such as X-band, thus the use of a temperature controlled crystal oscillator (TCXO) or light-weight and low cost atomic clock should be adequate for the first method. The second option can be achieved by having a direct communication link between the satellites, where the baseline distance is determined by the use of differential GPS [23]. Both satellites carry a GPS receiver as a routine element of the SSTL-100 guidance and navigation system, thus, this second method is proposed as being most appropriate.

For a 3.66 m antenna, operating at 1.28 GHz, the half-power beam width is ~4.5°, therefore a pointing accuracy of ~0.5° will be adequate – well within the capability of the SSTL-100 platform. However, the attitude stability requirements are severe – of the order of 5×10^{-4} degrees per second or less roll/yaw rates for 30 m resolution. This would probably require the use of control moment gyros (CMGs) in the attitude control system.

4.2 Waveform Consideration (PR or CW)

Classic pulse radar (PR) is being replaced by CW radars in several areas because of its unique features such as the lack of ambiguity, very low transmitted power and good electromagnetic compatibility with other radio devices [24]. CW transmissions employ 100% duty cycle, low continuous power instead of PR's high peak power for the same detection performance, and it is the average power that determines the detection characteristics [25]. Radar transmitters would be less complex and costly if they could simply operate in CW mode – like broadcast stations. Having to generate very high pulsed RF power leads to much higher operating voltages, energy storage problems, and the necessity for high power switching devices [26].

A fundamental limitation in CW radar systems is sensitivity to the transmitter power leaking into the receiver [26]. In order for a CW radar to work, the transmit and receive antennas must be separated to isolate the receiver from the transmitter. The bi-static configuration proposed offers this. None-the-less sensitivity to interference via the side-lobes of the antennas remains an issue. However, it is possible to use a feed through nulling technique to eliminate the residual transmitter leakage by using negative feedback at the receiver input to cancel a transmitter carrier when the two antennas cannot be separated enough to reduce transmitter leakage below the receiver noise level.

CW radars require some form of modulation to resolve targets in range. The well known and simplest modulation technique most often used is linear frequency modulation (FM) where the signal is "chirped" with triangular or saw-tooth waveforms controlling the oscillator and sweeping the carrier frequency. This type of modulation offers a significant advantage over a phase coded waveform (PCW) because it does not require Doppler correction [27]. Whether FMCW or PCW, it is average power that needs to be considered and the lower peak power associated with CW radars lends itself to implementation via solid state power amplifiers. These are generally smaller, cheaper and more robust than electron beam amplifiers (such as travelling wave tubes – TWTs) which are often necessary to provide the large peak powers for pulsed radars.

FMCW is relatively simple to implement and has already been demonstrated on unmanned aerial vehicles (UAVs) [28–30]. However, PCW – Pseudo Random Noise (PRN) coding/modulation is also appealing because of its use in GPS, (which operates in the same frequency band (L-band)), which means that some of the building blocks are available as commercial-off-the-shelf devices. This system also lends itself to multi-static systems as different PRN codes can be used to distinguish the different satellites. In the past, PRN coded radars were not practical due to digital technology and computational power limitations [24]. However, with recent advances in signal processing, digital processor speed and analogue to digital converters, PRN radars can now be implemented at lower cost [31], and this could potentially support a compact and cost effective micro SAR design. Therefore, both alternatives are being investigated and analysed along with improvements in power systems and energy storage.

Due to bi-static complexities, both FMCW and PRN systems rely heavily on signal processing techniques to produce the final image product. These are more complex in the bi-static case than for conventional mono-static SARs. The bi/multi-static backscatter coefficient and clutter resolution (which is geometry dependent) is more complex than in the mono-static case, and there is still much work to do in the development of these models.

The analysis of this bi-static LFMCW/PRN micro SAR system design and the development of special signal processing algorithms for imaging radar are still under active development within the Surrey Space Centre (SSC).

5 Conclusions

The feasibility of a low-cost L-band micro SAR satellite for forestry and flood monitoring has been studied. A CW radar in a bi-static configuration to provide modest (30 m) resolution over a 30 km swath is proposed. Most associated design requirements (mass, structure, power, antenna size and deployment) can be met, although synchronisation, platform stability and antenna sidelobe control remain critical.

Even though the bi/multi-static approach is complicated, and double the mass of a mono-static solution, it increases imaging flexibilities, reduces high peak power requirements, and supports a simplified SAR payload which paves the way for a low-cost constellation which helps fulfil coverage and shortens revisit time requirements.

We believe that such a configuration can lead to a (relatively) low-cost all-weather day/night imaging disaster monitoring constellation.

References

1. Sun, W., Hodgson, D., L. Boland, A. da Silva Curiel, C. Underwood, S. Mackin, M. Sweeting. "Operational Earth Observation – Continuity of the Disaster Monitoring Constellation", Paper presented at the International Astronautical Congress, Valencia, Spain, 2–6 October 2006. Published in the proceedings (IAC06-B5.1.02).
2. G. Keyte, A. Thornbury. Small satellite SAR. Journal of the British Interplanetary Society, 51, 381–386, 1998.
3. T. Algra, W.J. Looyen, M.D. Brouwer, et al. P-band SAR mission dedicated to global monitoring of forests. NLR 1997.
4. B. Hallberg, G. Smith, A. Olofsson, et al. "Performance simulation of spaceborne P-band SAR for global biomass retrieval," presented at Geoscience and Remote Sensing Symposium, 2004. IGARSS '04. Proceedings. 2004 IEEE International, 2004.
5. JPL, "LightSAR-Monitoring Floods," Jet Propulsion Laboratory, 1999.
6. JPL, "LightSAR Science Requirements and Mission Enhancements," vol. 2007, 1998, 1–22.
7. A.A. Kononov, K. Min-Ho. Model-associated forest parameter retrieval using VHF SAR data at the individual tree level. Geoscience and Remote Sensing, 46, 69–84, 2008.
8. H. Balzter, L. Skinner, A. Luckman, et al. Estimation of tree growth in a conifer plantation over 19 years from multi-satellite L-band SAR. Remote Sensing of Environment, 84, 184–191, 2003.
9. S.S.B. Committee on Earth Studies, National Research Council, "Development and Application of Small Spaceborne Synthetic Aperture Radars," vol. 2007: The National Academies Press, 1998.
10. Z.-W. Xu, J. Wu, Z.-S. Wu. "A survey of ionospheric effects on space-based radar," vol. 2007, 2004.
11. ITU, "Rec. ITU-R P.531-9 Ionospheric propagation data and prediction methods required for the design of satellite services and systems," vol. 2008, 2007.
12. ITU, "Rec. ITU-R P.618-9 Propagation data and prediction methods required for the design of Earth-space telecommunication systems," vol. 2008, 2007.
13. E. Chapin, S.F. Chan, B.D. Chapman, et al. "Impact of the ionosphere on an L-band space based radar," presented at Radar, 2006 IEEE Conference on, 2006.
14. F. Meyer, R. Bamler, N. Jakowski, et al. "The Potential of Low-Frequency SAR Systems for Mapping Ionospheric TEC Distributions," Geoscience and Remote Sensing Letters, 3, 560–564, 2006.
15. D.B. Marron. "Alternatives for Military Space Radar," The Congress of the United States – Congressional Budget Office 2007.
16. F.M. Henderson, A.J. Lewis. Principles and Applications of Imaging Radar, USA, John Wiley & Sons, Inc., 1998.
17. H.J. Kramer. "TecSAR (SAR Technology Demonstration Satellite)," 2006.
18. M. Alberto, K. Gerhard. "Spaceborne Synthetic Aperture Radar (SAR) Systems: State of the Art and Future Developments," 2003.
19. R.J. Sullivan. Radar Foundations for Imaging and Advanced Concepts. USA: Scitech Publishing, Inc., 2004.

20. SpotImage, "Spot Image – TerraSAR-X," vol. 2007, 2004.
21. W.N. Edelstein, C. Andricos, A. Moussessian, et al. "High-efficiency L-band transmit/receive module for synthetic aperture radar," presented at Radar Conference,2003.
22. N.J. Willis. Bistatic Radar. USA: Artech House, Inc., 1991.
23. A.M. Horne, G. Yates. "Bistatic synthetic aperture radar," presented at RADAR 2002, 2002.
24. K. Kulpa. "Continuous Wave Radars-Monostatic, Multistatic and Network". *NATO – Advanced Study Institute – Advances in Sensing with Security Applications*, 2005.
25. M. Jankiraman. Design of Multi-Frequency CW Radars, 2007.
26. M.I. Skolnik. Radar Handbook, 2nd Edition, USA, McGraw-Hill, 1990.
27. M. Cherniakov. Bistatic Radar – Principles and Practice. England, John Wiley & Sons Ltd., 2007.
28. J.J.M. de Wit, P. Hoogeboom. <misc>"SAR: A Novel Application for FM-CW Radars," presented at European Microwave Conference, 2001. 31st, 2001.
29. A. Meta, P. Hoogeboom. "High resolution airborne FM-CW SAR: digital signal processing aspects," presented at Geoscience and Remote Sensing Symposium, 2003.
30. J.J.M. de Wit, P. Hoogeboom. "High resolution FM-CW SAR performance analysis," presented at Geoscience and Remote Sensing Symposium, 2003.
31. E.J. Holder, D. Aalfs, B.M. Keel, et al. "A comparison of PRN and LFM waveforms and processing in terms of the impact on radar resources," presented at Radar, CIE International Conference, 2001.

Mission Design of the Dutch-Chinese FAST Micro-Satellite Mission

D. Maessen, J. Guo, E. Gill, E. Laan, S. Moon, and G.T. Zheng

Abstract The chapter treats the mission design for the Dutch-Chinese FAST (Formation for Atmospheric Science and Technology demonstration) mission. The space segment of the 2.5 year mission consists out of two formation flying micro-satellites. During the mission, new technologies will be demonstrated and, using spectropolarimeter and altimeter payloads on both spacecraft, observations will be performed characterizing atmospheric aerosols and seasonal variations of height profiles in the cryosphere. The mission is divided into four phases, each with a different orbital geometry. The rationale for and the orbital geometry during these phases as well as the transitions between the phases are treated in detail. A major complication to the mission design is the amount of data that can be sent to the ground. Since only two moderately capable ground stations, one in Delft and one in Beijing, are baselined, as much data processing as possible has to be performed onboard to allow high duty factors for the science payloads. When this is not possible, alternative payload operation modes have to be sought with which maximum scientific data return can be obtained through as little payload operation as possible. Both options are dealt with in the chapter.

1 Introduction

The Dutch-Chinese FAST (Formation for Atmospheric Science and Technology demonstration) mission is a collaboration between Delft University of Technology in the Netherlands and Tsinghua University in Beijing, China. During the 2.5 year mission, two micro-satellites flying in a loose formation will demonstrate new technologies and, using spectropolarimeter and altimeter payloads on both spacecraft, will perform observations characterizing atmospheric aerosols and seasonal variations of height profiles in the cryosphere [1–3].

S. Moon (✉)
cosine Research B.V., Niels Bohrweg 11, 2333 CA, Leiden, The Netherlands
e-mail: s.moon@cosine.nl

R. Sandau et al. (eds.), *Small Satellite Missions for Earth Observation*,
DOI 10.1007/978-3-642-03501-2_17, © Springer-Verlag Berlin Heidelberg 2010

The FAST mission is described in detail in the coming sections. First, the mission objectives and mission architecture are provided. These are followed by a detailed description of the various mission phases. The fifth section explores several ways to maximise science data return whilst minimizing the amount of data sent to the ground stations. Finally, several Conclusions are drawn.

2 Mission Objectives

The top level mission objectives for the FAST mission are threefold, namely: technology demonstration, science, and education. Each objective is discussed in the following sections.

2.1 Technology Demonstration

The technology demonstration objective is split into one primary and two secondary objectives. The primary objective is to demonstrate autonomous formation flying using various communication architectures with distributed propulsion systems and MEMS technology optimizing propellant consumption. The secondary objectives are to demonstrate distributed, fault tolerant, out-of-core computing using two spacecraft and to demonstrate real-time GNSS ephemerides updates for onboard navigation solution improvement.

2.2 Science

For the FAST mission, the opportunity exists to fly several advanced science payloads: SPEX (Spectropolarimeter for Planetary Exploration), SILAT (Stereo Imaging Laser Altimeter), and a modified Chinese altimeter earlier flown on Shenzou-4. This allows the formulation of the following scientific objective: Characterize atmospheric aerosols and monitor the seasonal variation of height profiles in the cryosphere and correlate data to visual and stereoscopic imagery for improved science return.

2.3 Education

In the field of education, the FAST mission has as objective to teach cutting-edge technology, broaden the international view of students and boost skills through exchange of students and staff members between Delft and Beijing.

3 Mission Architecture

The baseline orbit for both satellites, which are foreseen to be piggyback launched on a single Chinese launcher at the end of 2011, is a 650 km high sun synchronous orbit with an ascending node time of ~10:00 h. This orbit is a compromise between spatial resolution, coverage, ground station contact time, and expected launch opportunities. The ascending node time follows from the need to have a substantial angle between the orbital plane and the sun vector for optimal spectropolarimeter science output. The ground segment of the mission consists of a science data centre, a mission control centre, and two S-band university ground stations: one in Delft and one in Beijing, cf. Fig. 1.

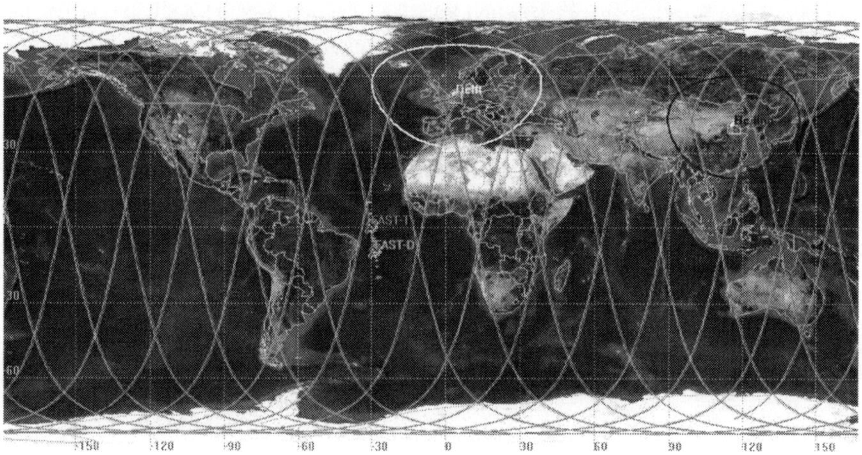

Fig. 1 FAST mission architecture

The 2.5 year mission is divided into four phases, each with a distinctly different orbital geometry. The rationale for and the orbital geometry during these phases as well as the transitions between the phases are treated in detail in the next section. As will be explained, the total velocity change (ΔV) per satellite is estimated to be around 5.5 m/s for the entire mission duration. However, taking into account the possible failure of the propulsion system of one of the satellites and allowing for operational anomalies and disturbances not taken into account in the calculations, a ΔV budget of 12 m/s is currently assigned to each satellite. This Results in approximately 1 kg of propellant mass in case of cold gas propulsion.

4 Mission Phases

Since FAST has both technological as well as scientific mission objectives, the mission design is a compromise between these two. As a result, the first weeks of the mission focus on technology demonstration while the remainder of the mission focuses on science. After the launch and early operations (LEOP) phase, the mission consists of two distinct one year phases, science mode A and science mode B. The aspect that sets these two phases apart is the line of sight between the satellites: in science mode A, there is a direct line of sight between the two satellites while in science mode B the distance between the satellites is such that no direct line of sight is existing. The coming sections detail the various mission phases.

4.1 Launch and Early Operations

The LEOP phase consists out of the launch, orbit insertion, detumbling, and check-out of the two FAST spacecraft, called FAST-T (Tsinghua) and FAST-D (Delft). The two spacecraft, which are not mated during launch, will be inserted into the same orbit with a small along-track separation. At any time during LEOP, the two spacecraft will not fly in formation. However, since collision of the two spacecraft has to be prevented, passive collision avoidance measures need to be implemented during this phase. The current approach to achieve this is to insert the spacecraft with the smallest ballistic coefficient, currently FAST-T, into orbit before the other spacecraft with sufficient Δt and ΔV with respect to the other spacecraft to rule out any chance for collision during LEOP. After LEOP, which will take approximately one month, a small manoeuvre will be performed to attain the correct orbital geometry for the science mode A1 phase.

4.2 Science Mode A1

During science mode A1, the two spacecraft will demonstrate the ability to autonomously maintain an along track separation of 1 ± 0.1 km by means of autonomous formation flying (AFF) using propellant optimized distributed propulsion, cf. Fig. 2. This capability requires a propulsion system on each spacecraft as well as an inter-satellite communication link.

Since the FAST mission is foreseen to be launched in 2011, the mission timeline coincides with a period of maximum solar activity. When assuming for FAST-D a

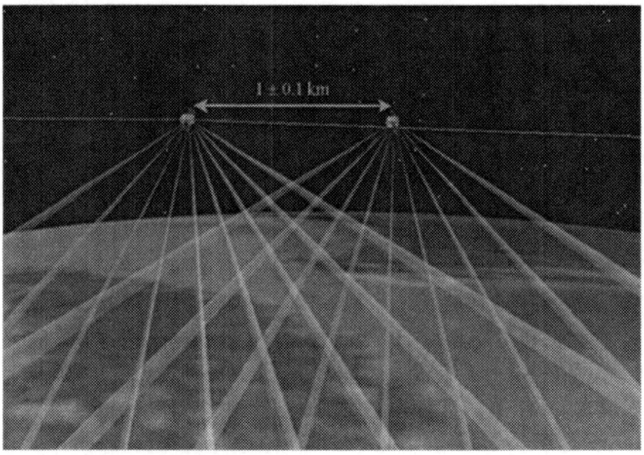

Fig. 2 Science mode A1

mass of 50 kg, a drag coefficient of 2.2, and a ballistic coefficient that differs 30% from the ballistic coefficient of FAST-T, the two spacecraft will rapidly drift apart in along track direction. In fact, the relative drift is such that a 1 km drift is achieved within 5 days for solar maximum conditions. Since LEOP takes approximately 1 month, a small manoeuvre will therefore be required to bring the inter-satellite distance back from approximately 60 to 1 km. An attractive option to accomplish this is to lower the perigee of FAST-D and to increase the apogee of FAST-T to let FAST-D "catch-up" with FAST-T. Once the distance between the satellites is as desired, the orbits are re-circularized and science mode A1 commences. When one day is chosen as the time frame for this manoeuvre, a ΔV of 0.1 m/s needs to be provided by each satellite.

For solar maximum conditions, it will take the satellites slightly over one day to traverse the 200 m control window of this mission phase. To maintain the desired formation geometry, taking into account differential drag only, a ΔV of 2 mm/s is then required to bring the inter-satellite distance back to the other extreme of the control window.

Although the orbital geometry is not optimized for it, some first science observations will be performed during this phase of the mission. Due to the relatively small inter-satellite distance, this phase lends itself well to perform distributed computing experiments between the two spacecraft and to perform cross-calibration between the instruments on the different spacecraft since they are essentially measuring the same properties at almost the same moment in time. The AFF demonstration will be performed for several weeks and will be followed by a transition to the required relative geometry for science mode A2.

4.3 Science Mode A2

In this mission mode, the along-track separation between the satellites will be 1,225 $\pm$ 5.7 km. This orbital geometry lends itself extremely well to perform synoptic and synergetic observations with the altimeters and especially the SPEX instruments on both spacecraft. The orbital geometry combined with the nine Earth looking fields of view (FOV) of the SPEX instrument [4] Results in many simultaneous intersections of SPEX FOVs and in several overlapping SPEX FOVs at the Earth's surface, cf. Fig. 3. This allows retrieval of aerosol characteristics at specific altitudes at a single moment in time and it allows making more observations of geolocations from various angles during a single pass (especially near the equator), which is highly desired for aerosol characterization. The $\pm$5.7 km accuracy in the inter-satellite range is driven by the pixel size of SPEX at 650 km altitude, which is approximately 11.4 km [4].

Changing the orbital geometry from mission mode A1 to mission mode A2 can be achieved simply by letting the two satellites drift apart due to differential drag and by correcting afterwards for relative orbital node drift. However, this will take roughly 100 days even for solar maximum conditions, which is too long. Therefore, the same approach as earlier is adopted only now the inter-satellite distance needs

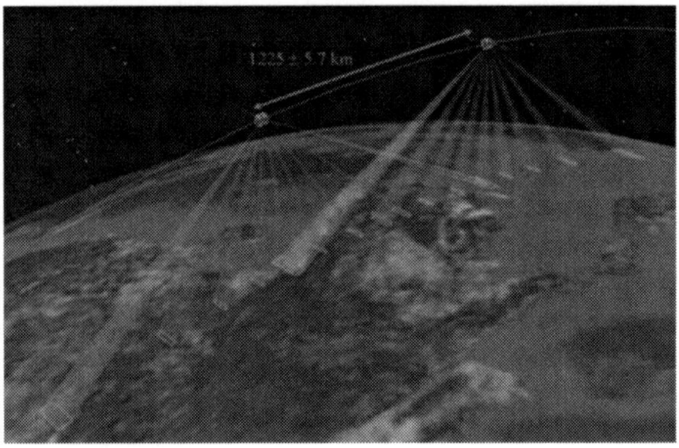

Fig. 3 Science mode A2

to increase. Thus, an increase in apogee for FAST-D and a decrease in perigee for FAST-T are required. When the change in semi-major axis for both satellites is set to 2 km, the manoeuvre requires a ΔV of 1 m/s for each satellite and will take a bit over two days. For one year of formation maintenance, again taking only differential drag into account, approximately 0.6 m/s of ΔV is required per satellite and a manoeuvre needs to be executed every 9 days.

4.4 Science Mode B

For science mode B a large temporal separation between the satellites is created, allowing valuable spectropolarimeter observations to be made of locations that exhibit a high temporal dependence on local aerosol characteristics such as cities and industrial sites. Ideally, the temporal separation is in the order of hours, allowing differences to be measured between e.g. morning and afternoon. Unfortunately, this requires a relative plane change to be performed by the satellites, which is very expensive in terms of propellant. For that reason, the current mission plan foresees in an orbital geometry where the satellites are separated 180° in the same orbit, creating a 49 min temporal separation.

The orbital geometry for this mission phase is achieved in the same way as for mission mode A2. When a transfer time of two weeks between the mission modes is taken as acceptable, a change in semi-major axis of roughly 5.5 km needs to be achieved by both satellites, resulting in a ΔV per satellite of 3 m/s. For one year of formation maintenance, again approximately 0.6 m/s of ΔV is required per satellite.

5 Increasing Scientific Data Return

For FAST, only two moderately capable S-band ground stations are baselined: one in Delft and one in Beijing. With an expected downlink data rate of 1–2 Mbps and raw payload data generation speeds of 1.2 Mbps (SPEX) and 26.2 Mbps (SILAT), low payload duty factors are inevitable for the FAST-D satellite. However, these can be improved upon using various strategies.

5.1 Strategies for SPEX

For SPEX, the allowable duty factor is much higher than for SILAT, but can easily be improved further by onboard deletion of pixels without significant information, reducing the amount of raw data produced by ~75%. The SPEX duty factor can even be drastically improved when by means of onboard processing a curve is fitted in a least squares sense through the recorded spectrum, resulting in merely six 32-bit floating-point numbers per imaged spectrum or less than 1% of the original raw data and allowing SPEX operation for the entire sunlit part of the orbit.

The curve fitting approach however requires significant onboard computing power. Even for a state of the art cell phone processor capable of speeds of 3 GFLOPS (floating-point operations per second) and a relatively low power consumption of 5 W (for the entire chipset), the energy required to do the curve fitting is roughly twice the energy needed to download the raw 10-bit SPEX data containing significant information for the same duty factor. Thus, advanced onboard data processing is in this case less efficient than downloading the raw data. However, keeping in mind that processor performance rapidly increases, performing the curve fitting onboard merely as an experiment to prepare for future capabilities is already valuable in itself. In addition, there is still the issue of a moderately capable ground segment, severely limiting the allowable duty factor of SILAT on FAST-D (currently less than 1%). Since processing the SPEX data onboard will free up a very considerable amount of bandwidth for extra SILAT data, thereby allowing a higher SILAT duty factor and increasing its science return, doubling the energy needed to download all SPEX data remains an option for the FAST mission. A final decision on this matter needs to be taken in a later stage of the design. Lastly, it is noted that to increase the tolerance of the non-space qualified processor to radiation effects, the FAST mission foresees to implement fault-detection algorithms that are guaranteed to produce a correct solution in case of e.g. bit flips for only a 10% computing overhead [5].

5.2 Strategies for SILAT

For SILAT, data reduction can be best achieved by reducing the data rate of its high resolution camera (HRC), which accounts for 95% of the raw data produced by SILAT. This can be done through technical means or through operational means. The first technical option is to bin the HRC detector pixels, resulting in lower resolution "macropixels". A second option is to reduce the size of the HRC detector, resulting in a smaller swath width. The third and last technical option is to reduce the number of colour filters used on the HRC from three to one, resulting in having to make only one picture of a scene instead of three, but coming at the price of much reduced science output.

Operationally, the duty factor of the HRC can be significantly reduced by using it only at locations with large slopes, such as edges of ice sheets or mountainous regions, since those are the places where HRC information greatly enhances the SILAT data quality. However, this requires meticulous operations planning and will therefore be costly.

Currently, it has not been decided yet which SILAT performance is acceptable. Since FAST is not a science mission, non-optimal performance is a viable option, but the right balance between performance and data amount still needs to be determined.

6 Conclusions

The chapter has introduced the bilateral FAST formation flying mission and presented its mission objectives. The mission architecture and a detailed mission description have been provided as well as several strategies to increase payload duty factors while maintaining an acceptable science level.

Since FAST focuses on technology demonstration as well as science, the mission design is a compromise between the two. In addition, the science return in the fields of aerosol characterization and cryosphere altimetry will not be able to compete with that of a dedicated science mission, but can be used to augment data from such missions. The formation flying capability of the FAST satellites allows the mission to be divided into several phases, each with its own orbital geometry and consequent technology demonstration and/or science benefits. Since the ground and space segments of the FAST mission only allow for moderate data rates between the ground stations and the satellites, several data processing activities are performed onboard the FAST satellites to enable higher payload duty factors and therefore increase the scientific output of the mission.

References

1. D.C. Maessen et al. Preliminary Design of the Dutch-Chinese FAST Micro-Satellite Mission. In Proceedings of the 4S Symposium, Rhodes, Greece, 26–30 May, 2008.
2. D.C. Maessen et al. Increasing System Performance and Flexibility: Distributed Computing and Routing of Data within the FAST Formation Flying Mission, IAC-08-D1.4.6, In Proc. of the 59th IAC, Glasgow, Scotland, September 29–October 3, 2008.
3. E. Gill et al. Atmospheric Aerosol Characterisation with the Dutch-Chinese FAST Formation Flying Mission, IAC-08-B1.I.1, In Proc. of the 59th IAC, Glasgow, Scotland, September 29–October 3, 2008.
4. D.C. Maessen et al. Conceptual Design of the FAST-D Formation Flying Spacecraft, To be published in Proc. of the 7th IAA SSEO Symposium, Berlin, Germany, May 4–8, 2009.
5. J.A. Gunnels, D.S. Katz, E.S. Quintana-Orti, R.A. van de Geijn. Fault-Tolerant High-Performance Matrix-Matrix Multiplication, FLAME Working Note #2, The University of Texas at Austin, Department of Computer Sciences, Technical Report TR-2000-34, December, 2000.

Part IV
University Satellites

SPRITE-SAT: A University Small Satellite for Observation of High-Altitude Luminous Events

Yukihiro Takahashi, Kazuya Yoshida, Yuji Sakamoto, and Takeshi Sakamoi

Abstract SPRITE-SAT is a micro satellite in the size of 50 cm cube and weighing 45-kg, designed and developed by Tohoku University. Its mission objective is to conduct scientific observation of atmospheric luminous emissions called "sprites" and terrestrial Gamma-ray flushes. Both are recently discovered phenomena and their mechanisms are still under the veil. SPRITE-SAT was developed to achieve significant observations to determine clear models of these mysterious phenomena. On January 23rd, 2009, SPRITE-SAT was successfully launched by JAXA's H-IIA rocket as a piggyback payload of Greenhouse Gas Observation Satellite (GOSAT). The spacecraft is now in a sun-synchronous polar orbit with 670 km altitude form the Earth's surface.

1 Introduction

SPRITE-SAT is an Earth science micro satellite designed and developed by the faculty and student members of Tohoku University, Japan. The objective is to monitor luminous emissions in the upper atmosphere, called "sprites" [1–3] (see also Fig. 1).

Sprites are a type of Transient Luminous Events (TLEs) that occur in the middle atmosphere (typically 40–90 km altitude, from upper Stratosphere to Mesosphere). They are frequently observed associated with lightning discharges in the lower atmosphere. Upward lightning discharges into clear air have been reported by pilots worldwide since the start of the early twentieth century. However, the first visual evidence of a TLE phenomenon was documented in 1989. An image of an unusual luminous electrical discharge over a thunderstorm had been obtained with a low-light-level television camera [4]. Such optical signatures of these events were later

K. Yoshida (✉)
Department of Aerospace Engineering, Graduate School of Engineering, Tohoku University, Aoba 6-6-01, Sendai 980-8579, Japan
e-mail: yoshida@astro.mech.tohoku.ac.jp

This chapter is also written by SPRITE-SAT Development Team.

R. Sandau et al. (eds.), *Small Satellite Missions for Earth Observation*,
DOI 10.1007/978-3-642-03501-2_18, © Springer-Verlag Berlin Heidelberg 2010

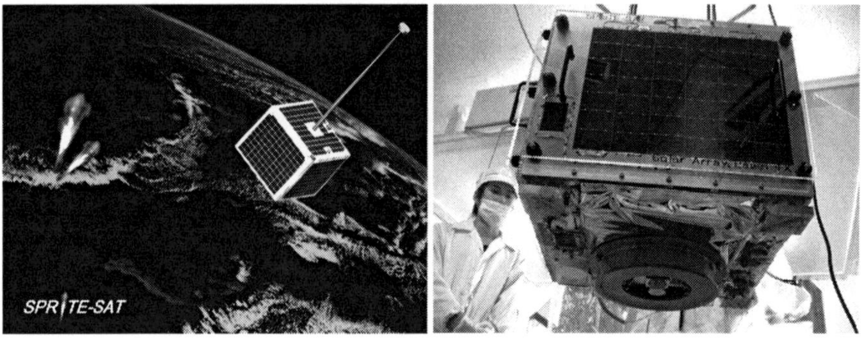

Fig. 1 Artist's view of SPRITE-SAT with the deployed gravity gradient boom (*left*) and a photo of the spacecraft during qualification tests

named "sprites" and "elves," both of which come from fairies in fantasy. Today, various types of the luminous events are reported and catalogued [5] (see also Fig. 2). Although extensive studies on sprites have been made since their first documentation, their characteristics and even the fundamental mechanisms cannot be fully explained so far.

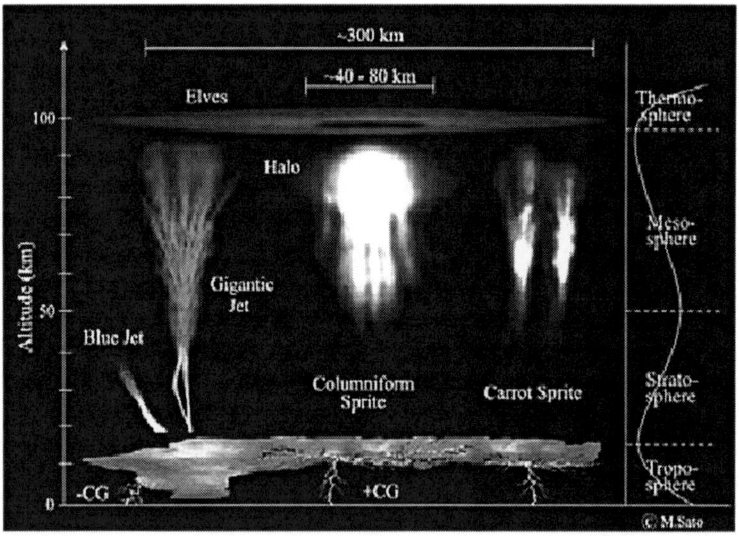

Fig. 2 Various TLE types in the upper atmosphere

Brief bursts of gamma-rays coming from outer space have been observed since the 1960s, but recent observations suggest that shorter bursts of gamma-rays occur in the Earth's atmosphere. Such bursts are called the Terrestrial Gamma-ray Flashes

(TGF). The flashes are considered generated by high-energy photons in the atmosphere and, therefore, a strong relationship is inferred with the thunderstorms and TLEs, such as sprites, as a possible source of these gamma-ray bursts.

SPRITE-SAT was designed and developed to achieve significant observations of these natural events in the upper atmosphere from the orbit. Nadir observations of sprites are particularly essential to identify the horizontal structure of the sprites. Also concurrent measurements of lower lightning discharges, TLEs and TGFs from space are important to understand the relationship and generation mechanisms of TLEs and TGFs.

SPRITE-SAT was manufactured and integrated by the faculty and student members in the Science and Engineering departments of Tohoku University, with technical supports from mentors (well-experienced experts). Students played leading roles in the assembling and testing processes. On January 23rd, 2009, the spacecraft was successfully launched as a piggyback payload of the H-IIA-15 rocket for the JAXA's Greenhouse Gas Observation Satellite (GOSAT). SPRITE-SAT is now in a sun-synchronous polar orbit with 670 km altitude form the Earth's surface.

2 Spacecraft Design

2.1 System Overview

The spacecraft structure is cubical with a side length of 50 cm. Its attitude stabilization is provided by a deployable gravity gradient boom which was newly developed for SPRITE-SAT [2]. The nominal attitude of the spacecraft is oriented in such a way that the bottom panel of the satellite (-Z panel) is always directed toward Earth. During the de-tumbling phase, two single-axis magnetic torquers are used to slow down the rotational motion of the spacecraft in prior to the boom deployment and to damp the liberation motion after the deployment.

The overall specification of the spacecraft is listed in the Table 1.

2.2 Onboard Science Instruments

As illustrated in Fig. 3, SPRITE-SAT equips with the following science instruments.

LSI-1 (Lightning Spectrum Imager-1): The objective is to detect lighting flashes. The camera features a CMOS detector with a format of 512 × 512 pixels. Observations are being made in the spectral band of 740–782 nm. LSI-1 and LSI-2 have a square FOV of 29° corresponding to a ground surface side length of 342 km.

LSI-2 (Lightning Spectrum Imager-2): The objective is to detect sprites. The camera features a CMOS detector with a format of 512 × 512 pixels. Observations are made in the spectral band at 762 nm. By subtracting simultaneous LSI-1 and LSI-2 images from each other, the image of sprites

Table 1 The overall specification of SPRITE-SAT

Spacecraft size	W 490 × D 490 × H 494 mm after mast deployment W 490 × D 490 × H 1354 mm
Spacecraft mass	45.3 kg
Attitude control	Type: gravity gradient stabilization Deployable boom: 860 mm in length, tip-mass: 3.680 kg, sleeves: 0.5 kg Sensors: 3-axis magnetometers in GAS (Geomagnetic Aspectmeter Unit) Actuators: 2-axis magnetic torquers (MTQ)
Attitude and orbit sensors	3-axis sun sensors (6 solar cells) 3-axis magnetometers (in GAS) 3-axis magnetometers, TAMU (Tohoku-AAC MEMS Unit) 3-axis gyro sensors (in TAMU) 3-axis accelerometers (in TAMU)
Power subsystem	Solar cells: P-type single crystalline silicon, 42 series/panel × 5 panels Batteries: 9-cell NiMH (total 3.3AH, 10.8 V) Power generation: 23.5 W (avg. in 62-min sunshine per orbit Power consumption: 22.0 W (observation mode), 12.0 W (communication mode), 9.5 W (at typical mode)
RF communications	Uplink: UHF, 1,200 bit/s at Sendai station, Japan Downlink: S-band, 0.1 W, 9.6 or 1.2 kbit/s Primary data acquisition: Sendai station, Japan Secondary data acquisition: Kiruna station, Sweden

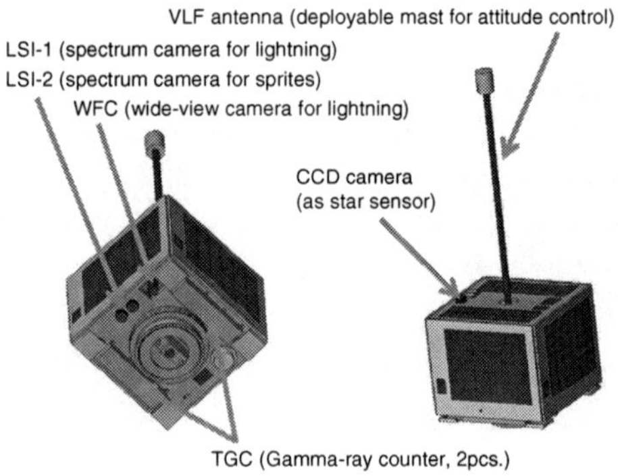

Fig. 3 Instrument locations on SPRITE-SAT

can be extracted from the background illumination of cloud-to-ground and inter-cloud lightning discharges.

WFC (Wide Field-of-view Camera): This camera is being used to determine the location of lightning flash which is relating to the TGF event. This high sensitivity panchromatic CCD camera has a fish-eye lens covering 140°.

TGC (Terrestrial Gamma-ray Counter): TGC is provided by JAXA/ISAS and the University of Tokyo. With a FOV (Field of View) = 134 × 180° detects gamma-ray flashes caused by lightning discharges with a time resolution of 0.25 ms.

HSS (High-sensitivity Star Sensor): The objective is to detect stars. This sensor features a high sensitivity panchromatic CCD imager used as a star tracker. Additional attitude information can be obtained by comparing the angle pattern of stars detected by the imager and that in a star-chart.

VLF receiver: The gravity boom is also used as a VLF antenna. The receiver electronics was developed by Stanford University (Stanford, CA). The objective of the VLF receiver is to measure the radio emissions from channels of lightning discharges not only of cloud-to-ground but also of intra-cloud discharges.

2.3 Attitude Sensors and Magnetic Torquers

SPRITE-SAT has multiple attitude sensors and an active damping control system. The significant sensor for the control system is 3-axis magnetometers (termed as Geomagnetic Aspect Sensor, GAS). The GAS was integrated by Tohoku University using a commercial-off-the-shelf three-axis magnetometer chip. The measurement sensitivity is 1 V/gauss, and the output range is +/– 2.5 gauss (2.5 V at 0 gauss). With an 8 bit AD converter, the GAS has the measurement range of magnetic field in +/– 0.6 gauss (= 60,000 nT) with the resolution of 470 nT.

In addition to GAS, the spacecraft equips SAS (Sun Aspect Sensor) using solar cells on 6 outer panels of the spacecraft, and MEMS gyros and accelerometers which was built in Tohoku-AAC MEMS Unit (TAMU) for off-line attitude determination. The 3-axis orientation of the spacecraft in the inertial space is obtained by the combination of GAS and SAS, and the time history of the attitude profile will be determined in a good precision by using gyros and accelerometers with a Kalman filter process.

In order to provide active damping capability, a couple of coreless-type magnetic torquers (MTQ, see Fig. 4) were developed and attached on the side panel of the spacecraft in X and Z axes. The specifications of the torqures are listed in Table 2.

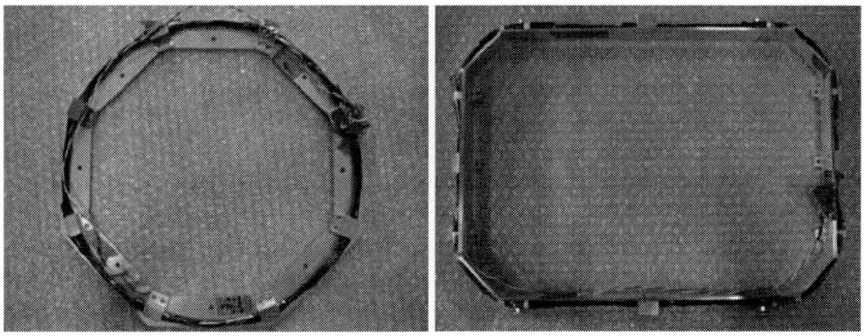

Fig. 4 The flight model of MTQ-X (*left*) and MTQ-Z (*right*)

Table 2 Specification of magnetic torqures

Device name	MTQ_X	MTQ_Z
Position	side panel (-X)	bottom panel (-Z)
Size (W × D × H)	315 × 315 × 10 mm	410 × 300 × 35 mm
Mass	420 g	668 g
Number of coil turn	62	11
Core area	0.0652 m^2	0.1039 m^2
Magnetic moment	0.22 Am^2	1.25 Am^2

2.4 Deployable Mast

The deployable mast (MST), which has a 3.7 kg mass at the end tip and in the length of 180 mm in the stored configuration (including the tip mass), can be deployed up to 1000 mm. A ribbon spring made of Be-Cu is used to provide the extension force (see Fig. 5). Tapered thin aluminum tubes were arranged to cover the ribbon spring. These tubes work as an extendable telescope to increase the stiffness of the structure and ensure the electrical conductivity as a VLF antenna after the deployment. The whole mast structure is electrically isolated from the spacecraft main body using glass-reinforced epoxy plates.

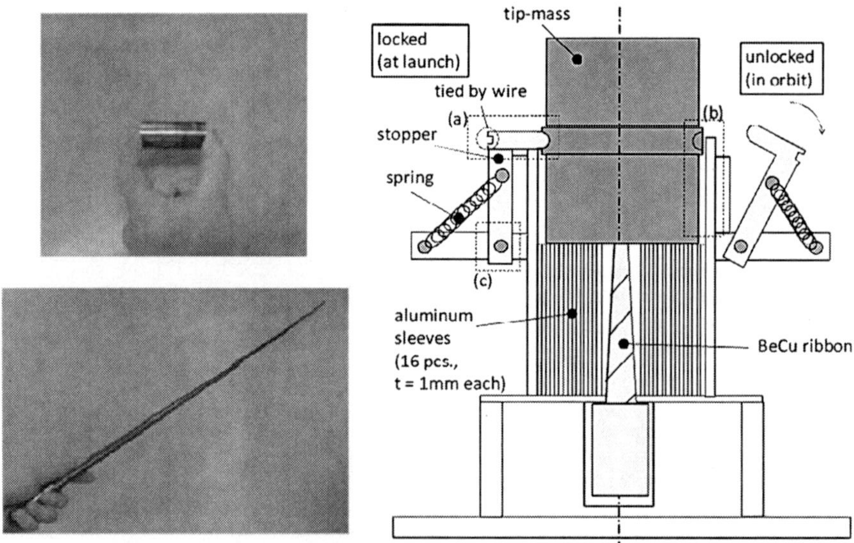

Fig. 5 Be-Cu ribbon spring (*left*) and the cress-section sketch of the mast mechanism

The tip mass is locked by four latching hooks, which are closed and tightened by nylon cables during the launch. The nylon cables are cut out by a burning command (using a Ni-Cr heater) in orbit, then the hooks are opened, so that the telescopic mast extends into the final configuration. Some pictures of the flight model of the mast are shown in Fig. 6.

Fig. 6 Flight model of the deployable mast (**A**: before deploy, **B**: after deploy, **C**: hooks closed, **D**: hooks opened)

2.5 Attitude Control

Onboard active damping control is conducted by a simple B-dot law using GAS and MTQ for the X and Z axes independently. A schematic timing chart is depicted in Fig. 7. Both MTQs are operated in a 4 second cycle such that they are active for 2 s after 2 s of interval, where the MTQ torques have only three states: positive

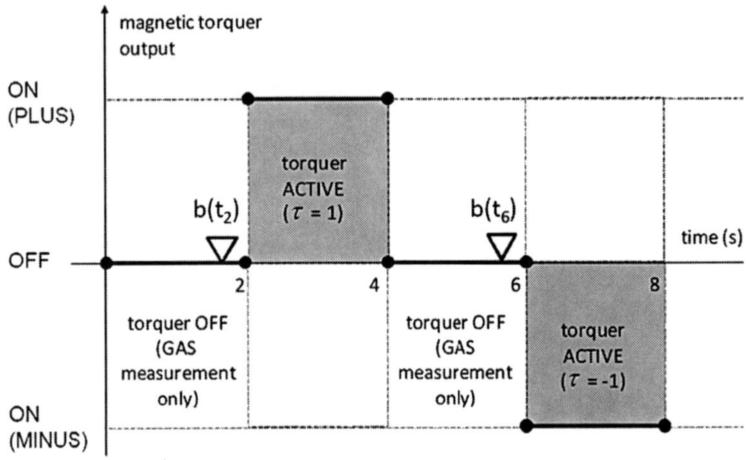

Fig. 7 A schematic time chart for the MTQ control

maximum, zero, and negative maximum. GAS measurement is conducted for the period when the MTQs are inactive.

The control law is relatively simple. Let $b_x(t)$, $b_z(t)$ be the GAS measurement values in x and z axes at time t, and τ_x, τ_z be the three-state torque command to each MTQ. Then τ_w ($w = x$ or z) is by the following rule:

if $b_w(t_i) - b_w(t_{i-4}) > 0$ then $\tau w = -1$ (negative maximum),
if $b_w(t_i) - b_w(t_{i-4}) < 0$ then $\tau w = +1$ (positive maximum),
else $\tau w = 0$

The control torque acting on the satellite can be simply computed by the following vector form equation:

$$N = M \times B,$$

where M is a torque vector generated by MTQ and B is a vector of the Earth's magnetic field.

2.6 Ground Station

A ground station that equips a twin Yagi antenna for UHF uplink and a 2.4 mϕ dish for S-band downlink was prepared in Tohoku University, see Fig. 8. Both antennas are mounted on a common remotely-controlled 2-axis gimbals table. Another downlink station with an S-band dish was prepared in Kiruna, Sweden to increase the amount of data transmission.

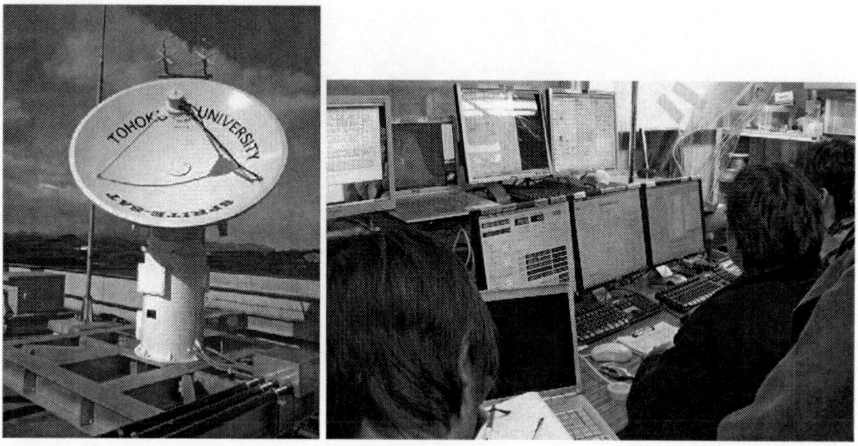

Fig. 8 UHF and S-band antennas on the roof of a science department building (*left*) and the operation room (*right*) in Tohoku University

3 Initial Operation and Results

After the successful launch and orbital insertion, the SPRITE-SAT has been successfully operated from the ground station in Tohoku University. The tracking of the satellite was based on the launch information of the rocket for the first few days, then has been done with the two-line elements (TLE) of the orbit. The international ID and catalog number of the SPRITE-SAT is 2009-002C 33494 [6].

Initial attitude operation was successful conducted. At the beginning, the spacecraft was in a tumbling motion at about 4–5°/s in all 3 axes. By means of magnetic torquers, the motion went into a single axis rotation, then a slow-rate tumbling at about 0.2–0.3°/s. Precise attitude determination is now under the process. The gravity boom/VLF antenna was also successfully deployed on February 4th, 2009, after the attitude rate went down around 0.1°/s and waiting for an appropriate timing so that the +Z panel, the boom side, goes opposite to the Earth.

In the initial operation period, TAMU (Tohoku-AAC MEMS Unit) was successfully tested. TAMU includes multiple commercial-off-the-shelf components, including 8051 microcontrollers, 4 Mbit Magnetic RAM, and CAN-bus interface, and all these components worked normally in the orbital environment. Attitude measurement devices were also worked well to record the de-tumbling and attitude damping sequence. The initial report was presented in [7].

A test image of the WFC, a high-sensitivity CCD imager with a fish-eye lens, was taken almost 12 h before the mast deployment. Figure 9 shows a night view of the Earth around the East China Sea, in which the airglow is seen on the Earth's limb and the location of the artificial lights of major cities are identified.

On the other hand, the onboard main controller had a problem after the mast deployment, and the satellite has been out-of-control, even though its battery and

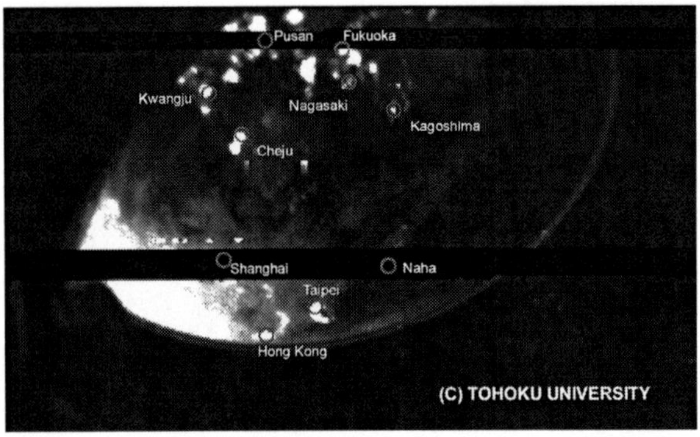

Fig. 9 An test image of night globe taken by WFC. Local Time = 2009/02/04 00:51:39 JST. Longitude = 128.972675 E (deg), Latitude = 33.397187 N (deg), Altitude = 667.994 (km)

power system is in a good condition. Scientific observation therefore has not yet started, as of May 1, 2009.

Acknowledgement This research is supported by Grant-in-Aid for Scientific Research No. 19002002, provided by Ministry of Education, Culture, Sports, Science and Technology (MEXT), Japan.

References

1. http://www.astro.mech.tohoku.ac.jp/SPRITE-SAT/index_e.html
2. Y. Sakamoto, K. Yoshida, E. Ujiie, Y. Takahashi, T. Sakanoi, K. Takiuchi, T. Sawakami, Y, Nakazato, S, Kondo. "Development of the Attitude Control System for Tohoku University SPRITE-SAT Using a New-Model Deployable Mast," Proceedings of the 59th IAC (International Astronautical Congress), Glasgow, Scotland, UK, Sept. 29–Oct. 3, 2008, IAC-08.C1.8.14.
3. E. Ujiie, K. Yoshida, Y. Takahashi, Y. Sakamoto, T. Sakanoi, Y. Masumoto, Y. Kasaba, S. Kondo. "Development of SPRITE-SAT for TLE and TGF measurements," IEICE (Institute of Electronics, Information and Communication Engineers), SANE2008-53, pp. 1–6.
4. R.C. Franz, R.J. Nemzek, J.R. Winckler. Television image of a large upward electrical discharge above a thunderstorm system. Science, 249(4964), 48–51, July 6, 1990.
5. http://www.spritesandjets.com/TLE-Introduction.htm
6. SPACEWARN Bulletin No. 663, NASA, Feb. 1, 2009.
7. F. Bruhn, E. Lamoureux, G. Chosson, J. Bergman, K. Yoshida, T. George, R. Thorslund, J. Köhler. "Bridging the Space Technology Valley of Death: Two Spaceflights in 2009 to Validate Advanced MEMS/Microtechnology Systems and Subsystems," CANEUS 2009 Workshop, NASA Ames, March 1–6, 2009.

SwissCube: The First Entirely-Built Swiss Student Satellite with an Earth Observation Payload

Maurice Borgeaud, Noémy Scheidegger, Muriel Noca, Guillaume Roethlisberger, Fabien Jordan, Ted Choueiri, and Nicolas Steiner

Abstract In 2006, Ecole Polytechnique Fédérale de Lausanne (EPFL) in cooperation with other Swiss academic and industrial partners started the SwissCube project based on the CubeSat program. The primary objective of the SwissCube project is to provide a dynamic and realistic learning environment for undergraduates, graduates and to foster the development of small satellite technologies. Besides the educational objectives, the SwissCube mission objective is to observe the airglow phenomenon (intensity) over selected latitudes and longitudes for a period of 4 months (possibly up to 1 year). The payload consists of a telescope which takes images of the airglow emissions. The telescope has a length of 50 mm. At one end, a CMOS detector captures images with a resolution of 188 × 120 pixels and a pixel size of 24 μm via a focusing optics. A bandpass filter centered at 767 nm, with a bandwidth of 20 nm, selects the desired wavelength of the airglow. At the other end, a baffle protects the optical system and the detector from straylight.

This chapter describes the mission of the satellite and provides information on the Earth observation payload.

1 Introduction

The SwissCube project was initiated in 2006 by the Space Center EPFL and the LMTS (Microsystems for Space Technologies Laboratory), parts of "Ecole Polytechnique Fédérale de Lausanne" in Switzerland. Its primary goal is educational, in the sense that students learn how to build a complex engineering system from A to Z. It is based on the CubeSat program [1] developed by Stanford University and the California Polytechnic State University (CalPoly). With this project, the Space Center EPFL was able to federate and attract students not only at EPFL but in several other Swiss academic institutions (e.g. Haute Ecole

M. Borgeaud (✉)
Space Center EPFL, Station 11, Ecole Polytechnique Fédérale de Lausanne, CH-1015 Lausanne, Switzerland
e-mail: maurice.borgeaud@epfl.ch

R. Sandau et al. (eds.), *Small Satellite Missions for Earth Observation*,
DOI 10.1007/978-3-642-03501-2_19, © Springer-Verlag Berlin Heidelberg 2010

Spécialisée de Suisse Occidentale, University of Neuchâtel, University of Applied Sciences Northwestern Switzerland) since the design, construction, and tests of such a complex system require expertise available from both types of these Swiss engineering universities. An additional benefit of this project was the excellent cooperation between academic partners and large Swiss space companies such as RUAG Aerospace and Oerlikon Space. SwissCube is scheduled to be launched during the first half of 2009 [2].

2 SwissCube

Though the satellite is rather small in size, a cube with sides of 10 cm for a weight of 1 kg, it contains all the critical sub-systems and functions present in larger satellites. SwissCube has also a scientific goal to image the night-glow luminescence created by the recombination of atoms of oxygen at 100 km of altitude and a technological aim to study a new generation of Earth-sensors for satellites. This photoluminescence of the atmosphere (more obviously seen at night and known as the nightglow) is principally due to the recombination of the diatomic oxygen molecule, which is dissociated during the day. The motivation for these observations is to demonstrate that the airglow emissions are strong enough to be measured by an off-the-shelf detector, thus validating the concept for a low-cost Earth sensor. An exploded view of the SwissCube satellite is shown in Fig. 1.

3 The Earth Observation Payload

The payload mission is to perform space-based observations of the airglow occurring in the upper atmosphere at an altitude of about 100 km. Figures 2 and 3 gives an example of the airglow phenomena. To study variations of the emissions as a function of time, the minimum science duration is 3 months, with an extended science mission of duration up to 1 year.

A number of studies exist on the temporal and latitudinal variability in the oxygen emissions in the 90–120 km region of the mesosphere and lower thermosphere. They provide a wide database on the airglow emission rates for different altitudes, local times and latitudes. Nevertheless, further observations would provide a more detailed characterization of the airglow intensity, especially:

- The minimum, maximum and mean airglow emissions at limb depending on altitude.
- The minimum, maximum and mean airglow emission at nadir and hence information about background radiation due to scattered sun- or moonlight.
- Mean tidal variations, with a particular attention on emission near the solar terminator.
- Aurora effects around the poles.

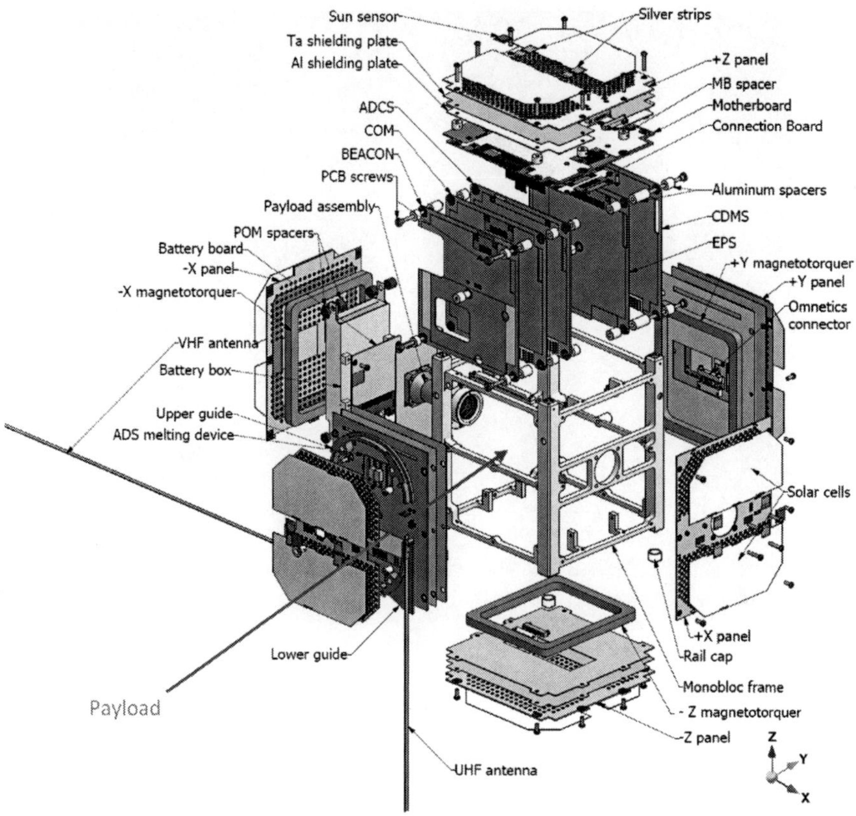

Fig. 1 Exploded view of SwissCube

The payload of the SwissCube satellite is a small telescope which takes images of the airglow. It satisfies the following conditions:

- The payload observes the airglow band at 762 nm, with a resolution between 10 and 40 nm.
- The payload has a spatial resolution of at least 0.3° and a field of view (FOV) of at least 20°.
- The payload survives a sun pointing attitude, with a permanent damage or performance degradation of less than 20% for an exposure time of at least 10 h.
- The payload is able to perform the science mission with the sun no closer than 30° from the sensor boresight.

The space which has been attributed to the payload has a volume of 30 (height) × 30 (width) × 65 (depth) mm^3 for the optical system and a volume of 80 (length) × 35 (width) × 15 (depth) mm^3 for the payload board. The total mass of the payload shall be less than 50 g. The payload consumes at maximum 450 mW (peak power)

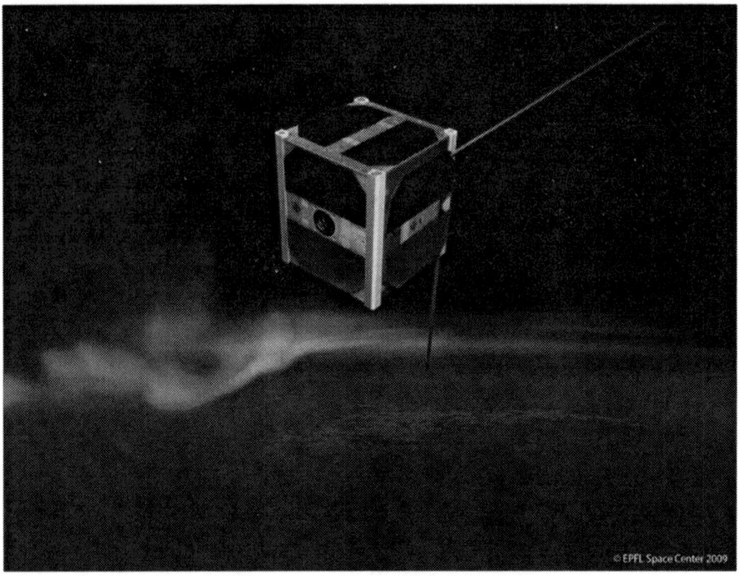

Fig. 2 Airglow phenomena with an artist view of SwissCube

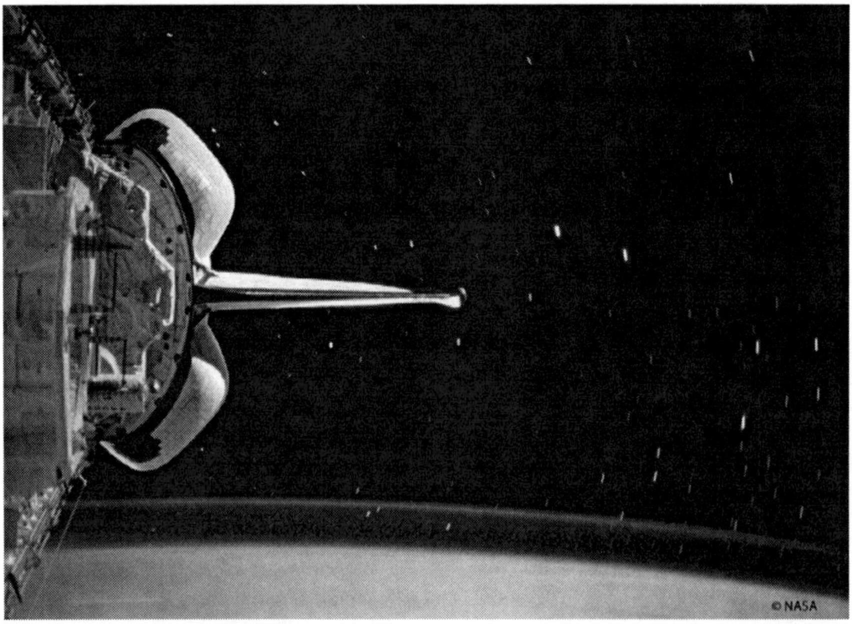

Fig. 3 Image of the airglow taken from the NASA Shuttle

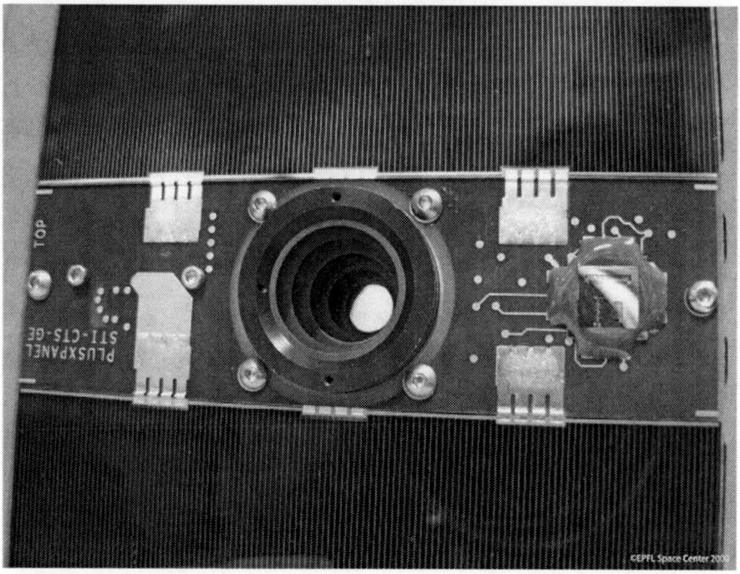

Fig. 4 Payload aperture with the baffle

during 30 s for each image capturing. In order to avoid straylight, the payload is mounted to one of the side of the satellite via a baffle as shown in Fig. 4.

Figures 5 and 6 show the overall design of the payload. It consists of a PCB, including a detector and its control electronics, and an optical system, comprising a baffle, the bandpass filter which selects the desired wavelength of the oxygen

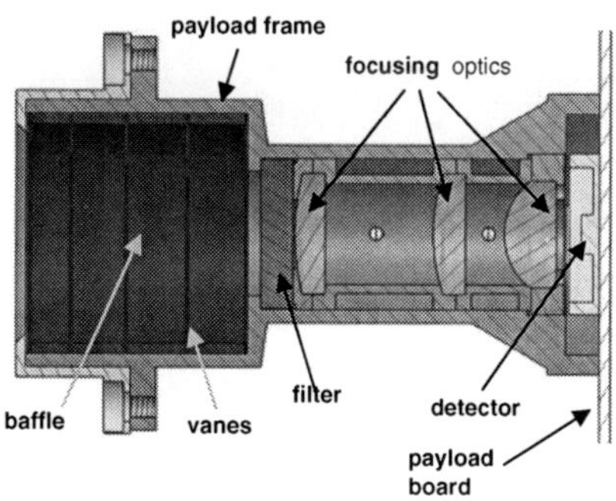

Mechanical design of the payload.

Element(s)	Material	Mass
Closing cap	Aluminium	3 g
Payload frame	Titanium	20 g
Vanes	Stainless steel	2 g
Spacers for the baffle	Aluminium	2 g
Spacers for the optics	Titanium	4 g
Lenses	Glass (BK 7)	1 g
Filter	Glass (RG-9)	1 g
Payload board	PCB	15 g
Screws	Stainless steel	2 g
Total mass:		**50 g**

Mass budget and material used for the payload.

Fig. 5 Mechanical design and information on the payload

emissions and focusing optics. The mass budget and a detailed description of the material used for the payload is also indicated.

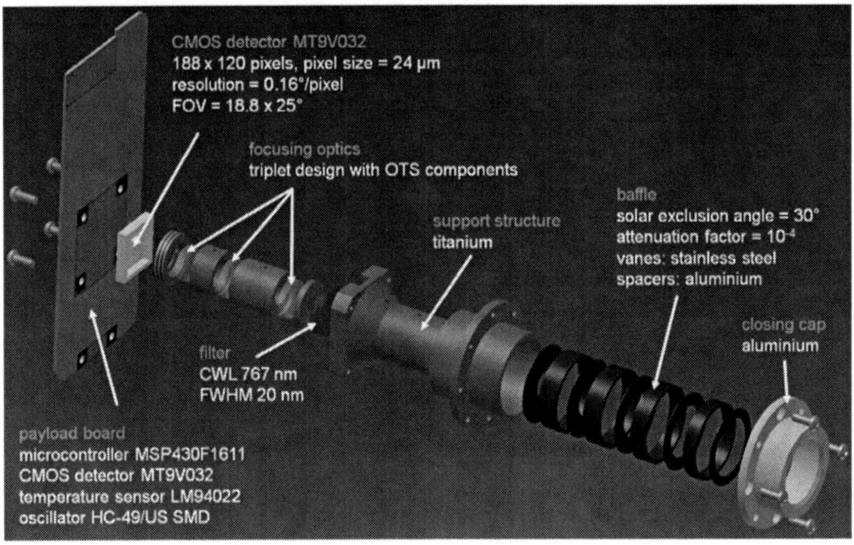

Fig. 6 Illustration of the SwissCube payload

The optical system provides a FOV of 18.8 × 25° and a resolution of 0.16°/pixel for the chosen CMOS detector. The payload is protected by a baffle so that no optical components are permitted to view any sunlit wall or vane edge directly. This means that at least two reflections from darkened surfaces of the baffle tube or vanes are required between the stray light sources and the optical elements. The targeted baffle attenuation factor of 10^{-4} guarantees that the signal perturbation due to the stray light noise which reaches the detector does not exceed the dark signal. A design using a commercial CMOS or CCD detector has been chosen for the payload. The micron model MT9V032 is a highly sensitive monochrome CMOS detector has been used for SwissCube.

A special care was taken during the development of the project. Several models were developed and the FM (flight model) during integration is illustrated in Fig. 7.

4 Conclusions

Designed by a group of universities and with a strong partnership with Swiss space industries, the first entirely-built Swiss student satellite is ready to be launched. Though the main objective is to show the students how to build a satellite, the mission has a scientific goal to observer the nightglow phenomena around the Earth.

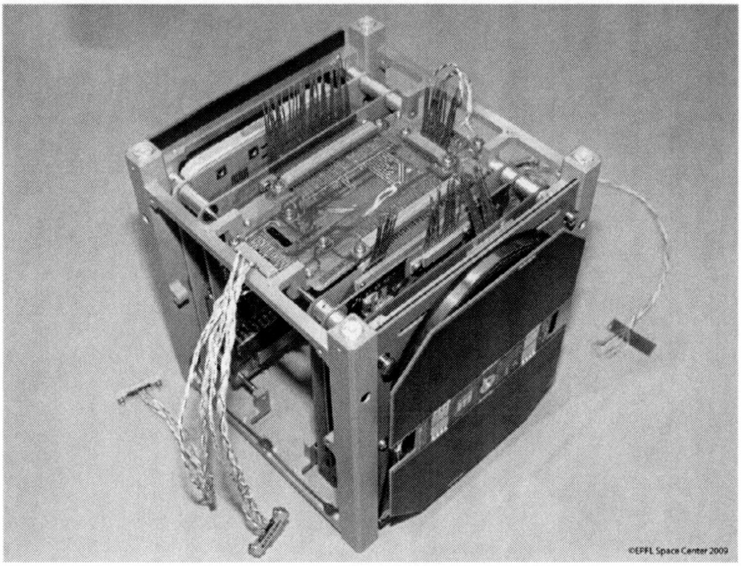

Fig. 7 Top view of the SwissCube flight model during integration

Furthermore, the technology developed could be used in the frame of cheap and small Earth detectors.

References

1. CubeSat Design Specification. Revision 10, Cal Poly, August 2007, http://cubesat.atl.calpoly.edu/
2. G. Roethlisberger, F. Jordan, A. Servonet, M. Borgeaud, R. Krpoun, H. R. Shea, "Advanced methods for structural machining and solar cell bonding allowing high system integration and their demonstration on a Pico-satellite", 22nd Annual AIAA/USU Conference on Small Satellites, Logan, USA, 11–14 August 2008

German Russian Education Satellite – Mission Outline and Objectives

D. Bindel, O. Khromov, M. Ovchinnikov, B. Rievers, J. Rodriguez Navarro, and A. Selivanov

Abstract International cooperation in space exploration and education can be a benefit but also a challenge to the accomplishment of a project. The German Russian Education Satellite (GREsat) is a cooperation between the German Center of Applied Space Technology and Microgravity (ZARM), the Russian Institute of Space Device Engineering and the Keldysh Institute of Applied Mathematics (KIAM). The main objective of the program is the education of students from both universities and the technology verification of the satellite subsystems. An annual exchange of students could be established with the support of the German Academic Exchange Service (DAAD).

The chapter will roughly outline the mission of GRESat and takes a deeper look onto the new technologies and concepts, that shall be demonstrated with this satellite. The cooperation of the partners also lead to a cooperation for the provision of components for the satellite bus. One of the main aspects will be a heterogeneous redundancy of the onboard data processing by using two different computers. Here the communication between the computers and the mode management are difficult topics to adress, when the cooperation of these units shall work without problems. The communication system for the command and data interface to the ground station is also planned as a redundant system, as Globalstar and ORBCOMM services shall be accessible from the satellite. That enables the test of both communication systems and provides a failure tolerant data interface. The chapter will outline what challenges the layout of the communication protocoll between both onboard computers has faced, and how the application of two independant satellite communication systems could be designed to benefit from the redundant hardware and services.

Another part of the on-orbit verification will be the usage of an AMR magnetometer and new micro torquers. These devices are acting as shared resources, so both onboard computers are allowed to access them for attitude control maneuvers. While the sensor delivers its data constantly to both computers, the magnetic torquers have to be managed for an exclusive access of only one computer at the same time. A

D. Bindel (✉)
Center of Applied Space Technology and Microgravity, University of Bremen, Bremen, Germany
e-mail: bindel@zarm.uni-bremen.de

R. Sandau et al. (eds.), *Small Satellite Missions for Earth Observation*,
DOI 10.1007/978-3-642-03501-2_20, © Springer-Verlag Berlin Heidelberg 2010

defined mode management and proper inter-computer communication shall ensure the proper function of these subsystems.

Due to the mission outline as a technology verification and student education project, another mission objectives is also the test of novel attitude control algorithms by using magnetic sensors and actuators. A fallback algorithm (B-dot) is used to stabilize the spacecraft in a non conform situation. This shall guarantee a good connectivity for the command and data interface with the ground station via Globalstar and ORBCOMM. Any further mode changes and failure handling can then be prepared offline on ground and uploaded to the satellite afterwards.

1 Mission Outline

The GRESat mission has multiple objectives during the preparation and operational phase. During the mission and spacecraft design process, the student education and the international cooperation between German and Russian universities shall be fostered. The technical aspects of GRESat can act as an example for student bachelor and master theses. The student exchange between the Russian Academy of Science in Moscow and the Center of Applied Space Technology and Microgravity in Bremen on one hand is a method to coordinate the work on the project and on the other hand gives the possibility of a cultural exchange for both student groups.

From the more hardware oriented view of the mission, it also acts as a demonstrator for technologies that are developed on both institutes. From the German side, a newly developed onboard computer shall be tested for its usability in space operations. There is also an AMR magnetometer and a smaller and compact version of magnetic torquers from the ZARM Technik AG which could be used for an on-orbit verification. Beside the pure hardware validation, GRESat can also provide a platform for attitude control experiments. The KIAM and ZARM institutes are researching and developing attitude and orbit control methods. New ideas for guidance and control algorithms are developed from time to time. Due to the sometimes experimental state of these methods, they are not considered as core algorithms for near term spacecraft. But after a successful demonstration of their capabilities in a simulation environment, they can be used onboard of GRESat as a kind of “software-payload”.

The basic instrumentation of the spacecraft consists in this early project phase of the planned magnetic field sensor and the onboard temperature sensors. So there is even a small contribution to a science aspect for the mission. The ZARM is working on advanced temperature simulations for spacecraft application. But for a verification of the software capabilities, a set of temperature data from a real spacecraft is required. The German onboard computer can attach up to 16 temperature sensors, to observe several spots inside the spacecraft.

2 System Outline

The overall design and the responsibility of the assembly of the spacecraft is done by the Russian Institute of Space Device Engineering. They have already build and launched a satellite of the TNS (Technology Nanosatellites, Tekhnologicesky Nanosputnik) family, which is also desired for the GRESat bus. The shape of the spacecraft is a six-sided cylinder with approximately 25 cm length and 17 cm diameter. A predecessor from the TNS family is shown in Fig. 1 (here still in round cylindrical shape). The top plate of the cylinder will be equipped with the antennas for the Globalstar and ORBCOMM modems. The lower end cap of the cylinder consists of a manual handle and a power switch to initialize and boot the spacecraft. That handle is required for the launch method, which is manual. The TNS family is designed to be launched from the International Space Station by a cosmonaut by simply throwing it away from the station.

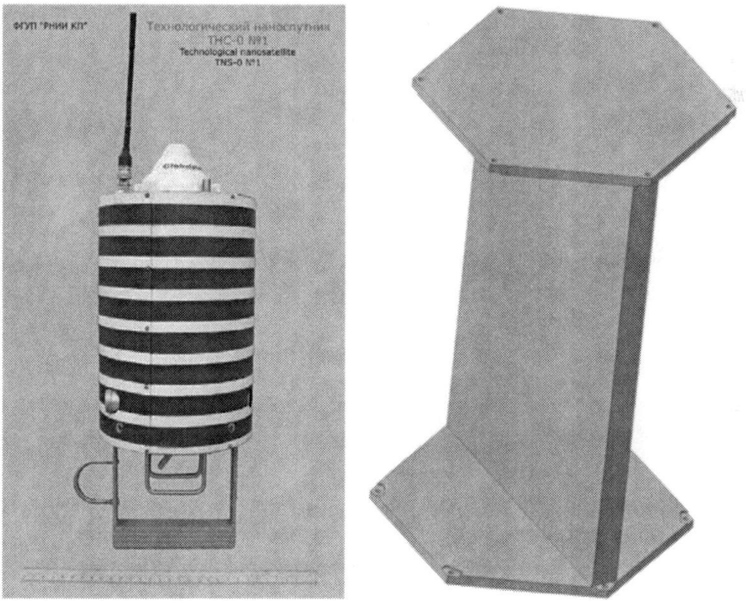

Fig. 1 TNS-0 spacecraft and bus structure of GRESat

Apart from the outer shape, the inside of the spacecraft is divided into two sections by a middle plate. This structure holds most of the electronic components, including the onboard computers, the modems and the energy distribution system. The six-sided surface will be equipped with solar cells for regenerative energy production.

Beside the spacecraft structure, the Russian participation to the mission will include the launch, one of the onboard computers, the Globalstar communication system and the energy system. The German participation will be the second onboard computer, the ORBCOMM communication system and the magnetic field sensor and actuators.

As there is only one unit for the magnetic field measurement, the magnetometer is connected to both computers via an RS422 line and a signal splitter. Also the magnetic torquers as momentum actuators are connected to both computers current drivers. A switch connects the coils to either one or the other computer, to prevent short circuits and power transfer from one amplifier unit to the other. So a concurrent electrical connection of both units is not possible. The magnetic torquer switch is operated by the Russian onboard computer.

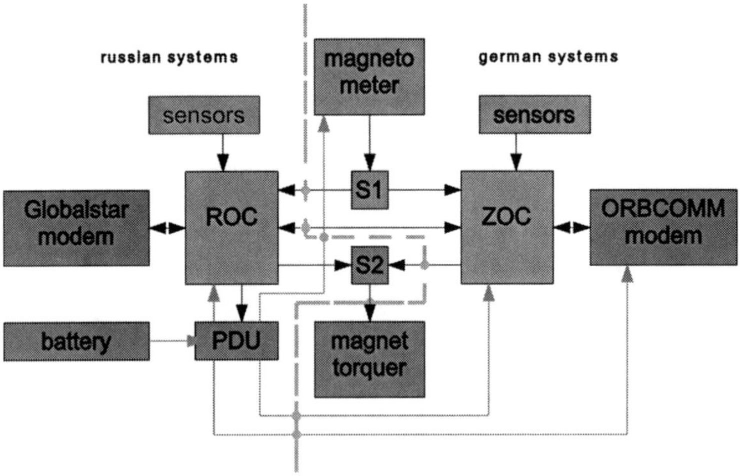

Fig. 2 GRESat system structure

By the outline of the mission concept, the Russian onboard computer (ROC) is the master unit of both computers and can command the German ZARM onboard computer (ZOC) to a passive state. Therefore a handover of the torquer control from ROC to ZOC is only done, when the ZOC unit is going to the attitude control mode afterwards. Although this solution incorporates several actions, the attitude control ability of both computers ensures, that the basic mission requirements can still be achieved, even if the switch of the magnetic torquers is failed and locked to one computer unintentionally.

3 Technology Demonstration

The presence of two onboard computers brings up the interesting combination of a heterogeneous computer system. For the German computer the mission is additionally used as a validation opportunity. But as an element of safety, both computers are able to handle the basic requirements of orbital communication and satellite attitude stabilisation. So a basic element of redundancy is introduced to the system.

The next demonstration elements are the two independent satellite communication systems. The main purpose is here not directly its operability in space, because that was already demonstrated by other missions. Due to the redundancy, a cross-usability for both computers shall be demonstrated. The goal is that one computer can use the modem of the other computer, by routing the messages through the inter-computer connection.

As mentioned above, also some components of the magnetic attitude sensor and actuation system shall be utilized in space. Here the AMR magnetometer from the ZARM Technik AG company can be taken to another demonstration opportunity. As there is no redundancy of the ACS components, each computer must be able to utilize these resources without conflicting the other computer. The GRESat system layout can be seen in Fig. 2. The magnetometer is configured to measure the surrounding magnetic field and delivers a set of measurements at a certain interval (10 Hz actually planned). A simple electrical unit provides this serial RS 422 signal to each of the computers simultaneously.

4 Communication Subsystem

The communication system of the GRESat consists of different elements, designed to serve specific purposes, that are based on the mission requirements. An interesting point is the data exchange and command interface between the ZOC and the ROC. While ROC is planned to be a small processor in the range of a Texas Instruments TMS320VC5 with 64 kB RAM and more than 100 MIPS processing power, the ZOC is choosen to be a low power and low cost ATMega128 microcontroller with around 4 kB RAM and 4 MIPS. So it shall be run temporarily to test out its ability for a full satellite attitude control and communication ability and to verify its operation in real space conditions.

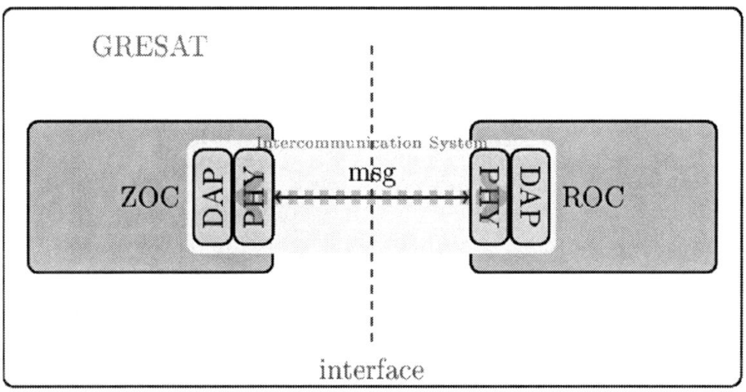

Fig. 3 ROC to ZOC communication interface

The connection between these two units is physically done by a simple half duplex RS232 serial line. As seen in Fig. 3, a protocol layer, the Data Architecture Protocol (DAP), is utilized as an interface to the user program layer of both computers. The DAP does not have to deal with network data stream methods, as the serial line is a point to point connection. But nevertheless, some effort had been done, to set up a full protocol that is able to establish and maintain a connection between the two units, no matter if the connection was lost or interrupted during the operation.

A set of functions was created to track the byte-wise synchronisation of the data stream. The USART interface of the ZOC is only able to hold eight bits of incoming data in a buffer, and has to dump its content to the RAM of the microcontroller. A second level ring-buffer collects all received bytes. By using a unique synchronisation byte order, the DAP can recognize the beginning of a valid data stream byte within the continuous bit-stream. Once that sync has been done, the next step of acknowledgement is done, to initialize the communication line.

All the data exchange between the two computers is done by a packet oriented approach with specified headers, used for protocol messages for the connection maintenance as well as data and command messages as payload information. Included in this package definition a Fletcher checksum ensures the data integrity. The protocol keeps track about the correct delivery of the packets to each of the two partners by using acknowledge messages. So the DAP can send lost messages again or resets and re-synchronizes the connection at all, if the number of lost package is to high.

Another requirement of the communication system is the data exchange with the satellite communication network ORBCOMM. A Q1000 modem is used to establish a connection from GRESat to one of the ORBCOMM satellites.

This part of the system is required to download the measurement and housekeeping data from GRESat to the groundstation in the ZARM institute in Bremen. The protocol is far less complicated, as the commands for the Q1000 modem are described and can be implemented in the ZOC microcontroller. But nethertheless, an acknowledgement system (see Fig. 4) keeps track of safe delivery of the packages. Unlike the DAP protocol between the two onboard computers, an additional logic monitors the amount of data packages to be send by the modem and that are stored temporarily in buffers of the modem and the RAM of ZOC. If a connection to

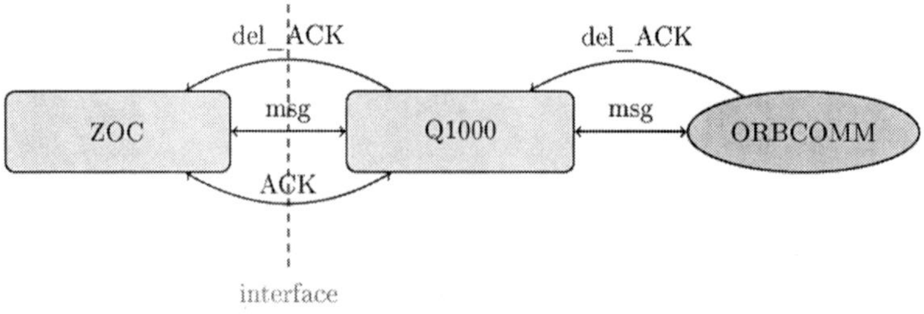

Fig. 4 Interface of ZOC, Q1000 and ORBCOMM network

the ORBCOMM network can not be established over a certain period of time, a data package overflow has to be avoided by the deletion of data packages of measurement data in a certain order. So the data rate of the measurement and house keeping data is artificially decreased, until the download by the ORBCOMM network can finally be achieved.

As both onboard computers possess the ability to contact a satellite data network (Globalstar and ORBCOMM), the DAP is designed to handle a bypass data transfer. So assuming a failure of the Globalstar communication of ROC, its downlink data packages can be transmitted by the DAP to ZOC and further by the ORBCOMM modem to the earth.

One of the most interesting Results of the GRESat mission will be the compare of the Globalstar and the ORBCOMM downlink capabilities under almost equal conditions in terms of usability and reliability.

5 International Cooperation

The GRESat project was established as a cooperation between the German ZARM institute in Bremen and the Russian KIAM institute and Russian Academy of Science in Moscow. This constellation gives the participating students an opportunity to explore development questions in the field of spacecraft design and production, but also to get into contact with students from other countries.

The German Academic Exchange Service (DAAD) is supporting the GRESat project with annual stipends for the Russian students from the Leonhard-Euler-Program. The stipend includes also a one-month stay of the Russian students in Bremen to concentrate their work with the German team. That visits can also be used for small reviews of the previous work.

During the development of the GRESat project, not only mechanical engineering or space-technology students have contributed but also electrical engineering and computer science students produced several theses for the onboard systems. So the GRESat mission does not only provide international cooperation between Bremen and Moscow, but also gives an interdisciplinary approach across the University of Bremen.

6 Conclusion

The German Russian Education Satellite project is an interesting example, how an international cooperation in the field of student education can be established. The resulting spacecraft will also be used for a technology demonstration of a heterogeneous onboard computer system and provides two independent satellite communication systems for redundancy and performance test purpose.

University Microsatellites Equipped with an Optical System for Space Debris Monitoring

Chantal Cappelletti and Fabrizio Paolillo

Abstract Currently, space missions must take into account a relatively new threat, which is represented by space debris, i.e. the orbital residuals of past artificial satellites and rockets. This problem has arisen in the last 25 years and requires specific strategies for mitigation, with the main intent of avoiding collisions between orbital debris and spacecrafts. Space debris monitoring and orbit determination is an essential premise to this task.

In the last decade, the Group of Astrodynamics of the University "La Sapienza" (GAUSS) of Rome has been involved in the optical surveillance of space debris. This activity was carried out using amateur astronomers' facilities and led to participating to the IADC (Inter-Agency Space Debris Coordination Committee) observation joint campaign. In 2006 GAUSS completed the first Italian observatory, fully dedicated to space debris monitoring. In this context, GAUSS students have analyzed the feasibility of two different formation flight missions, tailored to detecting space debris and, more specifically, focused on the detection of objects along MEO and LEO orbits. This paper deals with the possibility to carry out an optical surveillance of space objects by using a ground-based telescope and a microsatellites equipped with an optical payload. The feasibility of placing an optical payload (capable of detecting space debris) onboard the microsatellite UNISAT5 is investigated.

1 Introduction

Space debris are a central concern for space missions about the Earth. The accurate knowledge of the orbital debris environment is extremely relevant, in order to avoid possible collisions with operative spacecraft.

Space surveillance represents an important political and military concern and is related to a multitude of aspects, such as safety of spacecraft operations, safety for

C. Cappelletti (✉)
Scuola di Ingegneria Aerospaziale, "Sapienza" University of Rome, Rome, Italy
e-mail: chantal.cappelletti@gmail.com

R. Sandau et al. (eds.), *Small Satellite Missions for Earth Observation*,
DOI 10.1007/978-3-642-03501-2_21, © Springer-Verlag Berlin Heidelberg 2010

ground facilities (prediction of reentering objects and related hazards), basic information on the space debris population. In addition, for space missions an appropriate knowledge of debris environment allows realizing corrective maneuvers to prevent collisions.

A recent space debris-creation event was the Chinese anti-satellite weapon (ASAT) test, carried out on January 11, 2007. The impact was estimated to have generated about 35,000 pieces 1 cm or larger, and 1 million pieces 1 mm or larger. In June 2007, the NASA Terra environmental spacecraft was the first to be relocated in order to prevent impacts [1].

Space debris observation campaigns are aimed at preventing orbiting spacecraft against impacts. With regard to this issue, the Group of Astrodynamics of the University of Rome "La Sapienza" (GAUSS) started out the Italian space debris observation activity in 2002 [2] and promptly joined the Inter-Agency Space Debris Coordination Committee (IADC) coordinated campaigns addressed to the Geostationary orbit [3]. In 2003 GAUSS conducted a very successful test campaign to prove the effectiveness of a two sites observation and orbit determination of GEO objects [4]. Some observations were addressed towards Low Earth Orbits [5]. All these observations were performed using the facilities kindly made available by Associazione Astronomica Frusinate (Frosinone Astronomical Society) and by Observatorio Astronòmic de Mallorca (Mallorca Astronomical Observatory). Later, in 2006 GAUSS was appointed to design and build the first Italian optical observatory for space debris observation under a contract with the Italian Space Agency (ASI) [6].

Moreover since nineties, GAUSS has started the UNISAT program with the aim of designing, manufacturing and launching educational microsatellites, completely built and operated in orbit by students of the School of Aerospace Engineering [7]. In the framework of this program four satellites have already been launched, and a fifth satellite (UNISAT5) is nowadays under construction.

In this research, the feasibility of placing an optical payload (capable to detect space debris above the Earth's atmosphere) onboard UNISAT5 is analyzed. A short presentation of the different systems used for monitoring the debris population introduces the central topics of this paper, which is concerned with the main differences between observation campaigns performed by ground-based and orbital optical systems.

2 Systems for Space Debris Observation

Ground-based optical telescopes and radar can be used for the detection and orbit determination of space debris. The differences between the detection capability of these two systems is basically due to the fact that the power of the signal received by an optical system is proportional to $1/(d^2)$ (where d is the distance from the orbiting object) whereas for a radar it is proportionally to $1/(d^4)$.

For this reason radars are mainly used for the surveillance of LEO objects, whereas optical systems allow detecting more distant objects. With a ground-based radar system it is possible to detect LEO objects with size down to 1 cm; these objects cannot be seen with a ground-based optical system. On the other hand, optical systems have several advantages with respect to radars, such as the possibility of monitoring GEO regions, as well as the capability of identifying the shape and the attitude of an orbiting object. Another important feature of optical systems is the possibility to compare different images from distinct observation sites, thus allowing an improved orbital determination. In situ space debris monitoring systems are based on impact sensors and on the analysis of surfaces returned from space, but these methods do not allow preventing the possible collisions.

Otherwise in situ measuring of space debris can been exploited using a microsatellite equipped with an optical system or radar based system. This last system is not suitable for university microsatellite due to the large amount of required power. For these reasons the following sections describe the use of optical systems for debris detection.

3 Ground Based Optical Systems and Orbiting Systems Characteristics

The Earth's atmosphere hinders the monitoring of space debris environment. Clouds and pollution depend on the observation site; however, several undesired effects due to atmosphere remain even when the observations are performed from the best available site (as, for instance, from 4,200 m of altitude, in Mauna Kea, HI) [8]. Basically, the Earth's atmosphere filters selectively distinct electromagnetic wavelengths and this circumstance implies the existence of two specific narrow-band transparency windows available to optical and radiofrequency devices.

3.1 Atmospheric Effects in the Optical Band

- SEEING: the scattering effect caused by the passage of light through turbulent atmosphere.
- SKY GLOW: the atmosphere, even in the best Earth's observation site, glows due to atomic processes in the air. This light emitted by the sky, called sky glow, is a severe problem when observing faint objects, because the sky glow photons make extra noise which degrades the accuracy of measurements.
- ATMOSPHERIC EXTINCTION: the atmosphere absorbs and scatters some fraction of the light at all optical wavelengths. This causes objects to be dimmer than they would be without the atmosphere.
- ATMOSPHERIC REFRACTION: except when looking at the zenith, the atmosphere acts as a weak prism, spreading out light in a small spectrum along the line pointing to the zenith.

When compared to an identical telescope placed outside the atmosphere, these phenomena badly affect the observation campaigns. Seeing causes the lost of much of the detail that powerful (large) telescopes are able to provide. Seeing and sky glow severely limit the accuracy of the optical measurements regarding faint objects.

Moreover, also additional random disturbs can severely compromise the observations from the Earth, such as wind (which shakes the telescopes, degrading the image quality) or clouds (that interferes with light). Even high humidity (especially in presence of dust and pollution) severely affects the performance attainable by telescopes, due to the possible degradation of optical surfaces and rust metal parts.

3.2 Seeing and Angular Resolution

Stars and orbiting objects can be considered a point like source. However, even if the telescope is located outside the atmosphere, focusing a point source into a point image is impossible. The light can be viewed as a wave, and waves that come from different parts of the telescope mirror interfere with each other. In the image the result is a "bulls eye" pattern, termed "Airy disk".

The angular size of the Airy disk pattern on the sky only depends on the diameter (D) of the primary mirror or lens and is independent of its focal length. The linear size (in the focal plane) of the Airy disk in the image plane is determined by the angular size (related to D) and the image scale (related to the focal length) [8, 9].

The angular distance of the first dark ring radius (named the Dawes limit or the diffraction limit) is given by:

$$\Theta = 1.22 \cdot \frac{\lambda}{D} \text{ (rad)} ;$$

where λ is the wavelength of the radiation.

As aforementioned, the Earth's atmosphere spreads the light coming from a point source (seeing). The angular extension of the quasi-Gaussian profile associated to the seeing PSF (Point Spread Function) is determined by the atmosphere characteristics, and not by the telescope. In the best observation sites, seeing turns out to be responsible of a light scattering of 1 arcsec (measured at a level equal to half of the maximum, FWHM (Full Width at Half Maximum).However, these sites are hardly available for university purposes and funds [8].

In our University/ASI observatory GAUSS-SpaDE (Space DEbris) located in Collepardo (Fr), Italy, at 555 m of altitude [6]) seeing is responsible of a light scattering of about 8 arcsec. In Fig. 1 illustrates the Dawes limit angles as a function of the wavelength λ for different diameters D. The GAUSS-SpaDE CCD is a FLI PROLINE 16803, with a quantum efficiency peak of 60% at 550 nm [6]. For Example, boarding this CCD on a university microsatellite (equipped with a 20 cm diameter telescope) an angular resolution 16 times better than the ground-based one can be achieved. As shown in the plot of Fig. 1, if $\lambda = 0.5$ μm and D = 20 cm, $\theta = 0.5$ arcsec.

For this reason, outside the atmosphere the angular resolution limit is lower than the ground-based one, with the exception of very small telescopes (D < 15 cm), which have Airy patterns comparable to the seeing of the better ground observation sites (with regard to angular extension).

However, compared with the SpaDE observation site, also a 10 cm space based telescope improves 8 times the resolution limit. At the same time the weight and dimension are minimized.

3.3 The Ideal Resolution and Field of View

Pixels with sizes of about 1/3–1/2 the resolution limit represent the best solution to balance field of view and resolution [8].

With a 2.048 × 2.048 pixels CCD (pixel size of 5.8 μm), an optical system Field of View (FOV) set to 0.7° and a focal length of 1 m (f/ = 10, D = 10 cm), the angular dimension of the pixel FOV turns out to be equal to 1.2 arcsec, equal to the resolution limit. The 20 cm telescope gives a pixel size of 1/2 the resolution limit but produces a larger burden for UNISAT5 microsatellite.

Moreover, a large FOV is needed in order to obtain a comprehensive survey of the space debris environment. During an observation campaign, the problem of obtaining both a good resolution and a large FOV could be solved by carrying out joint campaigns through the use of the GAUSS-SpaDE ground-based optical system (equipped with a large FOV = 2.6°) and the microsatellite telescope.

3.4 Atmospheric Extinction

The atmosphere absorbs and scatters some fraction of the light at all optical wavelengths. This causes objects to be dimmer than they would be without the atmosphere. Astronomers call this phenomenon atmospheric extinction.

If this effect is taken into account, for a ground-based telescope the transmitted power is partially absorbed, and this circumstance implies that also the size of detectable debris reduces.

For this reason it is necessary to compare the transmitted power vs. debris dimensions for the two different systems with reference to different diameters. A CCD (Charge Coupled Device) with a peak quantum efficiency of 60% (in the range of commercial off the shelf CCD) has been considered in order to calculate the power noise threshold. In Fig. 2 it is possible to see that in the MEO case study we don't have a very important difference but in the LEO case study (Fig. 3) the difference is clearly visible.

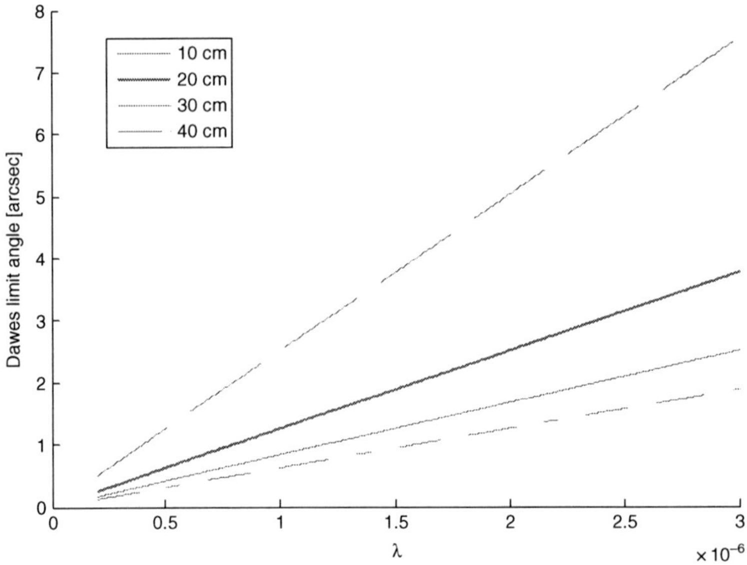

Fig. 1 Dawes limit angle (arcseconds) vs. wavelength lambda for different telescope diameters

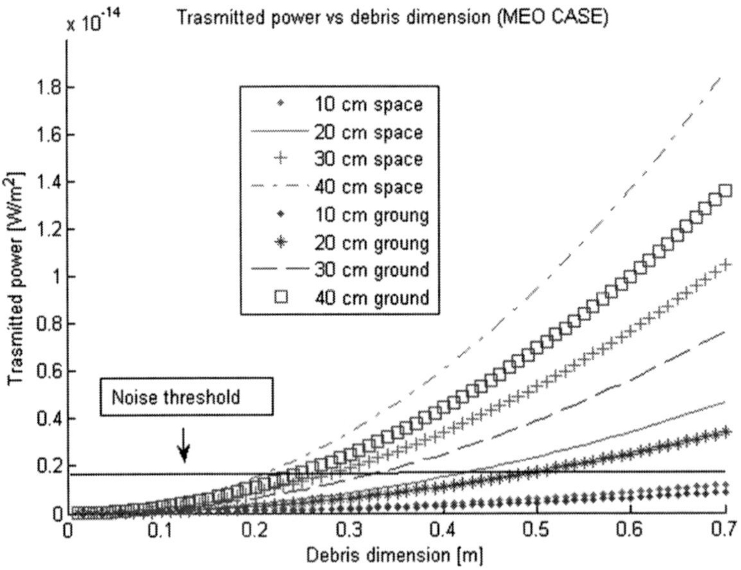

Fig. 2 Absorbed power vs. debris dimensions for different telescope diameters (MEO case)

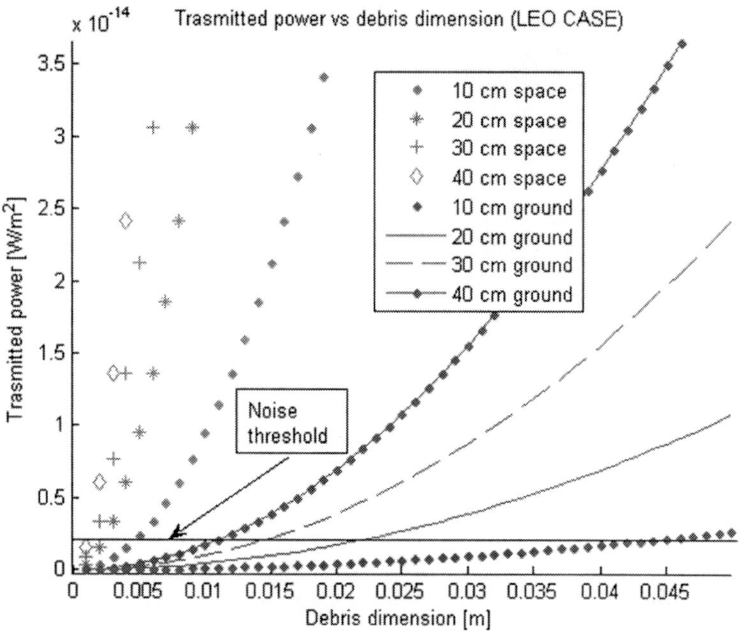

Fig. 3 Absorbed power vs. debris dimensions for different telescope diameters (LEO case)

In LEO case the difference between the two systems is not only due to the atmospheric extinction but mainly at the distance with the optical system. As matter of fact considering a LEO formation flight at 700 km to detect objects with an altitude of 800 km [10] we have a real advantage respect ground-based telescopes. Considering the MEO case [11], using microsatellites in a sun-synchronous dawn/dusk orbit, we don't have a significant difference on transmitted power but we can have others advantages which are illustrate in the following paragraphs. Another significant advantages in orbiting observations is the possibility to use thermal infrared which is absorbed by atmosphere.

3.5 Phase Angle and Baseline

An important difference between ground-based and orbiting optical systems is debris lighting condition. With a ground-based system it is impossible to observe orbiting objects in LEO with low phase angle because the objects are eclipsed by Earth. So it is possible to carry out observations only at dawn or at dusk with phase angle close to 90° [12]. Conversely, this circumstance does not occur if the observations are performed from orbiting telescopes, because different configurations (i.e. different mutual positions between the sensor and the observed object) can be encountered, thus allowing the use of different observation strategies

An accurate orbit determination requires the collection of observational data from a number of distinct locations. For this reason, joint space debris observational campaigns from different sites on the Earth's surface have been carried out. However, the baseline of these campaigns is limited.

Conversely, orbiting systems make possible the use of different baselines, which allow a more accurate orbit determination. In particular, in the MEO study case [11] the baseline has been fixed close to 14,000 km.

4 Space Based Telescope vs. Ground Based Telescope

Table 1 Optical system features of the ground based telescope and the space based telescope

Optical system features	Ground based	Space based
Diameter (m)	0.3	0.1
FOV (deg)	2.6	0.7
Number of pixels per row	2048	2048
Pixel size (width of square pixel) (μm)	18.0	5.8
Scale (pixel FOV) (arcsec/pixel)	4.4	1.2582
FWHM (full width at half max.) (pix)	2.0	2.0
Integration time (s)	2.0	0.01

This paragraph shows the differences between a 10 cm space based telescope and the SpaDE 30 cm ground based telescope, exploiting a 24 h ESA-PROOF2005 statistic simulation. In order to test these differences it has been chosen to monitoring LEO objects in the range from 700 to 1600 km.

The space telescope is placed in a 700 km dawn-dusk Sun Synchronous Orbit (SSO) in order to obtain an orbit node line orthogonal to the sun light direction. The launch date of the simulation is the 21st of December 2008. For this reason the telescope will point 90° of azimuth in order to obtain a phase angle close to zero during the whole orbital revolution. Also an elevation of 0° permits to detect orbiting objects in the selected range.

The Spade ground telescope is placed in Collepardo site (latitude 41.765°, longitude 13.368° and altitude 555 m). In Table 1 are showed the main features of the two optical systems.

4.1 Results

In the following figures (Figs. 4, 5 and 6) the Results of 24 h ESA-PROOF2005 statistic simulation are showed.

Figure 4 plots the number of crossing objects vs. range of the ground based telescope and the space based telescope. Due to the larger FOV, more objects cross the ground based sensor. In both case the crossing objects size is in a range from 0.01 to 10 m (Fig. 5-left side).

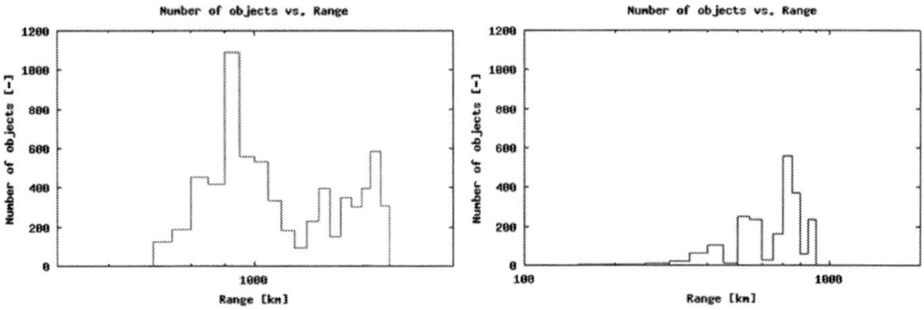

Fig. 4 Number of crossing objects vs. range (24 h ESA-PROOF2005 statistic simulation) of the ground based telescope (*left side*) and the space based telescope (*right side*)

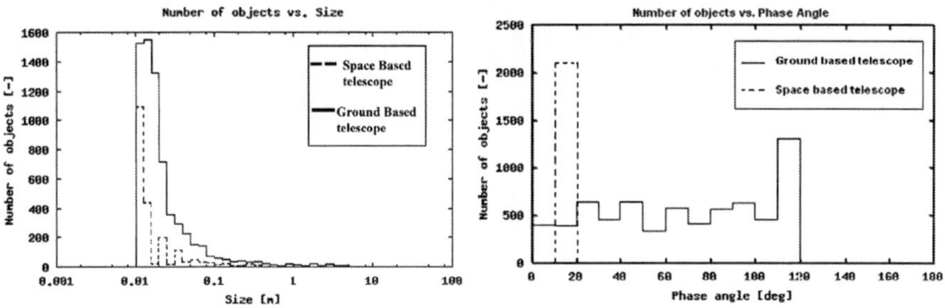

Fig. 5 *Left side*: Number of crossing objects vs. Size (24 h ESA-PROOF2005 statistic simulation) of the ground based telescope and the space based telescope. *Right side*: Number of crossing objects vs. Phase angle (24 h ESA-PROOF2005 statistic simulation) of the ground based telescope and the space based telescope

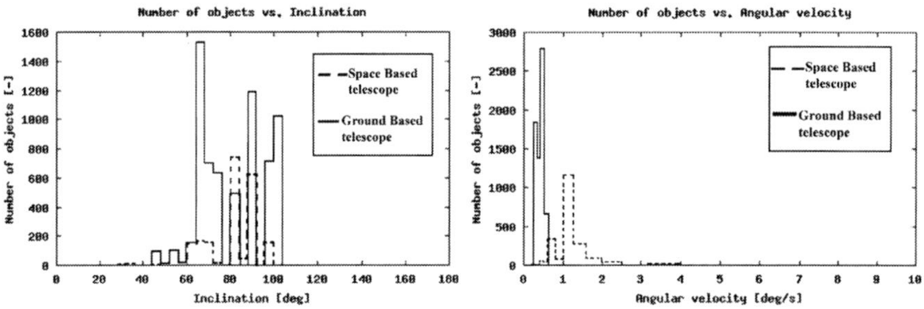

Fig. 6 *Left side*: Number of crossing objects vs. Inclination (24 h ESA-PROOF2005 statistic simulation) of the ground based telescope and the space based telescope. *Right side*: Number of crossing objects vs. Angular velocity (24 hours ESA-PROOF2005 statistic simulation) of the ground based telescope and the space based telescope

It is important to emphasize that with this mission configuration the orbiting telescope monitors LEO region with an optimal illumination condition (phase angle between 10° and 20° is due to the Earth inclination and orbit inclination).

In ground case the phase angle change during the day (only in dawn and dusk it is possible to observe with a good illumination condition).

The left side of Fig. 6 shows the number of crossing objects vs. inclination: Most of these are LEO satellites and debris with semi major axes in the range 7,000–8,000 km and inclination between 60° and 100°.

Another important result is plotted in the right side of Fig. 6. The space based FOV is crossed by faster objects than ground based one. In fact due to high SSO orbit velocity, the relative velocity between space sensor and target increases. So the short exposure time selected (0.01 s, see Table 1) improve the space telescope performance and allows to detect objects with 1°/s relative angular velocity. With this observation strategies these targets will be about 30 pixels trace shaped in the image. To process this kind of data a software for the on board images analysis has been already developed by GAUSS [10, 11].

The integration time comes from a trade off: more distance objects have a lower relative angular velocity and so an higher exposure time is needed.

In ground case the 2 s integration time allows to observe a 0.5°/s angular velocity target like a 400 pixels trailing trace in the image. The algorithms developed by GAUSS in order to process SpaDE data achieved in different observation strategy will permits to analyse these frames[13].

5 Conclusions

This study presents a preliminary selection of optical systems for debris observation. With an orbiting optical system it is possible to have a better resolution and to observe smaller objects in better illumination condition. For this reason GAUSS plans to board an optical payload on Unisat5 university microsatellite to carry on space debris observations. In particular, most likely an accurate orbit determination will be achieved by correlating the images captured by the ground-based GAUSS-SpaDE observatory with those taken by the UNISAT5 optical system. This solution appears not only very attractive under the scientific point of view, but also fully sustainable by educational institutions.

Acknowledgements The authors wish to thank Prof. F. Graziani, dean of the School of Aerospace Engineering of "Sapienza" University of Rome, for the suggestions and factual support provided. Moreover, the authors want to sincerely thank Prof. F. Santoni and F. Piergentili for the constant helpfulness and M. Pontani for support during the paper drafting phase.

References

1. Brian Berger, NASA's Terra Satellite Moved to Avoid Chinese ASAT Debris, Retrieved on 2007-07–06.

2. M. Porfilio, F. Piergentili, F. Graziani. First optical space debris detection campaign in Italy. Advances in Space Research, 34(5), 921–926, 2004
3. M. Porfilio, F. Piergentili, F. Graziani. The 2002 Italian optical observations of the geosynchronous region, Space-flight Mechanics 2003, Advances in the Astronautical Sciences, Vol. 114, pp. 1237–1252, Univelt, San Diego, USA, 2003, AAS paper n. AAS 03–186.
4. M. Porfilio, F. Piergentili, F. Graziani. Two-site orbit determination: The 2003 GEO observation campaign from Collepardo and Mallorca. Advances in Space Research, 38(9), 2084–2092, 2006.
5. F. Piergentili, M. Porfilio, F. Graziani. "Optical campaign for low Earth orbit satellites orbit determination", Proceed-ings of the Fourth European Confer-ence on Space Debris, ESA SP-587, pp. 689-692, ESA Publications Division, Noordwijk, The Netherlands, 2005.
6. F. Graziani, F. Piergentili, C. Cappelletti, L. Murrali, F. Paolillo, C. Marchiori, M. Porfilio. "The first Italian observatory for space debris observation", 58th International Astronautical Congress, Hyderabad 24-28 Settembre 2007, IAF paper IAC-07-A6.I.06.
7. F. Santoni, F. Piergentili., F. Graziani."In orbit performances of the UNISAT-3 solar arrays", 57th International Astronautical Congress, Valencia, Spain, 2–6th October 2006.
8. W. Romanishin. "An Introduction to Astronomical Photometry Using CCDs", University of Oklahoma, October 22, 2006.
9. M. De Petris, "Dispense del Corso di Ottica Applicata", University of Rome "Sapienza", September 13, 2006.
10. C. Cappelletti, F. Guarducci, F. Paolillo, L. Ridolfi, F. Graziani, M.L. Battagliere, F. Piergentili, F. Santoni. "Microsatellites Formation Flying for In-Situ Space Debris Detection", 5th International Workshop on constellations and formation flying, CRIMEA, Evpatoria, July 2–4, 2008.
11. F. Paolillo, F. Guarducci, C. Cappelletti, L. Ridolfi, L. Murrali. "Microsatellites Formation Flying For Optical Space Debris In-Orbit Observation", 58th International Astronautical Congress, 29th September – 3rd October 2008, Glasgow, Scotland, UK, IAF paper IAC-08-E2.3.3.
12. M. Porfilio. "Metodi per l'individuazione di detriti spaziali" PhD Thesis, Scuola di Ingegneria Aerospaziale, "Sapienza" University of Roma, 1999–2000
13. F. Paolillo, M. Porfilio, F. Piergentili. "First Italian space debris observatory: the image processing automation", 58th International Astronautical Congress, Hyderabad, 24–28 September 2007, IAF paper IAC-07-A6.I.05.

A First-MOVE in Satellite Development at the TU-München

Manuel Czech, Andreas Fleischner, and Ulrich Walter

Abstract MOVE (Munich Orbital Verification Experiment) is a program of the Institute of Astronautics (LRT) at the Technische Universität München (TUM), which aims on building pico-satellites with university students mainly for educational purposes. First-MOVE shall create a robust platform as a starting point for sophisticated satellite missions of the institute in the future. In the paper, the state of development is described, but emphasis is on the requirements for high reliability of the First-MOVE satellite and how robustness drives the actual design of the satellite.

1 Mission and Configuration of the First-MOVE Satellite

The First-MOVE satellite is mainly a verification experiment, which shall test the satellite platform and generate comprehensive data for future optimization. First-MOVE is a single-unit CubeSat [1] comprising not only standard subsystems such as power, communication and on-board data storage, but also a sophisticated Data Handling and Safety System as well as a structure with two deployable solar panels. These panels carry new generation solar cells which shall be precisely characterized in cooperation with an industrial partner.

In addition, the satellite carries a CCD camera which shall take pictures of the Earth. High quality pictures can be taken and be transferred back to the Earth despite the small bandwidth data channel given by the small size of the satellite, using a developed algorithm which filters the pictures due to content (space, Earth horizon), compresses them, fragments them to small data packages and generates thumbnails for preview.

M. Czech (✉)
Institute of Astronautics at the Technische Universität München, Boltzmannstraße 15, 85748 Garching, Germany
e-mail: m.czech@lrt.mw.tum.de

R. Sandau et al. (eds.), *Small Satellite Missions for Earth Observation*,
DOI 10.1007/978-3-642-03501-2_22, © Springer-Verlag Berlin Heidelberg 2010

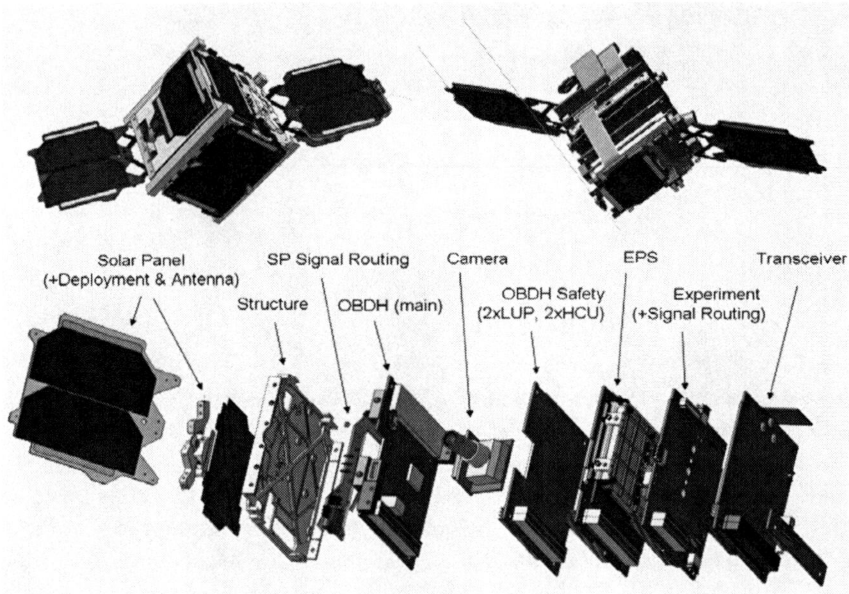

Fig. 1 Explosion drawing of the First-MOVE satellite

Table 1 Subsystems of the First-MOVE satellite

System	Short specification	Provider
Deployable solar panel	2 Deployables, carrying 4 GaAs [2] Solar Cells each	LRT
Antenna system	UHF/VHF, deployment with panels	LRT
Hold-down release mechanism	HDRM with redundant release system for each panel	LRT
Structure	Highly integrated lightweight structure	LRT
OBDH	– High performance, low power consuming MC Atmel SAM9260 [3] – Memories Flash, MRAM, SDRAM	LRT
Camera	1.3MP Static Camera	OES
Safety	– Latchup Protection – Hard Command Unit – Watch-Dog (all redundant)	LRT
EPS	– 12 GaAs Solar Cells [2]	EADS Astrium Clyde Space
	– 3 × MPPT [4] – Li-Ion Batteries 2 × 3.7 V/1,500 mAh [5]	
Experiment	4 GaAs Cells [2] + variable Load	EADS Astrium/LRT
Transceiver	UHF/VHF, 200 mW downlink [6]	ISIS

2 Progress and Timeline for the First-MOVE Satellite

The MOVE program was kicked-off in summer 2006. Since then many iterations have been performed to design a satellite which is coherent on all system levels.

Currently for every system there are verification models (form-fit-function) available which have been tested due to the interaction with other systems. About 70% of the subsystems are available in flight-configuration; the others are currently being transformed from prototype status to flight units.

Flight readiness of the two built flight-units of First-MOVE is expected to take place early June 2009. This leaves enough time for comprehensive acceptance testing and fine-adjustment of the thermal system (see also Sect. 3.2).

One of the satellites is supposed to be launched from India on a PSLV rocket [7] in late 2009; the other will stay at the LRT for the time being and has several functions:

1. Testbed for software enhancements after shipping the first satellite to the launch site.
2. Knowledge transfer to future project managers and team members.
3. Object of comparison for phenomenon observed on the flying satellite.
4. Initial object to apply lessons learned on and optionally to be launched shortly after First-MOVE as an improved MOVE-on satellite.

3 Innovative Concepts to Increase Reliability by Design

First-MOVE, as the name says, is the first mission of a series of satellites. It is therefore mainly for verification and educational purpose. For both purposes, it is important that the satellite lives long enough to acquire data from its space mission. The more the lessons learned through First-MOVE, the better can the following be optimized towards efficiency and scientific excellence.

Therefore the highest design goal for MOVE has been defined as:

> "Robustness rather than Efficiency"

For this reason the scientific purpose of the mission is intentionally kept on a reasonable level to focus on system aspects. As such small satellites are normally designed with a high amount of COTS components they are highly susceptible to a variety of failures. Reliability therefore cannot be guaranteed through the application of reliable components but has to be achieved by dedicated system concepts. The paper shows innovative concepts to increase reliability by design and testing in the following.

3.1 Failure and Risk Analysis as Prerequisites for Robust Design

The precondition to increase robustness is to understand failures and how they affect the overall system risk.

ECSS standards therefore define two basic methods:

- Fault Tree Analysis (FTA) [8]
- Failure Mode and Effects Analysis (FMEA) [9]

Both methods have been applied on the First-MOVE satellite, although FMEA has been emphasized. FMEA analyzes all failure sources of the satellite on component level. It can be displayed clearly and has therefore been chosen to be shown here.

Reliability of the satellite means lowering risk of failure. Due to ECSS risk is defined as following:

$$\text{Risk} = \text{Occurrence} \times \text{Severity} \times \text{Detection}$$

For the increase of system robustness the following driving failures have to be considered. For a clear understanding they are displayed in a non ECSS conform version, not regarding the factor Detection.

Table 2 Abstract of the failure analysis and countermeasures

	Occurrence w/o CM[a]	Severity w/o CM[a]	Main CM[a]	Occurrence with CM[a]	Severity with CM[a]
Failure of release detection	moderate	hazardous	Occurrence + Severity ↓ by Testing and Redundancy	low	moderate
Failure of Antenna deployment	moderate	hazardous	Occurrence + Severity ↓ by Testing and Redundancy	low	moderate
Software Crash	high	very high	Severity ↓ by Reset Strategies	high	moderate
Single Event Latch-up	moderate	very high	Severity ↓ by Latch-up Protection	moderate	moderate
Single Event Upsets	moderate	high	Severity ↓ by Software maintenance	moderate	low
Thermally miscalculated design	moderate	high	Severity ↓ by robust dimensioning Occurrence ↓ by test and adjustment	low	low

[a] CM = Counter Measure

Countermeasures depend on whether the affected component is a self developed component or off-the shelf. For self developed components, occurrence can be minimized by extensive testing and in-development improvements, whereas for COTS components the main countermeasure is to decrease severity by redundancy or more sophisticated concepts.

The countermeasures taken in First-MOVE to improve robustness are described in the following chapters.

3.2 Testing

Table 2 shows the need to test certain components, such as the switches for release detection and a safe deployment of the antennas with the solar panel.

Therefore these components were tested – in addition to extensive ground testing – under realistic conditions on the Rexus-4 sounding rocket [10]. Several switches and the deployment of a solar panel in space have been successfully verified on the rocket on 22nd October 2008. The testing conditions and results are described in [11].

Also the thermal system will be extensively tested. The design of the First-MOVE satellite leaves possibilities for specific adjustment of the overall thermal system by small changes on the hardware of the flight-unit of First-MOVE. The test results will be taken to correlate the thermal simulation models, the results from which are used to implement the changes.

3.3 Reset Strategies

The First-MOVE reset strategies deal with the occurrence of failures which can make a reset necessary.

One risk in using COTS components is their impairment by space radiation, especially microcontrollers and memories are affected by Single-Event-Latchups (SEL). In First-MOVE there are three separate power buses within the OBDH system which are supervised by a Latchup detector. A circuit breaker can do a complete satellite reset by cutting-off all power from the satellite for a short defined time.

Software crashes, especially those affecting the commanding line between ground and satellite are counter measured with a Hard Commanding Unit (HCU), which has been developed at the LRT. The HCU has an input for the Raw FSK signal from the transceiver and filters bitstreams to detect patterns of 80 bit length. In case such a pattern is detected, a signal can be output for resets on different levels.

Neither the transceiver nor the HCU use any software for triggering these resets. The transceiver demodulates the signal by hardware, and also the HCU uses

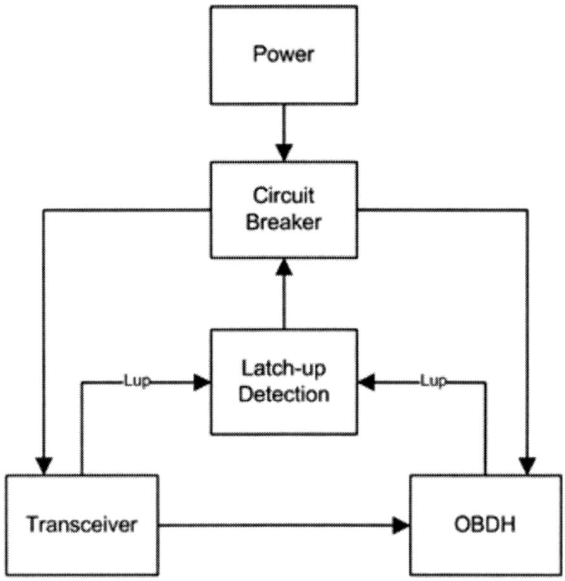

Fig. 2 Latchup CM scheme

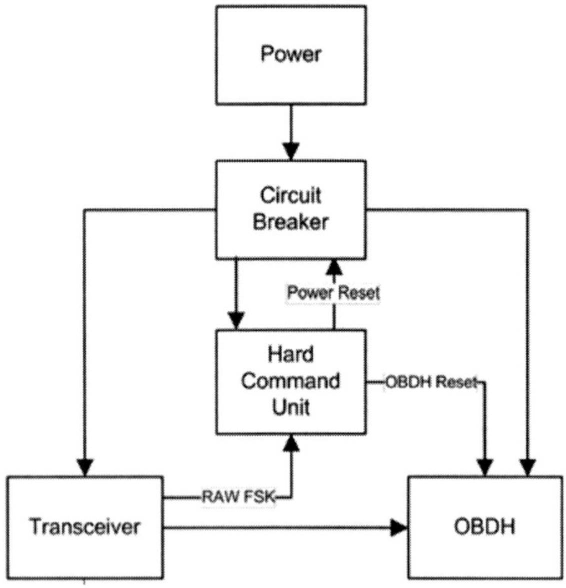

Fig. 3 HCU scheme

hardware filters. Therefore the satellite is accessible even in case of a complete software crash, and can be reset on highest system level triggering the circuit breaker which was described before.

In the First-MOVE satellite software crashes are additionally counter measured with Watch-Dog Timers (WDTs). One is integrated in the Microcontroller which it can reset; one performs a complete power reset on system level (using the circuit breaker) in case the user on ground does not send a specific command to reset this WDT. The latter therefore watches the user's ability to contact the satellite and ensures a link between ground and the satellite even if a problem with the transceivers Raw-FSK output occurs.

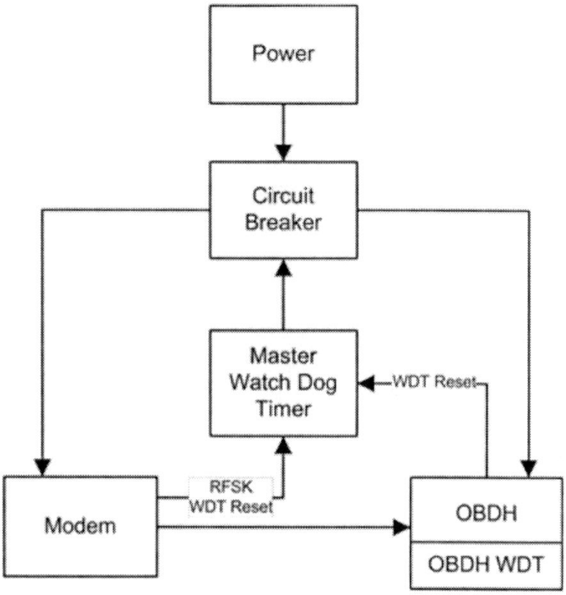

Fig. 4 Master WDT scheme

3.4 Software Maintenance

Software maintenance is essential for robustness, especially in combination with the reset strategies described above. It is expected that resets will occur during the mission, and it has to be ensured that the satellite has the right boot code available after a reset.

Radiation hard memories are expensive and are not available for the First-MOVE satellite. Therefore a combination of memories with an implemented software correction Procedure has been developed for software maintenance.

First MOVE uses a Magnetic RAM (MRAM) memory, which is very unlikely to have SELs and SEBs [12]. Therefore they are used as boot memory for the OBDH system.

However, MRAM is susceptible to Single Event Upsets (SEU), which might be hazardous for the mission if not counteracted. Therefore, the stored code is periodically compared to identical code within a flash memory device. Flash memory is big enough to contain the software several times, so that a concurrent corruption of code in several versions is very unlikely. With an implemented voting mechanism corrupted code can be found and corrected within the flash and the MRAM memories.

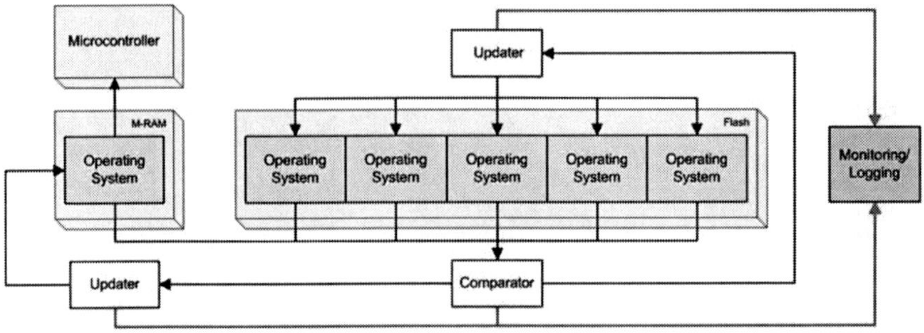

Fig. 5 Software error correction scheme

3.5 *Securing the Counter Measuring Technologies from Hazardous Behaviour*

All active countermeasures described in the previous chapters engage with high level systems of the satellite. Therefore, the risk of malfunction of one of the counter measures leading to hazardous interactions with the mission has to be kept low. In the First-MOVE mission this is done by using all countermeasures redundantly. The combination of the redundant systems and the interaction with the system is done with radiation tolerant components. Although these are expensive, the costs have to be beard, and they are still lower than designing all systems using radiation tolerant components.

4 How to Learn Lessons and React

The long term plan of MOVE is to verify the platform and to gain experience regarding realistic hazards for a satellite with First-MOVE. While in First-MOVE many systems are foreseen to improve robustness, it is intended to tailor down many of them to a more realistic risk scenario in future missions. The redundancies described in the previous chapter are most likely over-pessimistic. It is considered that some redundancies can be left out in following missions, making them less space consuming and less expensive, as the described radiation tolerant parts might be left away.

However, adequate evaluation is only possible if the behaviour of the safety systems can be adequately characterized over mission lifetime.

Table 3 Characterization of safety system behaviour

Action	Telemetry log-entry[a]
Deployment Detection	Signal of both redundant deployment switches
Antenna Deployment	–Activation of redundant melting units –Deployment Status –Antenna Quality
Microcontroller Reset	Reason for Reset
Latchup Detection	–Detected Latchups of all redundant detectors –Power Bus(es) showing a Latchup
Hard Command Detection	Detected Hard Command of all redundant HCUs
Circuit Break Trigger	–Trigger for all redundant circuit breakers –Source of trigger
Software Correction	–Location of the corrupted code –Counter for corrections
...	...

[a]All entries are logged with the exact time they occurred.

5 Future Prospects

The experience gained from First-MOVE can not only be applied on CubeSats. It is intended that it will be of use for all kinds of satellite missions.

However, development of MOVE satellites of the CubeSat standard is becoming a substantial part of the institute's research and education activities. Future MOVE satellites will be based on the MOVE platform, which in its optimized configuration will enable application of more sophisticated payloads.

Due to good compatibility of the developed MOVE systems with existing CubeSat systems, it is intended that MOVE components shall be also available for other CubeSat missions after being verified.

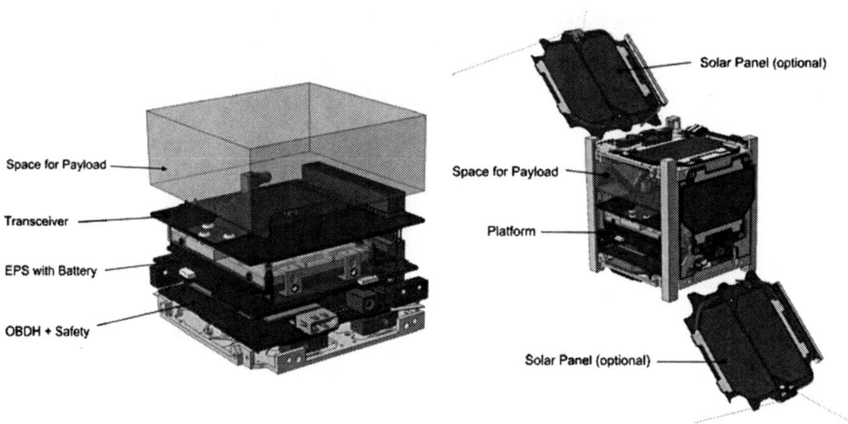

Fig. 6 Payload envelope for future missions using the MOVE platform

Table 4 Resulting configuration of a MOVE platform

Basic components	Development status
Clyde Space Power Subsystem	Several times flight verified by then
ISIS transceiver	Several times flight verified by then
On-Board Data Handling (LRT)	Flight-verified on First-MOVE
Safety Systems (LRT)	Flight-verified on First-MOVE
Software Operating System (LRT), Applications to be programmed in C++	Flight-verified on First-MOVE
Optional components	
Deployable Solar panels for high power needs (LRT)	Flight-verified on First-MOVE
Additional batteries (Clyde Space)	Several times flight verified by then
Light-weight structure for use with or without solar panels, incorporating Kill-Switches	Flight-verified on First-MOVE
Sun Sensors (DTU)	Flight-verified on First-MOVE
All components compatible with the CubeSat Kit standard	Experimental or verified
Interfaces of the Platform	
2 × SPI, 2 × UART, 2 × I^2C, 16 × GPIO (all on PC104 connector)	All flight-verified on First-MOVE
Power (Latch-up protected and resettable 3.3 V, 5 V and Battery Voltage Buses)	All flight-verified on First-MOVE
16 analog measuring channels	All flight-verified on First-MOVE
Programming interface (JTAG, Debug, USB, 3.3 V, Boot Mode Select) and battery charge interface, both in access port area	All flight-verified on First-MOVE
Space for experiment is 95 × 95 × 33 mm^3 (single cube)	TBC
Mass limit for experiments is approx. 250 g	TBC

References

1. http://cubesat.calpoly.edu/
2. AzurSpace, Datasheet 3G – 28%, http://www.azurspace.com/cache/88Data_Sheet_3G28.pdf
3. Atmel, Product Card SAM9260, http://www.atmel.com/dyn/products/product_card.asp?part_id=3870
4. Clyde Space, CubeSat Power Datasheet, http://www.clyde-space.com/resources/datasheets
5. Clyde Space, Battery Datasheet, http://www.clyde-space.com/resources/datasheets
6. Innovative Solutions in Space (ISIS), Small Satellite Transceiver, http://www.isispace.nl/media/products/TRANSCEIVER/Brochure_ISIS_Transceiver.pdf
7. Indian Space Research Organisation (ISRO), Polar Satellite Launch Vehicle, http://www.isro.org/pslv.htm
8. ECSS standards, ECSS Q-40–12A.
9. ECSS standards, ECSS-Q-30–02A.

10. A. Stamminger, *REXUS-4* Campaign Report.
11. C. Olthoff, R. Purschke, R. Winklmeier. Testing of Critical Pico-Satellite Systems on the Sounding Rocket Rexus-4. Proceedings of the 7th IAA Symposium on Small Satellites for Earth Observation.
12. M. Elghefari, S. McClure. Radiation Effects Assessment of MRAM Devices, JPL Publication 08-19 5/08.

Design of a Small Educational Satellite for the Italian High School Students: The EduSAT Project

Filippo Graziani, Giuseppina Pulcrano, Maria Libera Battagliere, Fabrizio Piergentili, Fabio Santoni, and Gabriele Mascetti

Abstract This chapter deals with the design of the EduSAT microsatellite: a small educational satellite developed by the Group of Astrodynamics of University of Roma "Sapienza" (GAUSS), on the basis of its previous experience. The UNISAT program (UNIversity SATellite) started at School of Aerospace Engineering of Roma in the early nineties. The EduSAT Project is funded and coordinated by Italian Space Agency with the aim to promote space education among high school students and to support the qualification and scientific careers of young people (university students, PhD students and young researchers). Another target of this program is to develop a small space mission for low cost scientific experiments and technological tests in orbit. The launch of EduSAT microsatellite is scheduled in 2010: a cluster launch in Low Earth Orbit, performed by Russian-Ukrainian DNEPR Launch Vehicle. This chapter synthesizes project motivations, program organization and describes system architecture and satellite main subsystems design.

1 Introduction

"The Educational Satellite EduSAT" is an educational and research project coordinated and funded by the Italian Space Agency and commissioned to Group of Astrodynamics of the Scuola di Ingegneria Aerospaziale of University of Roma "Sapienza" (GAUSS), in collaboration with an Italian company (IMT srl) active in the field of space systems. This project is organized in two main activities: design, building and launching in to orbit a small student-built satellite and laboratory activity with high school students, which are involved in a real space program. Satellite design and manufacturing are performed by the Group of Astrodynamics of the Scuola di Ingegneria Aerospaziale, on the basis of its previous experience in

F. Graziani (✉)
Scuola di Ingegneria Aerospaziale, "Sapienza" University of Roma, Via Eudossiana 16, 00184, Rome, Italy
e-mail: filippo.graziani@uniroma1.it

R. Sandau et al. (eds.), *Small Satellite Missions for Earth Observation*,
DOI 10.1007/978-3-642-03501-2_23, © Springer-Verlag Berlin Heidelberg 2010

this field [1, 2]. The Group of Astrodynamics is a research group which started its activity in the early nineties: PhD students, researchers and professors of the group are involved in many space activities and topics: among those the UNISAT program (UNIversity SATellite). In the framework of this program four satellites have been launched from the Baikonour Cosmodrome, in Kazakhstan, using DNEPR Launch Vehicle, starting in 2000 with UNISAT satellite. UNISAT-3 has been launched on 2004 and, currently, after 5 years is still operative [2].

EduSAT project gives to the university students the possibility to participate in an interdisciplinary experience, following the project life cycle and working in the field of systems engineering and project management techniques. Students have responsibilities to design mechanical devices, electronics board, software and coding, in cooperation with experts and university members. The project involves also high school students, coming from several institutes distributed on National territory, by a public competition and call for ideas and purposes for technological or scientific payload design and manufacturing. Winner payload will be hosted on board EduSAT satellite. This part of the project involving hands on education with high school students is carried out by an Italian company, IMT srl, in cooperation with the Second University of Roma “Tor Vergata”. In the framework of this activity, high school students use also a technological demonstrator to understand the complex architecture of a space system and get lectures about specific space related topics. EduSAT project policy is to provide Italian students with a practical opportunity to work in the space field with the main goal to stimulate their interest for scientific careers and to improve knowledge of scientific matters basic elements in the primary and high schools. Previous experiences with high school students and studies related to this field [3–5] demonstrated that students appreciate and prefer this learning methodology based on hands-on experience, obtaining better Results [5]. At the same time, this project provides an opportunity to perform technological tests in orbit in order to verify new ideas and realize small, low-cost and short time scientific experiments in orbit. The challenge for the future is to provide opportunities for low-cost or high risk investigations and innovative methods. As a matter of fact, the development of small university satellites became intensely active all over the world in the last twenty years and different space agencies announced programmes for the selection of piggyback passengers to small satellite projects performed by universities: two Examples in this field are given by the European Space Agency (ESA) [6] and Japan Aerospace Exploration Agency (JAXA) [7]. Instead, the role to create a “microsatellite market” and to develop low-cost “state of the art” small satellites can be pursued by professional spacecraft industry [8].

In the Sect. 2 of this chapter an Overview of EduSAT satellite system architecture is given, whereas in the Sect. 3 main satellite subsystems design is depicted. Currently Phases A/B have been concluded and Phases C/D are ongoing. EduSAT project started in December 2007 and the launch is scheduled on April 2010 by DNEPR launch vehicle.

2 EduSAT: System Architecture

EduSAT will be launched as a secondary payload by DNEPR LV. It is designed to fit different orbits in the range of 500–900 km. Nominal design orbit is polar with 500 km altitude. EduSAT's structure is a square prism with height size 26 cm and base size 31.5 cm. Its expected weight is about 10 kg. The internal space is divided by vertical panels, which host subsystems hardware and electronic boards. On the bottom plate are arranged mechanical interface with the launcher, antennas for S band transmission and for down link data transmission in UHF band (436 MHz), switches to turn on the satellite during the separation phase from the launcher (Fig. 1). On the upper plate are arranged antennas for up link data transmission in VHF band (145 MHz) and two solar cells strings. On the external wall of the first inner vertical panel (right side in Fig. 1) have been fixed radios for transmissions in S, UHF and VHF bands. Between these two inner panels the payload proposed by high school students will be located and will be fixed to the second vertical panel together with batteries. On the other side the magnetometer and electronic board of the OBDH subsystem will be arranged. Solar arrays are body mounted and photovoltaic system is based upon triple junction solar cells which are glued to a non conductive support. This panelization Procedure has been used on board UNISAT-3 and UNISAT-4 and it has been extensively tested in orbit [2].

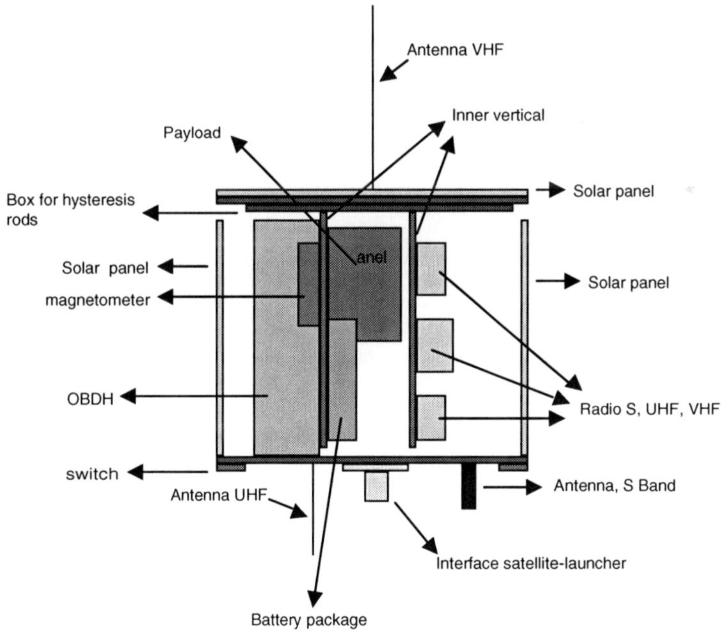

Fig. 1 EduSAT LAY-OUT

Commands and telemetry data will be received in the UHF-VHF bands using the Ground Station SPIV (San Pietro in Vincoli) located at the Scuola di Ingegneria Aerospaziale of Roma. This station is also able to receive and transmit data in S-Band, mainly used to receive video images and pictures.

Due to the educational objectives of this mission, spacecraft has to be simple, commercial off-the shelf components and parts have to be employed and project time schedule is strongly fixed.

3 Main Subsystems

3.1 Structure

Structure is easy and fast to assemble and disassemble thus permitting a good flexibility even in substituting a complete sect whenever it is needed and giving to the students the possibility to work, at the same time, on different subsystems on separate sects. Lateral panels, inner vertical panels, upper and bottom plates have been manufactured by using Aluminium honeycomb (AI 3003-3/8-0.0015, density 83 kg/m^3) with face sheets in Aluminium AI 5251 H16 (thickness 0.5 mm) [9]. The vented Aluminium honeycomb has already been employed for the manufacturing of the previous satellites UNISAT [9] and it has been selected also to minimize weight.

There are not electrical connections between launcher and satellite; during ascending phase satellite is turned off and at the separation it is switched on. EduSAT preliminary mass budget is given in Table 1 [10, 11].

Table 1 EduSAT mass budget

EduSAT mass budget	
Structure (lateral panels, inner vertical panels, adaptator, secondary structures, mechanisms)	4 kg
Electrical Power System (battery package, solar arrays)	2 kg
Wiring, cables, screws, glue	2 kg
Payload	1 kg
Electronics (PCB, magnetometer, antennas, radios)	1 kg
Margin (underestimation, program additions)	2 kg
Total expected mass	12 kg

EduSAT moments of inertia have been calculated on satellite digital mock-up with respect to main axis of inertia (green axis = x, red axis = y, blue axis = z, Fig. 2); they are $I_x = 0.094\ kg{\cdot}m^2$, $I_y = 0.11\ kg{\cdot}m^2$, $I_z = 0.12\ kg{\cdot}m^2$ [10, 11].

In order to have an idea of the structure vibration modes, preliminary analysis FEM using specific software has been assessed in the design phase (Fig. 2). A launch-vehicle structure has natural frequencies responding to forces from both internal (engine oscillations) and external (aerodynamic effects) sources [12]. For DNEPR launch vehicle requirements are: natural frequencies larger then 10 Hz for lateral axis and larger then 20 Hz for longitudinal axis [13]. The Results of

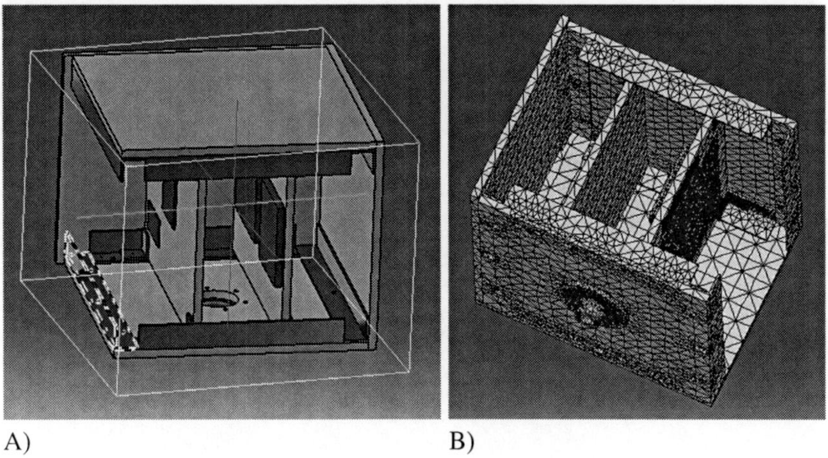

Fig. 2 (a) EduSAT digital mock-up employed for preliminary FEM analysis; (b) model mesh

Table 2 EduSAT natural vibration mode

Vibration mode	Frequency (Hz)
1	97
2	100
3	264
4	493
5	829
6	1,440
7	1,802
8	2,080
9	2,393
10	2,618

the preliminary analysis showed the first ten satellite natural vibration frequencies (Table 2). First natural vibration frequency is 97 Hz, greater then launcher requirements. The first vibration frequency is related to satellite oscillations on pyrotechnic which works as cantilevered element [10, 11].

3.2 Electrical Power Subsystem

Four body mounted solar panels provide power required for the spacecraft. In order to evaluate the solar-array area it has been considered a low Earth orbit with 500 km of height. Since EduSAT is stabilized using a passive magnetic system, the worst illumination condition in this case can be evaluated considering only one solar-array in the Sun direction (Fig. 3, [11]).

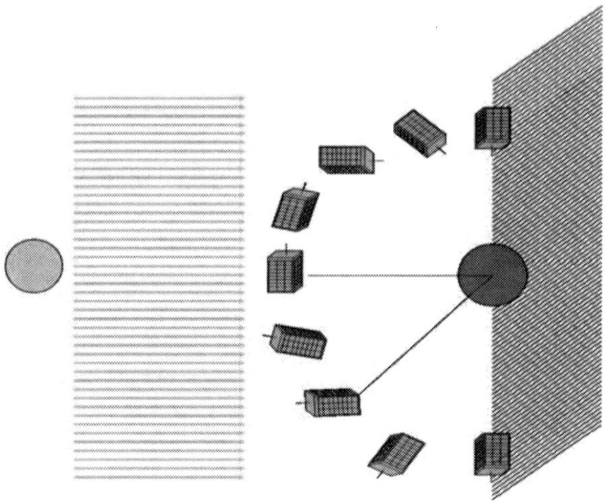

Fig. 3 Sun exposition in orbit for EduSAT satellite

Table 3 Satellite power loads

P(W)	t/T	Subsystem	Required mean power (W)
0.5	0.5	Attitude determination (sensors)	0.3
4.0	0.125	Data transmission (down link)	0.5
2.0	0.5	Payload	1.0
2.0	1.0	CPU (OBDH)	2.0
8.5	–	Total	3.75

To evaluate the minimum solar array size it is necessary to compare the satellite power consumption during the whole orbit with the photovoltaic system power generation during daylight (P_{sat}) [12], Satellite power loads are sketched in Table 3. P is the required power, t is the operation time and T is the orbital period. On the basis of these requirements the estimated solar array power is 7 W.

The effective solar array area toward Sun direction, considering attitude motion shown in Fig. 3, is the 63% of the solar array geometrical area. Thus, the power provided by a single solar array is 63% of the solar array power when the Sun direction is perpendicular to the solar array [10, 11]. Minimum estimated area for each solar panel is 400 cm^2, considering triple junction solar cells not space qualified, with efficiency 21% [2, 12]. Effective satellite sizes satisfy widely this requirement. Results in orbit for the selected kind of solar cells demonstrated an efficiency loss of 20% per year [2] . On the basis of this consideration, a solar panel with 20 solar cells has been designed, choosing 5 solar cells strings in parallel. Solar cell voltage is namely 2.3 V; satellite bus voltage is non regulated 8 V. EduSAT's solar arrays are manufactured using the same Procedure exploited for UNISAT solar arrays. In Fig. 4 EduSAT's solar cells characterization is shown. Nichel-Cadmium batteries will be used for energy storage.

Fig. 4 Solar cells characterization at GAUSS Laboratory

3.3 Attitude Control Subsystem

Satellite attitude control system is based upon a passive magnetic stabilization system which exploits a permanent magnet to provide a restoring torque and permeable rods for oscillation damping [11, 14]. Permanent magnetic dipole sizing has been assessed by estimating worst-case disturbance torques (Table 4 [11, 12]). It is on the order of 1 A·m^2 [11].

Table 4 Worst case disturbance torques estimate for EduSAT

Disturbance	Gravity gradient	Solar radiation	Magnetic field	Aerodynamic	Worst case
Value (N·m)	$4.7\cdot10^{-8}$	$2.8\cdot10^{-8}$	$9.8\cdot10^{-7}$	$7.6\cdot10^{-7}$	$1.8\cdot10^{-6}$

Passive magnetic attitude stabilization techniques require damping devices to dissipate rotational kinetic energy and to ensure asymptotic stability of the attitude motion: magnetic permeable rods have been designed and manufactured with this aim for EduSAT microsatellite. These rods have been manufactured using commercial soft magnetic material (Mumetal) with density 8,700 kg/m^3 and chemical composition in weight %: 77% Ni, 4.4% Cu, 2.67% Mo, 0.29% Mn and Fe balance. Permeable rods 20 cm-long have been selected to be boarded on EduSAT; the rod volume is 0.2 cm^3. In design phase following configuration has been proposed for rods arrangement on board the satellite: 5 rods per axis on two perpendicular directions in a plane normal to on board permanent dipole axis. The permeable rods magnetic parameters have been measured in order to compare them with the nominal values obtained from the reference literature and to improve accuracy of the satellite attitude dynamics numerical simulations, as shown in [11].

3.4 Communication and OBDH Systems

EduSAT communication system is based on UNISAT satellites communication system; it will use data transmission in VHF band (145 MHz) for up link and in UHF band (436 MHz) for down link. Bit rate is 9,600 bps for these channels. Payload data transmission can use a buck-up channel in S band (2.2–2.4 GHz) with a bit rate of 38800 bps. On board computer is an integrated system RCM3400 of Z-World based on a microcontroller Rabbit 3000 [9].

4 Conclusions

In this chapter the motivations, the mission requirements, the design drivers and the main subsystems design of the EduSAT satellite are summarized and depicted. This is an university student-built satellite which involves also high school students coming from several Italian schools in the proposal, design and manufacturing of small technological and scientific payloads. Payload selected on the basis of a public competition, coordinated by Italian Space Agency, will fly on board EduSAT satellite.

EduSAT is currently under realization and its launch is scheduled in April 2010.

References

1. F. Santoni, F. Piergentili, F. Bulgarelli, F. Graziani. “The UNISAT program: lessons learned and achieved Results”, 57th IAC, International Astronautical Congress 2006, Valencia, Spagna, Ottobre 2006, IAC-06-E1.1.10.
2. F. Santoni, F. Piergentili. Analysis of the UNISAT-3 Solar Arrays In-Orbit Performance. Journal of Spacecraft and Rockets, 45(1), Jan–Feb 2008.
3. F. Graziani, F. Santoni, C. Marchiori, F. Piergentili, M.L. Battagliere. An educational microsatellite for high school students, XIX National Congress A.I.D.A.A. 17–21 Sept. 2007, Forlì, Italy.
4. C. Cappelletti, M.L. Battagliere, F. Piergentili, F. Santoni, F. Graziani. “A new educational program: Astronautics in High School”, 58° International Astronautical Congress – IAC 2008, 24–28 Sept. 2007, Hydebarad, India.
5. F. Antinucci, R. Nesler. “Imparo giocando: videogiochi e apprendimento. Rapporto di ricerca sul quadriennio di sperimentazione”, Research Thecnical Report, IPRASE Trentino Region, August 2007, Italy.
6. http://www.esa.int
7. Y. Tsuruda, T. Hanada, J.C. van der Ha. “QSAT: A Low-Cost Design for 50 kg Class Piggyback Satellite”, 2008, www.senkyo.co.jp/ists2008/pdf/2008-f-20.pdf
8. M.A. Swartwout. “The role of Universities in small satellite research”, http://aa.stanford.edu/~ssdl/papers/SSDL9705.pdf, August 1997.

9. F. Graziani, F. Santoni, F. Piergentili, F. Bulgarelli, M. Sgubini, M. Ronzitti, M.L. Battagliere. "UNISAT microsatellites: an affordable way to test in orbit innovative technologies and ideas", 23rd AIAA International Communications Satellite Systems Conference (ICSSC-2005), 25–28 Sept., Rome, Italy.
10. M.L. Battagliere, F. Piergentili, F. Santoni, F. Graziani and others, Technical Report ASI_RA1_V1.0, "Satellite Didattico EduSAT", Scuola di Ingegneria Aerospaziale, University of Roma "La Sapienza", Sept. 2008.
11. M.L. Battagliere. "Design and Manufacturing of small satellites with scientific and educational purposes", Ph.D Thesis, Scuola di Ingegneria Aerospaziale, University of Roma "Sapienza", February 2009.
12. W.J. Larson, J.R. Wertz. "Space Mission Analysis and Design", Second Edition, Space Technology Series, 1995.
13. DNEPR user's guide, ISC Kosmotras.
14. M.L. Battagliere, F. Piergentili, F. Santoni. "Libration Damping System in the Attitude Stabilization of University Microsatellites", Paper AS-08-013, Automatic Control in Aerospace, on line Journal, Vol. 2, Luglio 2008, ISSN 1974-5168.

Testing of Critical Pico-Satellite Systems on the Sounding Rocket Rexus-4

Claas Olthoff, Ralf Purschke, Roland Winklmeier, and Manuel Czech

Abstract On October 22nd 2008, the VERTICAL (VERification and Test of the Initiation of CubeSats After Launch) experiment was flown on the REXUS 4 sounding rocket mission at Esrange in Kiruna, Sweden. The experiment's objective was to verify critical hardware to be used on the MOVE *CubeSat* in a space environment. The items to be verified were multiple micro switches from different manufacturers and a *solar panel deployment mechanism* developed at TUM. The deployment mechanism is triggered by a melt wire. During launch, the switches are depressed by a plate which is retracted once the rocket is near its apogee. This simulates the satellite's ejection from the launch vehicle. The verification sequence was executed as planned during the 10-min flight and the experiment was safely recovered. The acquired data suggests that the deployment mechanism can be used as is and COTS of verified quality micro switches will survive LEOP conditions and are suitable for further testing, addressing their long-term reliability.

1 Introduction

A picosatellite verification platform called MOVE (Munich Orbital Verification Experiment), which is based on the CubeSat standard, is being developed at the Institute of Astronautics at the Technische Universität München [1]. MOVE shall be a low-cost verification possibility for MEMS (Micro Electro Mechanical System) technology, helping to bridge the so-called "Valley of Death" which prevented operational aerospace application of such new technologies often in the last years.

More important than this technical goal however, was MOVE's educational aspect. From the beginning the project was considered to be a great opportunity for students to participate in a real space project using tools and methods that are commonly used in industrial projects. All participants in the VERTICAL project are

C. Olthoff (✉)
Institute of Astronautics at the Technische Universität München (TUM), Boltzmannstraße 15, 85748, Garching, Germany
e-mail: office@lrt.mw.tum.de

R. Sandau et al. (eds.), *Small Satellite Missions for Earth Observation*,
DOI 10.1007/978-3-642-03501-2_24, © Springer-Verlag Berlin Heidelberg 2010

also MOVE team members who took their knowledge from MOVE into their own project, which they executed with minimal intervention by members of the institute and the MOVE project manager [2].

2 Background

Regarding CubeSat missions one of the most critical events is the startup of the internal systems after separation from the launcher. CubeSats are by definition not allowed to be active during launch; the power circuits have to be mechanically disconnected. After deployment from the launch vehicle, micro switches close these circuits and the systems have to startup autonomously. The reliability of MOVE verification platforms, but also other CubeSats can be significantly increased by a qualified autonomous system executing startup operations.

Sounding rockets are a very sufficient option for verification of these operations, as mission characteristics are very similar to those of an orbital mission until deployment. A flight on a sounding rocket simulates the following conditions very close to those on a CubeSat orbital launcher:

- Load characteristics during ascent phase
- Decrease of pressure to vacuum during ascent phase
- Microgravity environment during experiment operations.

The REXUS program is realized under a bilateral Agency Agreement between the German Aerospace Center (DLR) and the Swedish National Space Board (SNSB). The Swedish share of the payload has been made available to students from other European countries through a collaboration with the European Space Agency (ESA). Experts from ESA, the Swedish Space Coporation (SSC) and DLR provide technical support to the student teams throughout the project [3].

The 10-day REXUS 4 campaign took place at Esrange in Kiruna, Sweden, beginning October 13th 2008. During this time the experiment went through final testing with the REXUS service system in conjunction with ground systems and the other payloads. In a fairly short timeframe, the experiment had to be readied for launch.

3 Method

The first concept for the VERTICAL experiment included deploying a fully functional CubeSat from the launcher. The short time in microgravity would have been used to simulate the first few seconds after the satellite deploys on orbit. Soon it was realized, that this verification method was unnecessarily complex. It would have involved the development of ground stations for TM/TC, a deployment mechanism for the satellite and numerous smaller items. Additionally, most of the MOVE subsystems needed for this test would not have been ready for flight. To reduce complexity it was decided, not to deploy the satellite outside of the rocket, but inside. This would have eliminated the need for complex communication systems, plus the hardware could be recovered for post-flight analysis. Unfortunately, the CubeSat would have been floating free within the rocket and the uncertain mechanical loads during reentry and landing made it necessary to capture the released CubeSat, again increasing complexity and error-proneness. This resulted in a concept where the CubeSat itself is not moved, but a plate is retracted from the satellite structure, simulating it's deployment. While further developing this option, it became apparent, that the limited volume inside a REXUS module would make it difficult to have a full size mock up of the satellite on board. This analogy was technically possible, but not necessary from a testing standpoint. So it was decided to verify only those items critical for deployment and system start up. These so called Verification Items (VI) were to be tested independently to maximize the scientific return, even if one of the components failed.

The VIs that where considered most valuable for MOVE are:

VI1 Deployment Switches (DS): Launch service providers require CubeSats to be powered off during launch until deployment. Usually micro switches are used for deployment detection, by sensing the loss of contact to the launch vehicle. All following operations, including power up, require feedback from these switches to be initiated. On VERTICAL, a total of 16 commercial switches are verified to cover a broad spectrum of manufacturers, the CubeSat standard only requires one switch on board [4].

VI2 Solar Array Deployment Mechanism (SADM): MOVE's solar panels have to be stowed during launch and cannot be deployed before or during the ejection of the CubeSat from the launch vehicle. The SADM, consisting of a set of springs and a Hold Down Release Mechanism (HDRM) was newly developed at TUM and has no space heritage. The HDRM is activated by melting a nylon wire and the solar panel is deployed by a set of springs.

3.1 Components

Figure 1 shows the three basic components of the VERTICAL experiment and the bulkhead it was mounted to inside the launch vehicle. The components are: the

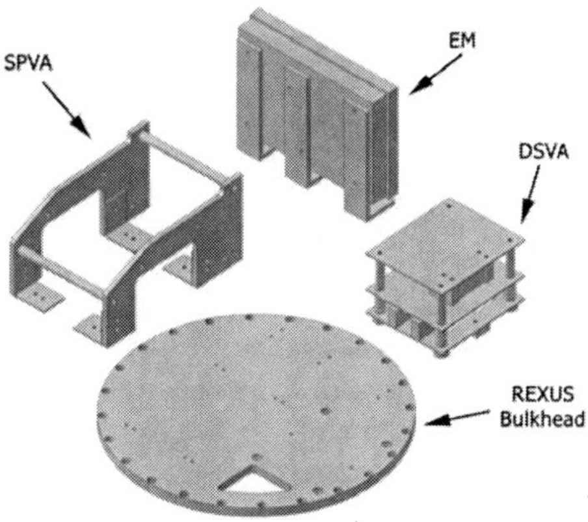

Fig. 1 Overview

Solar Panel Verification Assembly (SPVA), the Deployment Switch Verification Assembly (DSVA) and the Electronics Module (EM). They are mounted on the bulkhead using standard M4 screws, the bulkhead itself is secured to the structure of the payload module by screws along its perimeter.

3.1.1 Deployment Switch Verification Assembly

The function of the DSVA is to verify the deployment switches. A total of 16 micro switches are mounted on two brackets in a parallel configuration. During the launch the switches have to be depressed. This is accomplished by a plate that covers all the switches. This so-called retraction plate is actively pushed towards the switches by an actuator. After launch the switches have to be triggered, so the actuator is commanded to turn 90° and the plate is pushed away from the switches by four springs. The actuator turns at 250°/s making the retraction speed of the plate slightly slower than an actual CubeSat deployment.

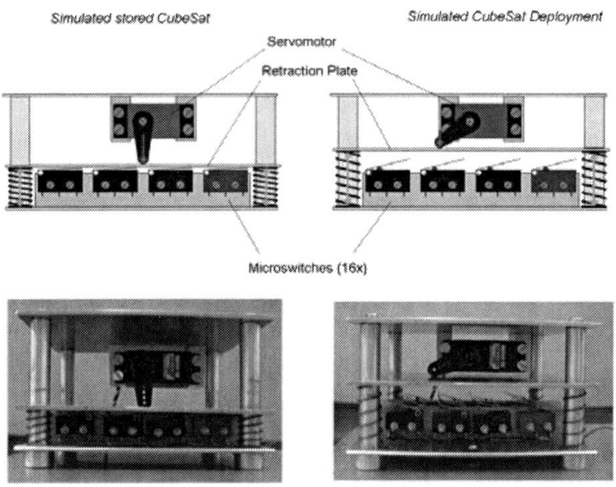

Fig. 2 CubeSat deployment simulation with DSVA

3.1.2 Solar Panel Verification Assembly

The SPVA is designed to verify the HDRM and deployment mechanism of the solar panel. The SPVA achieves two objectives, it detects the deployment of the solar panel, confirming the HDRM's function and it also measures the panel's motion during deployment. To accomplish this, photo diodes and LEDs are mounted opposite each other on two plates located on both sides of the solar panel. When the panel deploys, the sensors are shaded for a very short time. This time is sufficient for the drop in electric voltage in the diodes to be recorded by the on-board electronics.

The HDRM consists of two clamps that enclose two cones, to one of which the solar panel is attached; the other is mounted on the satellite structure [5].

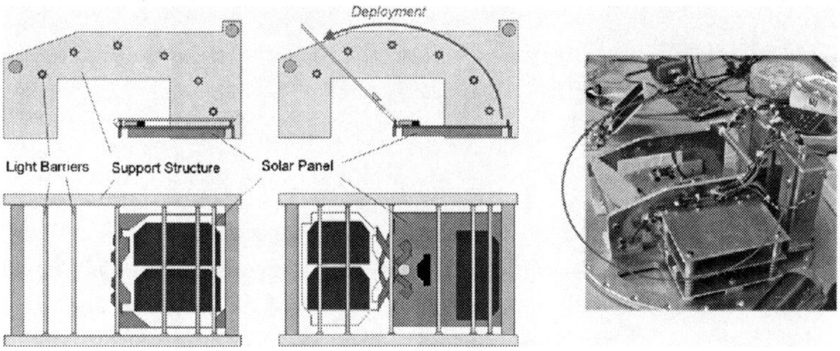

Fig. 3 Solar panel verification assembly

Fig. 4 HDRM

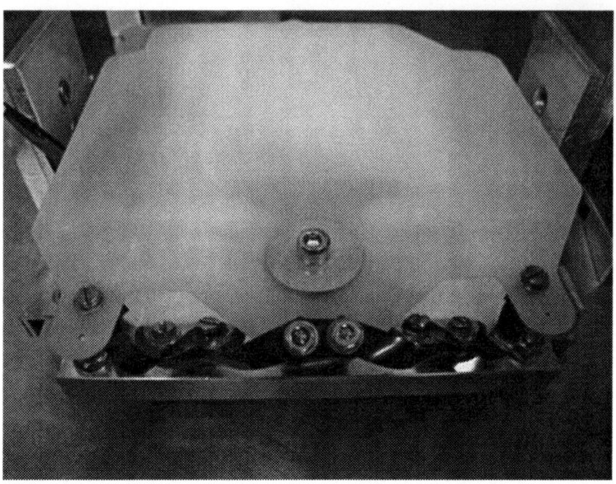

Fig. 5 Solar panel

The solar panel itself is made of a thin glass-fiber epoxy panel with a total of four cells glued to it. It is connected to the CubeSat's structure by a set of springs, which also generate the deployment forces for the panel.

3.1.3 Electronics Module

The EM contains a single electronics board, which handles all of the on board commanding, data acquisition, storage and forwarding. It draws power from the REXUS service system through which it also downlinks part of the acquired data. All of the data is stored on an on-board flash memory device.

None of the electrical components had to be made radiation hard due to the relatively low apogee and short time spent in space. Most parts were SMD packages because they are small and resistant to vibrations. To reduce the risk of the electronics board being hit by loose components within the experiment module and causing a short circuit, an enclosure box was manufactured by bending sheet metal [6].

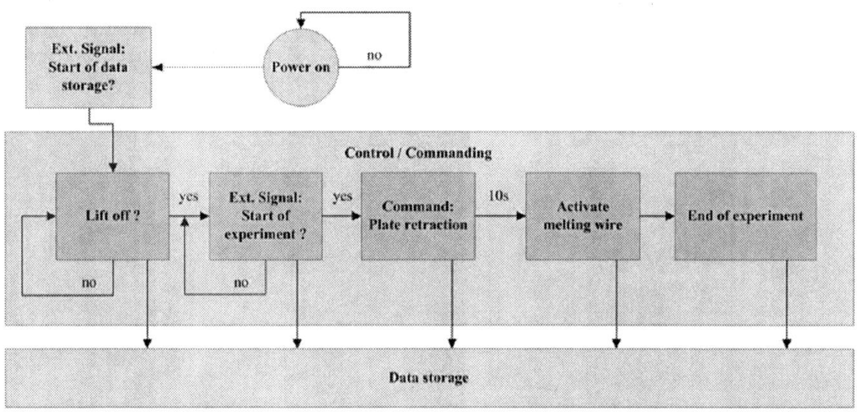

Fig. 6 Software flow diagram

4 Launch Campaign

The REXUS launch vehicle used for the REXUS-4 mission had two solid rocket motor stages. The first stage was a Nike motor, which burned for about 3.5 s, the second stage motor was an improved Orion and it's burn time was about 25 s. Both motors came from German military surplus. Figure 6 shows different views of the REXUS-4 launch vehicle and VERTICAL's position within the experiment module.

The rocket launch itself was flawless, both stages performed nominally and lifted the payload section to an apogee of 176 km. The maximum altitude was reached 200 s after lift-off. Maximum accelerations during launch were 21.6 g during the first stage burn and 15.12 g with the second stage. Figure 8 shows a graph of the accelerations during the entire flight. The temperature in the payload compartment

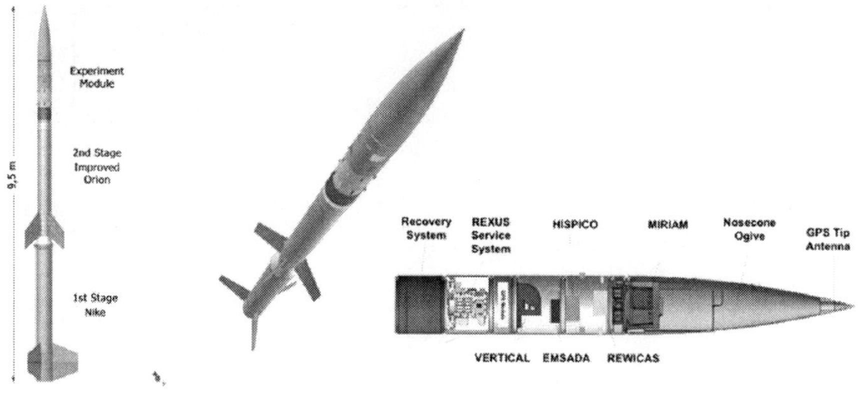

Fig. 7 REXUS 4 launch vehicle (Image credit: DLR)

was 7.2°C prior to launch and it rose to 45.85°C during the ascent. As Fig. 9 shows, the temperatures during reentry were not as high. After 600 s the power supply was cut by the REXUS service module [7].

During the verification sequence there was a ground system data drop out, so mission success could not be verified live during the flight. Fortunately, all of the spacecraft data was recorded on the ground and during a data replay a few minutes after touch-down, all of the experiment data was delivered to the terminal computer.

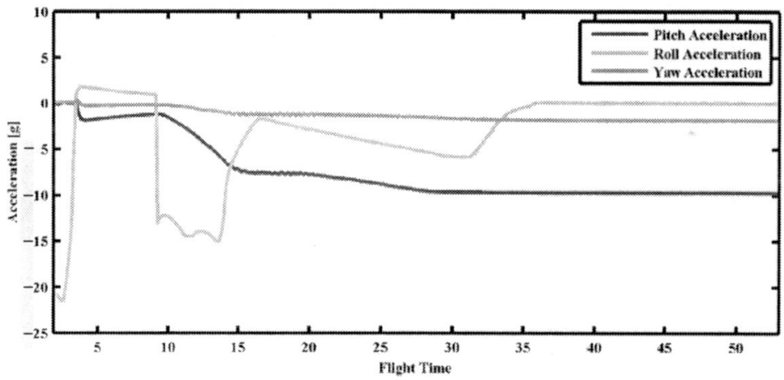

Fig. 8 Accelerations

5 Results

Out of the 16 deployment switches, 15 worked as expected, the one that did not had already been damaged during ground testing. All working switches remained open during the ascent. When the retraction plate was commanded to retract, all switches closed within 73 ms of each other (see Fig. 10). This was consistent with ground

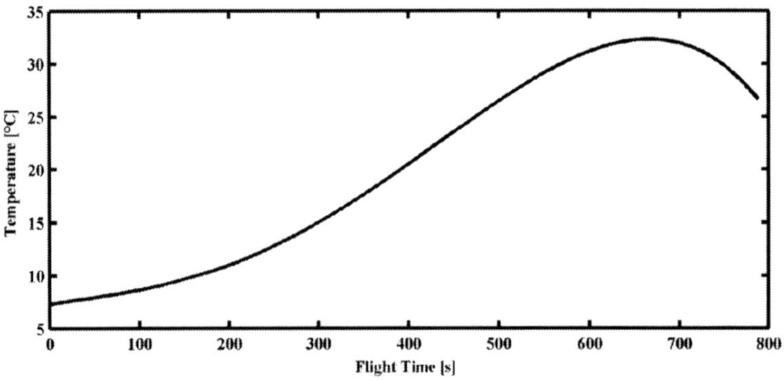

Fig. 9 Temperature within the experiment module

test results and is due to the mechanical configuration of the individual switches. The switches remained open until the power supply was turned off by the REXUS service module prior to touchdown.

The HDRM released the solar panel 1.48 s after the melting wire was activated. This was the same duration that was observed during ground tests. The maximum panel velocity observed was 1.37 m/s, this is also consistent with ground test results. The EM also performed as expected and did not show any off-nominal behavior.

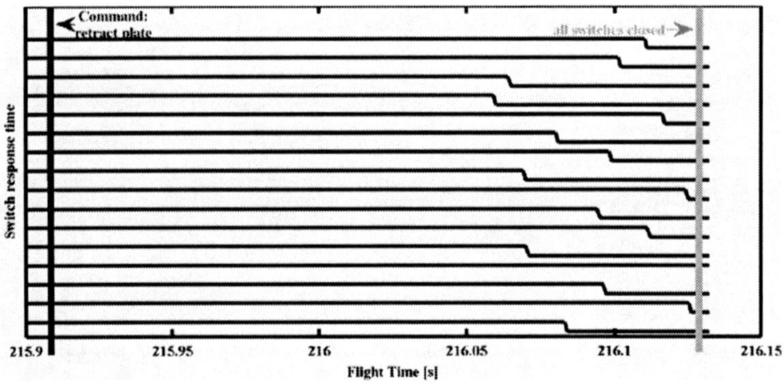

Fig. 10 Switch response time

6 Discussion

Most importantly it is safe to say that the SADM in its current configuration is a working concept to be used on the upcoming First-MOVE mission. Since the primary payload of the First-MOVE mission are solar cells, the successful deployment of the solar panels is critical for achieving the mission objective. Furthermore, off the shelf micro switches of verified quality can survive a rocket launch and the space

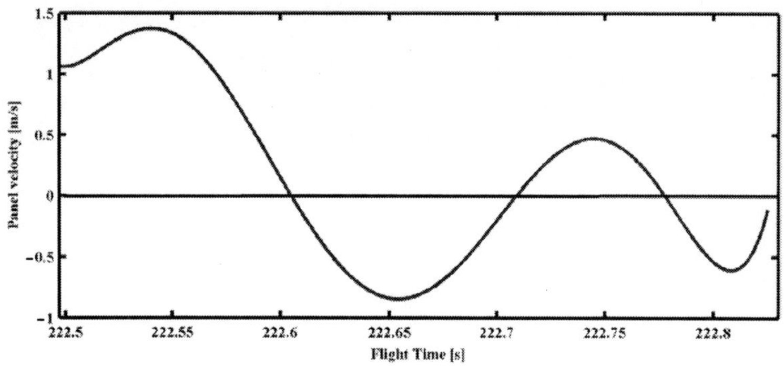

Fig. 11 Panel velocity

environment. This is an important result for the MOVE team because it ensures a successful start up of the CubeSat after deployment.

However, there are some limitations to the comparability between the REXUS flight and an orbital mission. These are shown in Table 1.

Table 1 Comparability to orbital flight

	Good comparability	Moderate comparability	Bad comparability
Launch loads	X		
Vacuum		X	
Temperature			X
Microgravity	X		

The main factor that influences the comparability in a negative way is the short duration of the flight compared to the mission of a satellite. The low altitude also influences the quality of the vacuum, as well as the unknown outgassing of the payload compartment. The launch loads are also shorter than on a launch to orbit, albeit at a higher magnitude.

This shows that further testing of the components is necessary, especially in the areas in which the REXUS environment does not compare well to an orbital mission. These tests would mainly be thermal vacuum and radiation tests and should focus on the long duration nature of an orbital mission.

In conclusion it can be said, that regarding certain environmental factors, sounding rockets are an adequate way to prove the functionality of components during a space flight.

References

1. M. Czech, A. Fleischner, U. Walter. *A First-MOVE in Satellite Development at the TU-München*, Proceedings of the 7th IAA Symposium on Small Satellites for Earth Observation.

2. C. Olthoff, R. Purschke. *Verification and Test of the Initiation of CubeSats after Launch*, Semester Thesis, Technische Universität München, RT SA 2008/32.
3. DLR/MoRaBa, *REXUS User's Manual*, Version 3 (2007).
4. *CubeSat Design Specification*, Revision 9 (2005).
5. W. Rackl. *Multi Purpose Deployment Mechanism for the MOVE Satellite*, Semester Thesis, Technische Universität München, RT SA 2009/1.
6. R. Winklmeier. *Data Acquisition and Control Unit for VERTICAL*, Semester Thesis, Technische Universität München, RT SA (2009).
7. A. Stamminger. *REXUS-4 Campaign Report*, DLR (2009).

Laboratory Facility for Simulation and Verification of Formation Motion Control Algorithms

D. Ivanov, Z. Ming, T. Walter, and I. Zaramenskikh

Abstract The laboratory facility named LuVeX is developed for motion control algorithms verification of the formation flying purposed for the Earth remote sensing. The mock-up description, the parameter determination of its individual elements, description of the realized algorithms for its thruster engine control are given in the chapter. The mathematical simulation results of a mock-up motion controlled with the algorithms which provide the required mock-up formation motion in given trajectories are presented. Several configurations of the formation and formation's elements coupling methods are shown. Required motion precision and its dependence on formation parameters are examined.

1 Introduction

The miniature spacecraft advent caused the appearance of a new view on space system configuration forming. In particular, now it is possible to create a group of satellites which move in orbit in a certain distance one from another and accomplishing a common task, for example, the Earth remote sensing one. The fundamental feature of the satellite group named *formation flying* is autonomous navigation of a single spacecraft within the group and their mutual relative intragroup position control. The formation merit is a possibility of concurrent realization of observation, measurement and other experiments at different points under predetermined and operated mutual position of the satellites.

Prior to the launch campaign satellites are put through multistage verification and test procedure. The tests provide important data about its future behaviour in orbit, allow us to reveal errors committed during its elaboration and development, help to find the application restrictions either one or another control method and others contributing to exact and reliable satellite functioning in the orbit.

D. Ivanov (✉)
Moscow Institute of Physics and Technology, State University, Moscow, Russia
e-mail: danilivanov@mail.ru

R. Sandau et al. (eds.), *Small Satellite Missions for Earth Observation*,
DOI 10.1007/978-3-642-03501-2_25, © Springer-Verlag Berlin Heidelberg 2010

Mostly for this purpose, the laboratory facility called LuVeX (**Lu**ftkissen **V**ehikel **Ex**periment) (Fig. 1) is developed in ZARM, the Center of Space Technology and Microgravity of the Bremen University (Germany) with contribution of graduating students from KIAM and MIPT (Russia). It allows us to explore the satellite dynamics, to give a work-out to different types of satellite formation cooperation and to verify the satellite group motion control algorithms [1]. At the moment there are two mock-ups which enable to move on an air-cushion and set in motion by compressed-air thruster engines. Elaborating the engine control algorithms provides the experiments with control of the mock-up's position and orientation.

Fig. 1 The LuVeX mock-ups

The LuVeX facility can be used also as a training base for students and graduates in space technology development. It makes possible new hypotheses examination and the sensor measurements processing. The facility control algorithms can also be verified and problems of the mock-up group autonomous navigation can be solved.

The mock-up design, the mock-up cooperation means, relative autonomous navigation modes are to be considered in the chapter. One of the realized control algorithms will be cited as an example and examined.

2 Mock-Up Design

Because of the gravity the mock-up mass weighs upon the supports represented by three small plates (legs). For friction exclusion so called air-cushions are used. Actually, it is an air layer between the supporting plates and the bench which is produced through air transmission under some pressure through porous material of the supports (Fig. 2). Due to the air-cushion the vehicle moves freely over the plane actually imitating orbital motion but in projection onto a two-dimensional plane [2].

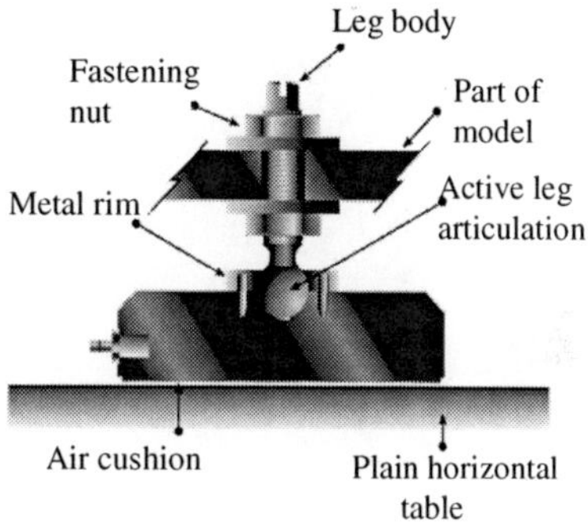

Fig. 2 Sketch of the model leg

The thruster engines installed at the mock-up are powered by compressed air arriving from the tank assembled by air-cylinders. The thruster engine (Fig. 3) consists of a valve shutting the compressed air tank and a tube directing the gas stream. Since the engine is commanded the valve opens and the compressed air stream runs out transmitting thereby an impulse to the mock-up.

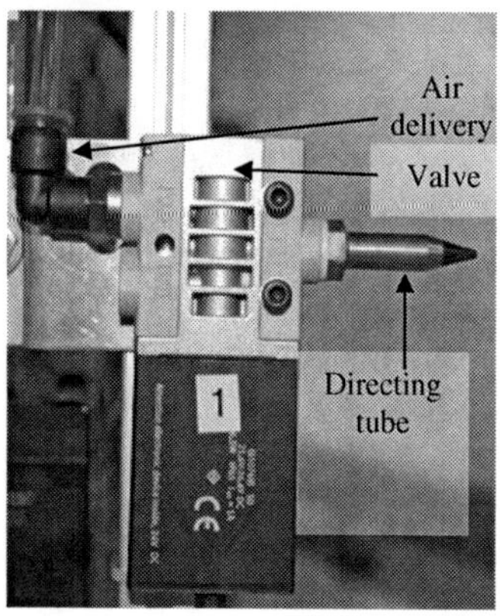

Fig. 3 Thruster engine

The air delivery system is supplied by four compressed air cylinders (Fig. 4). Their inside pressure equals to 200 Bar. However, the pressure $P = 2\text{Bar}$ at the engines inlet is necessary for maximum impulse. For this purpose a pressure regulator was installed. It guarantees constancy of the air pressure delivered to the engines and air-cushion.

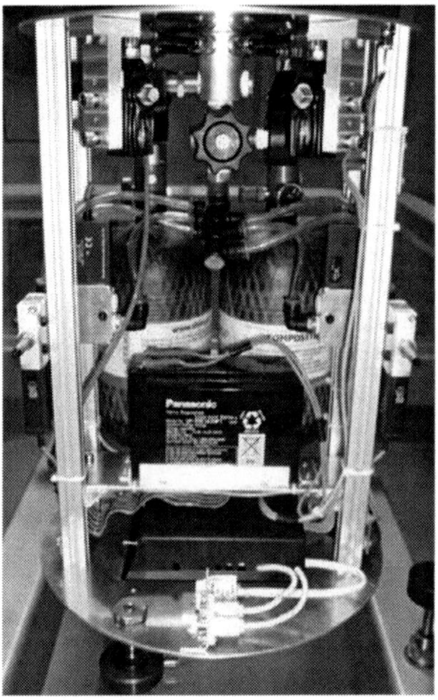

Fig. 4 Mock-up appearance

The power supply system is made up of an accumulator and voltage transducer. The mock-up orientation and position control system consists of the on-board computer, microcontroller and thruster engine system.

Motion determination system consists of an angular velocity sensor, two-axis accelerometer, a star tracker and a web-camera, which imitates distance finder work, and a calculator, which processes the data received from the sensors.

A flat glass bench is also a part of the facility. The mock-ups move over its surface (Fig. 5). The bench is enclosed with dark cloth for unlit space imitation. There is a light source – the Sun imitator – on one side of the bench and a starry sky imitator (luminous LEDs ceiling) above it.

Fig. 5 The mock-up environment

3 The Mock-Up Cooperation Gear and Relative Navigation Modes

When the satellite group solves a unified task, for example, the Earth remote sensing one, it must act in coordination, for instance, turn in order to point a camera to a sub-satellite point on the Earth's surface. At that the data about the satellites position in the inertial frame or the autonomous mutual measurements of their positions should be available. Consider such variants of cooperation gear which are realized by means of the mock-ups.

Name the mock-ups at our disposal the mock-up #1 and the mock-up #2. If an angular velocity sensor and a two-axis accelerometer are installed on board of the mock-up #2 and the mock-up's initial state vector is determined, the on-board computer can calculate the mock-up state vector at each time by the Kalman filter [3]. The calculated values can be transmitted via WLAN connection to the leading mock-up. This mock-up (#1), which is familiar with own position and supporting trajectory, can calculate the dependent mock-up control parameters and transmit them to the mock-up's #2 on-board computer also via WLAN.

The disadvantage of such a realization is the necessity to determine the mock-up initial motion conditions that is a rather difficult task. It can be solved using position sensors such as starry-camera or direct observation of the leading mock-up using a web-camera.

Consider the scheme the position sensors are used in (Fig. 6). If the leading mock-up is provided with the distant light source (Fig. 7) (the Sun imitator), one can determine the relative position and relative orientation after predetermined calibration with the web-camera picture from the second mock-up. These data can be used as the positional ones which imitate, for example, the distance meter indications [4]. Such method of the camera pictures use for navigation was applied in [5].

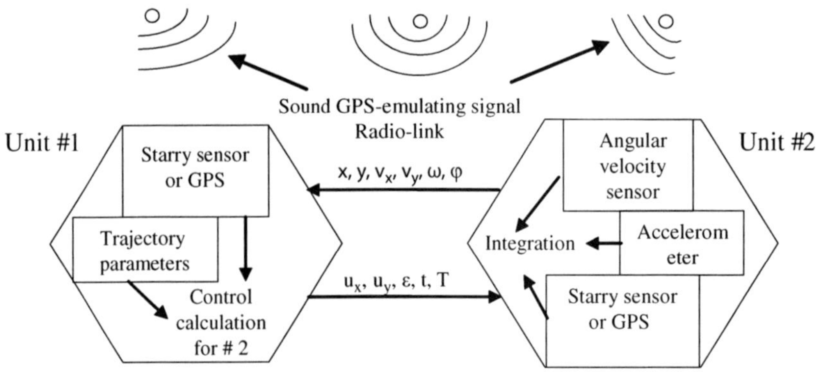

Fig. 6 The mock-ups' cooperation mechanism using position sensors

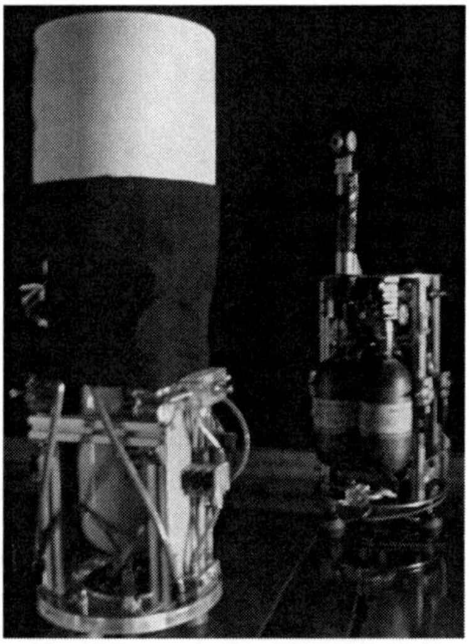

Fig. 7 The web-camera shooting for relative position determination

The set of sensors avoids the problem of initial conditions definition because the position sensors output is the mock-up position on the testbench. However, the disadvantage of the starry-camera use is a low rate of image processing. It does not allow us to get the position data at each step of the control software processing loop so such a position sensor can only correct the values obtained during integration.

4 The Mock-Up Configuration Maintenance Algorithm

It is important for the image sharpness that the mobile camera follows the chosen point on the Earth surface during the shooting. Otherwise, the image will be out of focus and in the case of the delay lag the camera can miss its goal and shoot another area of the Earth. For these camera turns guarantee it is required that the satellite rotates about the centre of mass along the reference trajectory. When observing a point of the Earth from several satellites at the same time, satellites' relative motion along the reference trajectory must be also guaranteed. Consider as an example the realized mock-up algorithm which guarantees the mock-ups' relative motion along the certain reference trajectory with some given accuracy.

4.1 Control Task Statement

Consider the mock-up motion control task. Develop the control so that it would move following the defined trajectory with given accuracy. The control task is to return the mock-up to the defined trajectory as fast as possible in case of deviation. Notice that such a motion scheme approaches configurational state of the satellites during the formation motion.

As the mock-up has three degrees of freedom (two translational degrees and one rotational one), it is possible to control three coordinates independently. One coordinate defines the rotational angle on the bench around the vertical line and two other coordinates define the position of the mock-up centre of mass on the bench (Fig. 8).

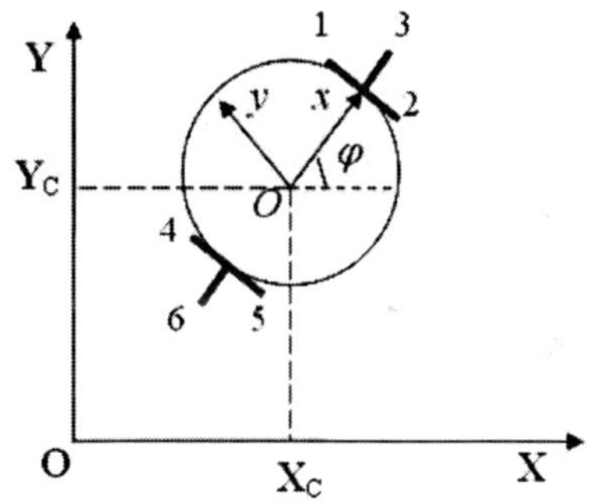

Fig. 8 Engines location on mock-up and reference frames

Consider the following scheme of two mock-ups' motion (Fig. 9). Let the mock-up #2 (the *dependent* one) accomplishes circular motion around the mock-up #1 (the *leading* one) with a constant velocity ω keeping the circle radius R constant. During that the dependent mock-up must be orientated at the leading one in a given direction, for example, by the axis oy of its reference system as shown in Fig. 9. Such a motion is realized only when the angular velocity of circular motion velocity ω equals to the spin velocity ω.

The configuration considered is very similar to the orbital configuration and, at the same time, it is the complex in the view of a control. The fundamental feature is that it is necessary to execute the dependent mock-ups' motion control already basing on the idea that they move around a mobile point, which motion is not known. At this juncture the control is developed on the basis of that this point stays immobile

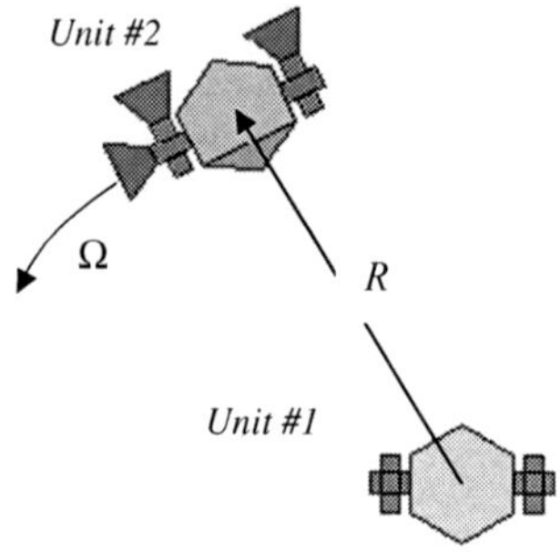

Fig. 9 Pattern of motion

till the end of the dependent mock-up control session. This leads to errors of the whole formation motion.

Consider the classical control method which is optimal in time and the purpose of which is to restore the body to the reference trajectory. According to the Pontryagin maximum principle the solution of such an equation is double-pulse control action. Having the mathematical model of the object motion helps to find out the switch curves for the double-pulse manoeuvre. As the mock-up can move along the bench thanks to the air-cushion almost without friction, the law of motion, for example, in the x coordinate can be written as

$$\dot{x}_1(t) = x_2(t),\ \dot{x}_2(t) = u(t),$$

where x_1 is a coordinate, x_2 is a velocity, u is a control acceleration. The body phase path which passes through the point of origin is given by

$$x = \pm\frac{\dot{x}^2}{2}.$$

Having found the shuffle function [6], get the following control law

$$u = \begin{cases} +u^{\max},\ \textit{if}\ \dot{x}^2\mathrm{sign}(\dot{x}) < -2x\ \textit{or}\ \dot{x}^2\mathrm{sign}(\dot{x}) = -2x,\ x > 0, \\ -u^{\max},\ \textit{if}\ \dot{x}^2\mathrm{sign}(\dot{x}) > -2x\ \textit{or}\ \dot{x}^2\mathrm{sign}(\dot{x}) = -2x,\ x < 0. \end{cases}$$

Thus, knowing the system point location on the phase plane at start time one can learn which control "sign" should be applied to the system for optimal motion to the given point (Fig. 10). It is necessary to monitor at that the system real position at the plane at every computing cycle. In case of the curve switch-over reaching one should reverse control sign and move further to the given point along the switch-over trajectory. When the body position is in the region of acceptability, the control has to be switched off.

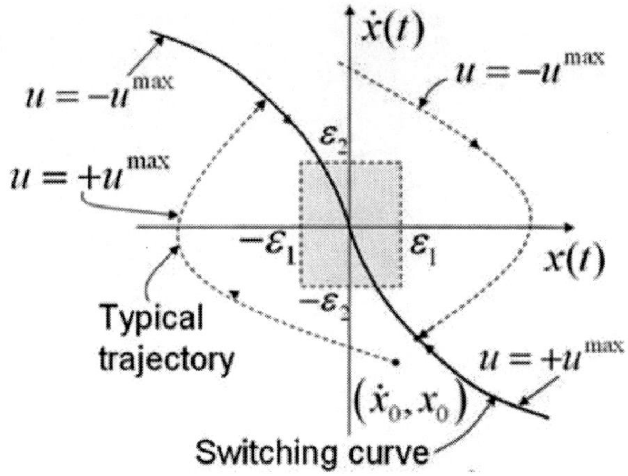

Fig. 10 The phase trajectories and the switch curves

4.2 The Mock-Up Control Algorithm

The routine realizes the close-loop principle of control. The feedback channel is organized using the imitation of two sensors (the angular velocity sensor one and the two-axis accelerometer) work. Thus, three values enter the routine input. Those are the body angular velocity and the accelerations which are measured in the direction of the body coordinates Ox and Oy axes (Fig. 8). Then, transfer the measured accelerations into the OXY static system. The transformed data from the sensors enter the Kalman filter input. At the filter output the six-dimension vector of the body phase state value is obtained. First, the phase components of the body state vector are checked for the body location in the area of acceptability. If it is in the area about the reference trajectory, the orientation phase elements are checked. If the check is positive, the orientation control parameters are calculated. Thereafter, the control is realized. But if the check result of the body location phase elements turned out to be positive, the rotation is stopped and only when the rotation stops, the position control parameters are calculated.

It is intuitively obvious that the formation control is effective for slow motion because in such a case the leading mock-up position changes in the course of time slightly and the dependent mock-ups' control copes with a task of the reference trajectory motion (Fig. 11). Consider how the average position deviation for the modelling time depends on the formation angular velocity in case of two dependent moving around mock-ups (Fig. 12). In the diagram it is seen that control is effective with given trajectory deviation $\varepsilon = 0.05\,m$ since the angular velocity satisfies to inequality $\Omega \leq 7°/\mathrm{s}$. Above this value of the velocity the error of mutual position oversteps the legitimate area.

Consider how the average angular orientation deviation depends on the angular velocity. As in case of position deviation the orientation control is effective by the angular velocity of $\Omega \leq 7°/\mathrm{s}$ (Fig. 13).

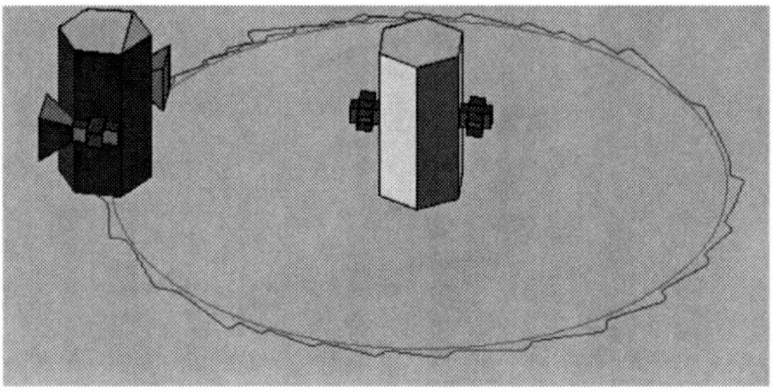

Fig. 11 Relative motion trajectory: the circular *line* is required trajectory, the jagged *line* is real trajectory

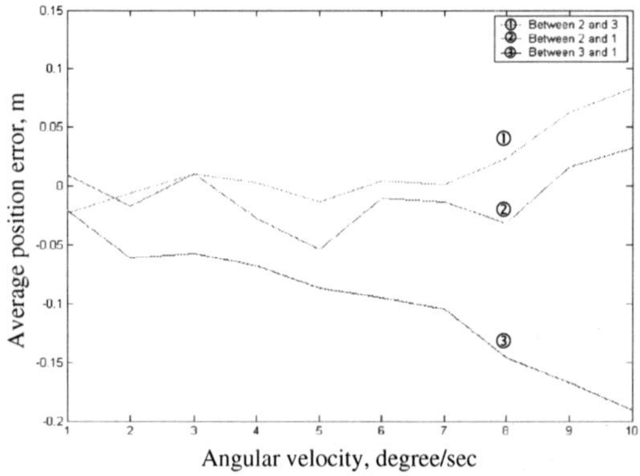

Fig. 12 The plot of mutual position error against angular velocity

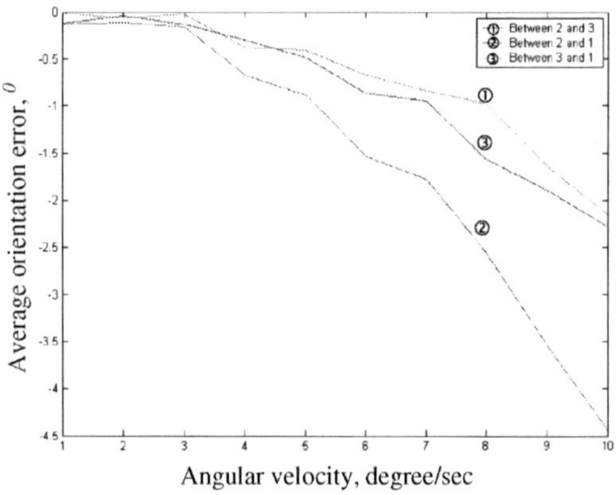

Fig. 13 The plot of mutual orientation error against angular velocity

5 Conclusion

The mathematical and laboratory simulations of mock-up group configuration motion have been executed. For configurations relevant for the Earth remote sensing the motion parameter limits within which the considered control is effective are obtained. Various cooperation schemes between mock-ups vs. hardware available are considered.

Later on the authors intend to continue development of the mock-up control algorithms, to verify them at the laboratory facility in order to transfer the results obtained to the industry for the TNS-0 satellite group use [7, 8].

The research is carried out under support by DAAD (Leonhard Euler Program, Unit 325), Russian Foundation for Basic Research (Grants 09-01-00431 and 09-01-09210), and Russian Federal Agency of Science and Innovations (Contract № 02.514.11.4068).

References

1. S. Theil, R. Stanislowski, A. Schleicher, S. Scheidhauer. Entwurf und Entwicklung eines Luftkissenfahrzeugs für einen Satellitendynamiksimulator / Projektbericht, Universität Bremen, Bremen, 2003, 38p.
2. D. Ivanov, T. Walter, D. Bindel, M. Ovchinnikov. Laboratory Facility to Verify Motion Control Algorithms for Multi-element Systems, Preprint of KIAM RAS, M.: 2008, N 56, 32p. (http://www.keldysh.ru/papers/2008/prep56/ prep2008_56.html)
3. D. Ivanov, M. Ovchinnikov. Determination of the Orientation of a Suspended Mock-up Using One-Axis Gyroscope, Preprint of KIAM RAS, Moscow, 2008, N 11, 32p. (http://www.keldysh.ru/papers/2008/prep11/prep2008_11.html)
4. C. Mehlen, C. Peyrotte, S. Furlan, N. Wilhelm, R. Frenkiel, N. Perriault, L. Pirson, M. Sghedoni. RF Metrology and Data-link Instrument for Formation-flying Missions, Paper on 30th Annual AAS Guidance and Control Conference, AAS 07-067, pp. 567–583.
5. C. Kuehl, M. Mehl, H. Diehl, E. Gottzein. Micro-Tech.-Sensor for Attitude and Orbit Determination, Paper on 26th Annual AAS Guidance and Control Conference, AAS 03-001, pp. 1–18.
6. F.P. Vasilyev. Numerical Methods for Extremal Tasks Solving, Moscow. Nauka, 1988, 552p.
7. M. Ovchinnikov, A. Ilyin, N. Kupriynova, V. Penkov, A. Selivanov. Attitude dynamics of the first Russian nanosatellite TNS-0, Acta Astronautica, 2007, Vol. 61, Issue 1–6, pp. 277–285.
8. I. Zaramenskikh, M. Ovchinnikov, I. Ritus. Nullifying feasibility of Slow Gravitational Perturbation Effect on the Formation Flying Configuration, Paper on 5th IAA Workshop on Constellations and Formation Flying, Evpatoria, Ukraina, 1–4 July, 2008, 16p.

Asynchronous Parallel Reactive System for Intelligent Small Satellite on-Board Computing Systems

Toshinori Kuwahara, Claas Ziemke, Michael Fritz, Jens Eickhoff, and Hans-Peter Röser

Abstract This chapter describes the development of satellite control algorithms implemented into FPGA (Field Programmable Gate Array) hardware logic, which is gathering a great interest in applying for central computing systems of small satellites with high computational demands. In order to maximize the parallel processing capability of the FPGA, the satellite control algorithms are classified into subsystems and are implemented as a parallel running multi-agent system. The developed asynchronous parallel reactive system is capable of real-time processing and reasoning, which enables the implementation of the multi-agent system in FPGAs. This system introduces a reactive subsumption system, which has been ever proven its potential for autonomous systems, and is exemplary applied for the small satellite *Flying Laptop*. This concept offers further possibilities to be extended as an on-board autonomous system based on a Belief-Desire-Intention architecture. The performance of the implemented control algorithm is evaluated and its potential for autonomous space systems is investigated.

1 Introduction

Small satellites have cultivated their application fields and are nowadays playing very important roles in the field of Earth observation, e.g., in disaster monitoring. A new field of simultaneous mulTipoint remote sensing by distributed small satellites is also emerging [1]. Responding to their growing mission requirements, the capabilities of payload instruments are significantly improved. The amount of data by latest Earth observation sensors is becoming enormous due to high spectral, spatial, temporal, and radiometric resolutions. Furthermore, the required autonomy level for small satellites is increasing to achieve a higher availability, to fulfill real-time demands and to expand the scope of their application fields. Indeed, autonomy is the key technology in missions of recent interest, such as for autonomous relative

T. Kuwahara (✉)
Institute of Space Systems, Universität Stuttgart, Pfaffenwaldring 31, 70569 Stuttgart, Germany
e-mail: kuwahara@irs.uni-stuttgart.de

R. Sandau et al. (eds.), *Small Satellite Missions for Earth Observation*,
DOI 10.1007/978-3-642-03501-2_26, © Springer-Verlag Berlin Heidelberg 2010

attitude and orbit control of constellation satellites, for low-cost small satellites, and for their operations based on a small number of ground stations. Consequently, despite of the limitation of available size, mass and power aboard small satellites, there are great needs on advanced on-board computing systems with high computational throughputs and higher degrees of autonomy.

The Institute of Space Systems at the Universität Stuttgart is now developing a series of small satellites including *Flying Laptop*, *PERSEUS*, and *Lunar Mission BW1*. The *Flying Laptop* is a testbed for an On-Board Computer (OBC) based on FPGAs (Field Programmable Gate Arrays), which shall be applied for the successive missions. Together with the flexibility based on the reconfigurable feature, FPGAs are especially very attractive to small satellites in terms of their extremely high throughput (based on internal parallel processing mechanisms), dense interface implementation capability, low power consumption (due to the lower clock frequency), low cost (commercial off-the-shelf components), and speed of technology advancement. Indeed, there has been a tremendous interest in reconfigurable FPGA's for the realization of reconfigurable satellites [2]. The focus of this chapter is to describe the development of the control algorithm implemented into the FPGA hardware logic, which not only maximizes the FPGA's real-time parallel processing capability but also enables the realization of on-board autonomy. This developed system provides innovative solutions to recent and future advanced small satellite missions.

2 Hardware Platform

The central processing node of the OBC of the *Flying Laptop* utilizes reconfigurable SRAM-FPGAs (Fig. 1). The high computational capability of FPGAs is based on internal parallel processing and precise timing. The best way to maximize this capability is to implement processes asynchronous. In order to implement whole satellite control algorithms as asynchronous parallel real-time processes, the usual

Fig. 1 FPGA-based on-board computer

programming approach cannot be applied and thus new control algorithm architectures must be developed. This chapter describes an implementation method of asynchronous parallel reactive systems, based on a reactive subsumption architecture [3, 4]. The OBC of the *Flying Laptop* is a centralized system connected with all satellite peripheral electronics, which are classified into subsystems. This OBC performs control functions of a traditional satellite bus controller as well as all tasks of the payload data processing. The unification of different computing functions aboard a small satellite onto a single computing system Results in a clear-cut system structure and provides a high degree of fault tolerance with minimal resources. The speed gain over a software based system is significant; typically a speed gain of 40–100 is achieved [5]. The complete control algorithms are modeled by means of Handel-C by Mentor Graphics, a C-based high level hardware description language.

3 Control Algorithm Architecture

The developed control algorithm is divided vertically in four layers and horizontally in domains. In Fig. 2, the highest two layers are illustrated which are relevant to this section. The Platform Abstraction Layer Core (PAL-Core) serves as an abstraction layer to provide generic accesses to the platform specific peripheral devices. The Application Layer is the highest layer, in which actual satellite control algorithms are implemented. This layered structure assures a higher portability of the application algorithms. The satellite control algorithms are classified into subsystems through the layers and are implemented as a parallel running multi-agent system in different domains.

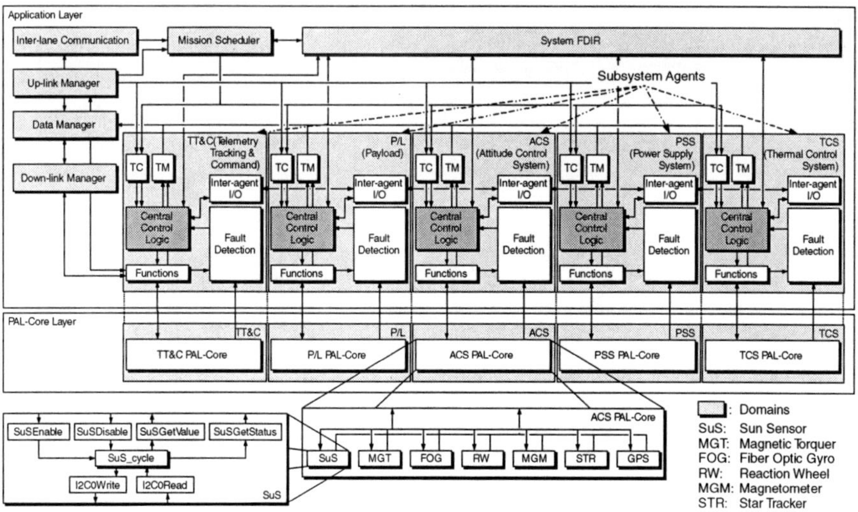

Fig. 2 Control algorithm architecture

3.1 Multi-Agent System

Each subsystem control algorithm is implemented as an agent. Within the scope of this chapter, an agent is a unified complex of processes that is situated in a certain environment and is capable of conducting actions in it in order to meet design objectives. The important characteristic of the multi-agent system is that the system level algorithms can be decentralized, letting each agent be responsible to a part of system tasks, which can be implemented as parallel processes inside an FPGA. Each agent possesses a local view and communicates, interacts, and collaborates with other agents. This implementation concept achieves simplicity of over-all system level algorithms, clear functional partitioning, and generic buildup of agents. Furthermore, this encapsulation of functional units allows easy extension to multi-instance redundancy in a functional grain level for critical missions with high reliability requirements.

The behavioural rules of agents are arranged into individual state-machines so that no inconsistency between agents arises. The exemplary one of the attitude control system is illustrated in Fig. 3. The details of the implementation method of these state-machines are beyond the scope of this chapter, and will be the subject of a separate publication. What shall be emphasized is that there is no governing central process unlike conventional CPU-type OBC's. All agents and domains as well as all functions in lower level layers are running in parallel. The system level

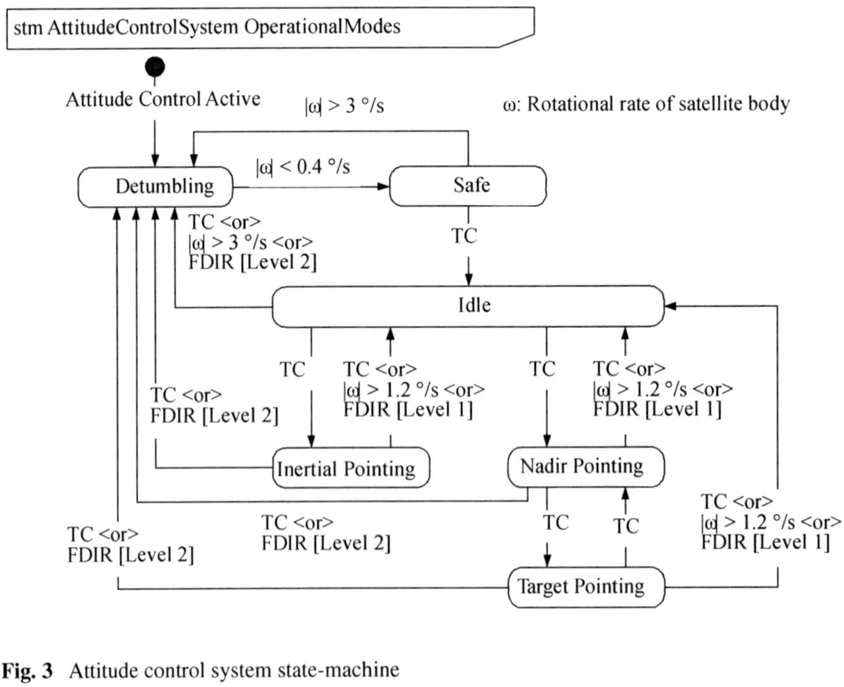

Fig. 3 Attitude control system state-machine

FDIR (Failure Detection, Isolation and Recovery) function is designed to deal with potential risks of dead/live-locks using watch-dog functions moving the system into safe mode. The communication interfaces between agents are realized as hard-wired signal lines inside the FPGA. The bit information is exchanged through signal lines, which can be simply connected to mulTiple receiving agents. These signals can be delivered within one clock cycle without any delay. Certain changes in one subsystem, for Example mode changes or component on/off status changes, result in reactions in other subsystems.

3.2 Asynchronous Parallel Reactive System

The above mentioned behaviours of agents are implemented in individual Central Control Logic (CCLs) as illustrated in Fig. 2. A CCL receives inputs from inter-agent communication (I/O), Fault Detection, and TC (Telecommand) modules, and gives output to Functions. A CCL is the main logic governing the behaviour of one agent and consequently the combination of the CCLs govern the behaviour of the satellite system. Interactions between CCLs in agents ensure that combinations of states and the knowledge databases of all agents are always consistent. A CCL shall be able to process all parallel inputs and generate decision outputs in real-time, in order to keep state transitions and corresponding interactions between agents stable. The implementation of CCLs is the key technology to realize the multi-agent system and can be mathematically represented as follows. The *environment states* of an agent can be described as a set $S = \{s_1, s_2, \ldots\}$, and the *actions* of an agent as a set $A = \{a_1, a_2, \ldots\}$. Let P be a set of *perception* capability, D be a set of internal *database*, and I be the set of all internal *states* of the agent. Then the agent can observe its environment through

$$\text{perception: } S^* \rightarrow P, \tag{1}$$

which "maps" sequences of environment states to perceptions, where S^* denotes the set of sequences of elements of S. The database and the state can be updated as

$$\text{database: } D \times P \rightarrow D, \tag{2}$$

$$\text{state: } I \times P \rightarrow I, \tag{3}$$

respectively, and the agent selects the next action based on

$$\text{action: } I \times D \rightarrow A. \tag{4}$$

Figure 4 illustrates the internal architecture of a CCL. It is structured in a layered manner. The input information is prioritized starting from Low Priority TC as the lowest one. In order to process parallel inputs in real-time, each input is processed by a corresponding interpreter. Interpreters "react" on the input information in each

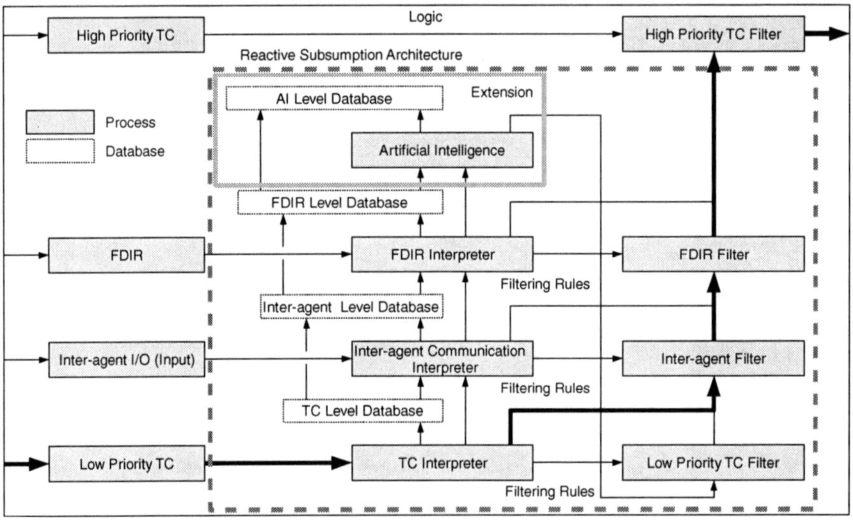

Fig. 4 Central control logic architecture

priority level by generating outputs. It is arranged in such a way that the output information of an interpreter is input information for the higher level interpreter. The processed information is also stored in the database, which is shared by the higher level interpreter. In this way higher level interpreters can make higher level decisions based on the input information and knowledge over the lower level environment. Parallel to this, interpreters set the filtering rules of corresponding filters, which shall filter out or modify the output signals from the lower level interpreter. For Example, an output signal interpreted by Low Priority TC Interpreter shall go through Inter-agent Filter and FDIR Filter, as indicated with thick lines (Fig. 4), so that it can be authorized as the final decision by the reactive subsumption system. This process takes only several clock cycles. The output of this system is finally filtered by High Priority TC Filter. Due to its architecture, this proposed new concept can be referred to as asynchronous parallel reactive system.

This proposed system is able to be extended to an on-board autonomous system, by adding higher level layers for different types of artificial intelligence (AI) technologies. It is planned to demonstrate this functionality on the *Flying Laptop* in orbit by introducing a single AI-layer (Fig. 4), based on a Belief-Desire-Intention (BDI) architecture, in which agent's deliberation process is represented in a filter function as:

$$\text{filter: } \wp(\text{Belief}) \times \wp(\text{Desire}) \times \wp(\text{Intention}) \times \wp(\text{Intention}), \tag{5}$$

while the state of an agent at any given moment is a triple (B, D, I), where $B \subseteq \textit{Belief}$, $D \subseteq \textit{Desire}$, and $I \subseteq \textit{Intention}$ [6]. This layer generates practically low priority command equivalent output signals for safety and demonstration reasons.

Generally, the degree of autonomy can be classified in following three levels [7]:

- Level 0: Autonomous monitoring of parameters, autonomous navigation, and autonomous failure management based on redundancy and switching.
- Level 1: Detailed reaction to failures, failure diagnosis through combinational logic, and operational autonomy for predictable contingency events.
- Level 2: Knowledge based planning of actions, reactive planning, autonomous optimization of Procedures and flexible knowledge based diagnosis.

Consequently, the autonomy level of the designed system without the final AI-layer can be regarded as Level 1 and this can be extended to Level 2 by adding the final AI-layer. The decision making in this layer can be implemented based on the knowledge on the available FDIR level database. For Example, planning of a series of Earth's image acquisition processes based on the achieved attitude pointing accuracy, available storage capacity, and meteorological condition of the target regions can be implemented.

4 Simulation and Verification

The developed control algorithms were tested with the system simulation infrastructure MDVE (Model-based Development and Verification Environment), which has been developed by EADS Astrium and is applied at the Institute of Space Systems, utilizing hardware-in-the-loop configurations (Fig. 5) [8]. This real-time simulation environment is capable of simulating spacecraft attitude, power consumption, component temperatures, communication, and space environment at the same time. The applied Mission Control System software is based on SCOS-2000 by ESA. MOIS from Rhea is used as Procedure Execution Engine enabling automated tests.

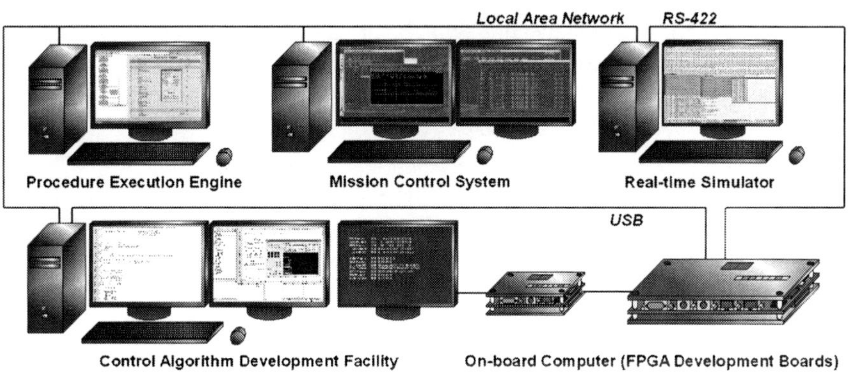

Fig. 5 Control algorithm simulation and verification environment

The developed control algorithms implemented inside the FPGA are operated and monitored from SCOS-2000.

SCOS-2000 enables live-monitoring of variables and conditions. The connection between development board and Mission Control System is established by a proxy in the real-time simulator. In Fig. 6, a graphical live visualization of the satellite angular velocity is illustrated together with a numerical display of recently received telemetry packets and communication logs between the real-time simulator and the OBC.

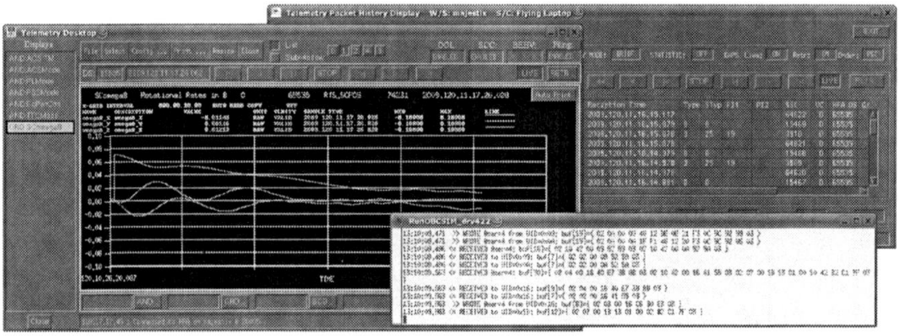

Fig. 6 Live visualizations of SCOS-2000 and communication log

The purpose of the following simulation of implemented algorithms of *Flying Laptop* is to demonstrate the correct functionality of the developed asynchronous parallel reactive system inside an FPGA. Because the reaction time of the implemented processes are too short (typically few clock cycles), it is not feasible to follow every single processing result. The simulation scenario of the mode transition of the subsystems and of a few components is summarized in Table 1. In Fig. 7, the mode transition Results, satellite's rotational rate, and summary of power balancing are illustrated. As summarized in Table 1, the *Flying Laptop* performs an Earth observation in nadir with the payload camera instruments at first, and then a high-speed Ka-band communication with the ground station, followed by malfunction of one of the four reaction wheels (RWs), which Results in a high rotational rate of more than 4°/s. In Fig. 7, it can be seen that the FDIR function of the attitude control subsystem automatically makes a decision to move back into the detumbling mode from the target pointing mode through idle mode immediately and turns corresponding components off according to the state transition rules defined in Fig. 3. After the rotational rate has been reduced to 0.3°/s, the system moves to safe mode again and stabilizes the attitude. The total power consumption illustrated in Fig. 7 proves that the components operation (on/off) took place correctly controlled by each subsystem agent. As the result of the interaction of the agents, the satellite system fulfils the desired missions. With this simulation, the correct functionality of the implemented control algorithm of the asynchronous parallel reactive system is validated.

Table 1 Simulation scenario of *Flying Laptop* satellite operation (mode numbers in parenthesis)

Time (min)	Sys. event	ACS event	P/L event	TT&C event
0	Safe mode (4)	Safe mode (1)	Off mode (0)	Stand-by mode (0)
110	Idle mode (5)	Idle mode (2)	–	–
115	–	–	Operation mode (1)	–
120	Active op. (6)	Nadir P. mode (4)	–	–
125	–	–	Camera Idle (1)	–
130	–	Target P. mode (4)	Camera Operation (2)	–
145	–	Nadir P. mode (4)	Camera Idle (1)	–
150	–	–	Camera Off (0)	–
155	Idle mode (5)	Idle mode (2)	Off mode (0)	–
210	Active op. (6)	Nadir P. mode (4)	–	Standard contact mode (1)
215	–	–	–	S-Band LG On (1)
220	–	–	–	High-speed com. Mode (3)
–	–	–	–	S-Band LG Off (0)
–	–	–	–	Ka-Band On (1)
225	–	Target P. mode (5)	–	
240	–	–	–	Stand-by mode (0)
–	–	–	–	Ka-Band Off (0)
242	Safe mode (4)	RW malfunction	–	–

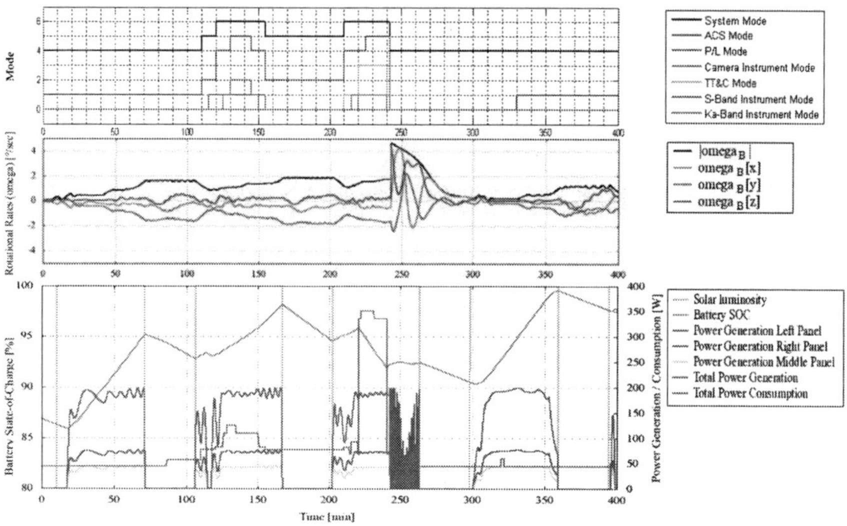

Fig. 7 Hardware-in-the-loop simulation Results: subsystem and component mode transitions (*above*), satellite rotational rate (*middle*), and power balance (*below*)

5 Conclusions and Future Work

The developed asynchronous parallel reactive system enables the implementation of satellite control algorithms as a parallel running multi-agent system. This system is based on the reactive subsumption architecture and realizes real-time processing and reasoning of agents. The conceptualized control algorithm is implemented in an FPGA-based on-board computer and its functionality is validated in a hardware-in-the-loop simulation environment. As the next work, the formal verification of the designed hardware logic is planned. This system can be extended to on-board autonomous systems for future small satellites as well as for autonomous robotic and interplanetary missions. This asynchronous parallel reactive system will be one of the key technologies for the *Lunar Mission BW1* and will be demonstrated with the *Flying Laptop* satellite.

References

1. D.J. Barnhart, T. Vladimirova, M.N. Sweeting. Very-small-satellite design for distributed space missions. Journal of Spacecraft and Rockets, 44(6), 1294–1306 (2007).
2. J.J. Wang, R.B. Katz, J.S. Sun, B.E. Cronquist, J.L. McCollum, T.M. Speers, W.C. Plants. SRAM based re-configurable FPGA for space application. IEEE Transactions on Nuclear Science, 46(6), 1728–1735 (1999).
3. R.A. Brooks. A robust layered control system for a mobile robot. IEEE Journal of Robotics and Automation, RA-2(1), 14–23 (1986).
4. R.A. Brooks, A.M. Elynn. Fast, cheap and out of control: a robot invasion of the solar system. Journal of the British Interplanetary Society, 42, 478–485 (1981).
5. F. Huber, P. Behr, H.-P. Röser, S. Pletner. FPGA based on-board computer system fort he Flying Laptop micro-satellite. Proceedings of the Data System in Aerospace Conference, SP-638, ESA, Naples (2007).
6. G. Weiss (ed.) Multiagent Systems: A Modern Approach To Distributed Artificial Intelligence. The MIT Press, USA (2000).
7. T. Kuwahara, A. Falke, C. Ziemke, Y. Muhammad, J. Eickhoff, H.-P. Röser. "Development of a hardware-in-the-loop simulation environment on a MDVE for FPGA-based on-board computing systems. Transactions of the Japan Society for Aeronautical and Space Sciences, Space Technology Japan, 7(ists 26), Pf_1-Pf_9 (2009).
8. J. Eickhoff. Systemsimulation in der Satellitenentwicklung I&II – (System simulation in Satellite Engineering I&II), Annual lectures at Institute of Space systems, Universität Stuttgart, Germany.

NanoSiGN – Nanosatellite for scientific interpretation of GNSS dual-frequency signals in the low Earth orbit

Fabian Pacholke, Huu Quan Vu, and Götz Kornemann

Abstract NanoSiGN will be the first nanosatellite with permanent 3-axes stabilization dedicated to the scientific interpretation of GNSS (Global Navigation Satellite System) dual-frequency signals in the low Earth orbit. It will carry and operate a complex GNSS receiver and antenna system for Precise Orbit Determination (POD) and GNSS ionospheric remote sensing based on measurements with a dual-frequency GNSS receiver for space applications whose design is based on Commercial-of-the-shelf (COTS) technologies and which is therefore free of International Traffic in Arms Regulations (ITAR) limitations. The NanoSiGN design builds on the know-how of the longtime TUBSAT series and on volume- and power-saving technologies from the recent faulttolerant BeeSat design by the TU Berlin. It is a robust design set up to be adaptable to other payloads. Despite its comparatively high volume of $350 \times 350 \times 350$ mm^3 NanoSiGN will be a nanosatellite of not more than 20 kg. It takes advantage of new and innovative technologies sized for nanosatellites such as micro-reaction wheels and a star tracker for precise attitude determination and control.

1 Introduction

Due to the fact that current spaceborne GPS (Global Positioning System) receivers are very expensive and under ITAR (International Traffic in Arms Regulations) agreements the German Research Center for Geosciences (GFZ) at Potsdam in cooperation with the German Aerospace Center (DLR) and other German and European partners is developing a new dual-frequency GNSS receiver for space applications based on Commercial-of-the-shelf (COTS) technologies [1]. This new GNSS receiver is designed to be compatible not only with GPS but with GLONASS and GALILEO as well, able to perform Precise Orbit Determination (POD) and

F. Pacholke (✉)
Aerospace Institute (ILR), Technical University of Berlin (TU Berlin), Marchstrasse 12, 10587 Berlin, Germany
e-mail: fabian.pacholke@ilr.tu-berlin.de

R. Sandau et al. (eds.), *Small Satellite Missions for Earth Observation*, 289
DOI 10.1007/978-3-642-03501-2_27, © Springer-Verlag Berlin Heidelberg 2010

GNSS ionospheric remote sensing measurements of the Total Electron Content (TEC) above the satellite's orbit.

1.1 Scientific Background

The ionosphere is a in general electrical neutral layer of the atmosphere (altitudes between about 50 and 1,000 km) and consists of neutral particles, ionized atoms, molecules and free electrons. The charged particles are low in number compared to the neutral ones. The TEC or free electron density of the ionosphere between the transmitter and receiver of radio waves can be measured by the use of two widely spaced frequencies and allows the description of the structure of the ionosphere. Hence its influence on the phase of electromagnetic waves can be corrected.

The POD will be achieved by the precise analysis of the GNSS satellite-satellite tracking (SST) data, which can be validated with the Satellite Laser Ranging (SLR) and Very Long Baseline Interferometry (VLBI) data.

1.2 Primary Objectives

On orbit verification of the new GNSS receiver on-board a low Earth orbiting small satellite.

Technology demonstration of a nanosatellite capable of permanent 3-axes stabilization, carrying and operating a complex GNSS receiver and antenna system as a baseline design adaptable to other payloads.

1.3 Technical Requirements

Table 1 gives an Overview of the major technical requirements for the NanoSiGN nanosatellite to achieve the stated objectives.

Table 1 Major technical requirements

Mission lifetime	about 2 years
Orbit	500–1,000 km, (near) polar
Attitude control	3-axes stabilized (+/–5°), continuous nadir pointing
Attitude determination	accuracy of about 0.5° in 3-axes
Payload operation	continuous GNSS data recording
Payload data downlink	1 MBit/s
Ground station	near polar

2 NanoSiGN Design

The NanoSiGN concept aims to build on the know-how of the TU Berlin developed with the longtime experience of the TUBSAT series (by Prof. Dr.-Ing. U. Renner) and on new miniaturized technologies already developed by the TU Berlin for the more recently BeeSat picosatellite (Berlin Experimental & Educational Satellite). The NanoSiGN spacecraft bus is based on a robust design (avoiding mechanisms where possible) which is set up to be adaptable to other payloads with challenging requirements for volume and power consumption. It provides capacity in volume for the integration of a propulsion system in behalf of orbit maneuvers or spacecraft deorbiting at the end of its lifetime.

Besides the requirement for a nanosatellite with not more than 20 kg there are three primary design drivers: At first the attitude determination and control requirements, secondly a sufficient electrical power supply and thirdly the dimensions of the large payload antenna system. Hence the NanoSiGN baseline design consists of a cube shaped spacecraft bus of 350 × 350 × 350 mm^3 with the POD GNSS antenna in zenith direction (as shown in Fig. 1 and 2) and body-mounted solar arrays in flight and both cross-flight directions.

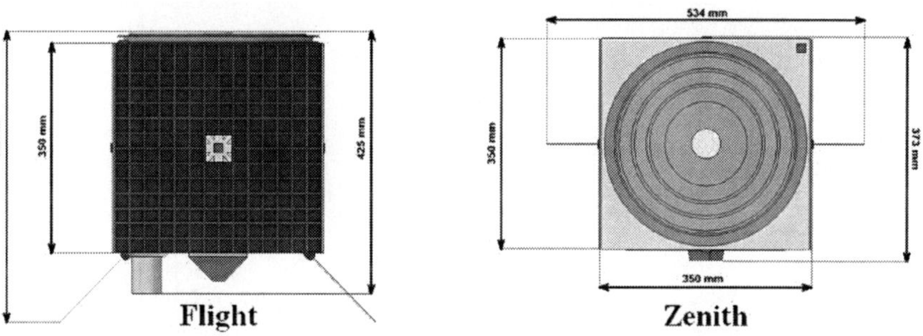

Fig. 1 View of the payload antenna in NanoSiGN CAD model

Fig. 2 3D-View of NanoSiGN CAD model and one of the authors with mockup (**X** points in flight, **Y** in cross-flight and **Z** in nadir direction)

2.1 Functional System Overview

Figure 3 illustrates the schematic functional system Overview of NanoSiGN. The major functional blocks are the electrical power supply (EPS), the on-board data handling (OBDH) including the power control (PCU), the communication (TM/TC and S-Band Downlink), the attitude determination and control (ACS), the payload (GNSS, SLR and VLBI) and the payload control (PLCU) as well as the thermal control (TS).

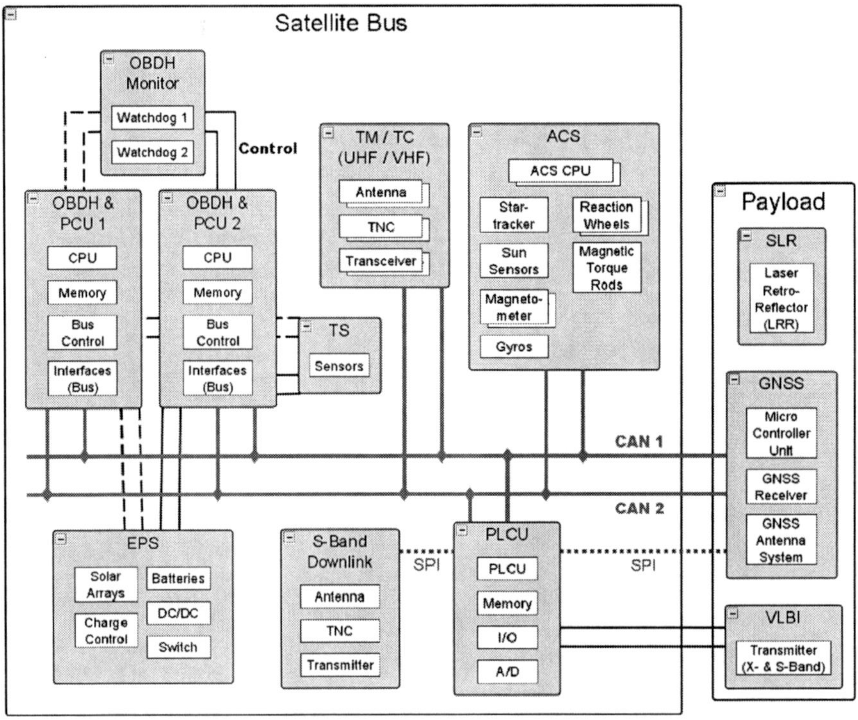

Fig. 3 NanoSiGN functional system Overview

The data link between components will be carried out mainly by a redundant bus system of two separate CAN-Buses. Merely the payload data stream will take place via a SPI data link to prevent the CAN-Buses from overloading.

2.2 Attitude Determination and Control

NanoSiGN will include a combination of sun sensors located on each face of the satellite and redundant three-axes magnetometers for attitude determination of about +/–2° (also in use on-board BeeSat), a star tracker for the accuracy of 10^{-2} degrees and a system of gyros. The gyros will only be operated infrequent in case of tumbling motions with high rotation speed (after satellite separation or during retrieval from safemode) to provide information for implicit attitude determination of low accuracy. The star tracker is an internal development by Prof. Dr.-Ing. U. Renner and will have a mass of about 600 g, a dimension of about 60 × 60 × 160 mm (depending on baffle) and a maximum power consumption of 3 W.

For attitude control NanoSiGN will use micro-reaction wheels suited for a nanosatellite of up to 20 kg total mass. NanoSiGN will carry four micro-reaction wheels – three perpendicular to each other and one in a 45° angle to them for redundancy. Magnetic torque rods will be applied for desaturation of the wheels and momentum dumping. Considered are already available reaction wheels and a yet to be developed upgrade to the micro-reaction wheel RW1 by Astro- und Feinwerktechnik Adlershof GmbH which evolved from the micro-reaction wheel RW-B developed in cooperation with TU Berlin for BeeSat. Tables 2 and 3 are representing some Results of the investigations for the reaction wheel sizing of NanoSiGN. The worst case torque from the reaction wheel for disturbance rejection in Table 2 equates to the sum of the expected disturbance torques (gravity gradient, solar radiation pressure, magnetic field and atmosphere of the Earth) including a margin factor of 1.5. The momentum storage in a reaction wheel is the outcome of the accumulation of the disturbances within an entire orbit. Using the Example of the available RW-0.060-28 reaction wheel for nanosatellites by Sinclair Interplanetary NanoSiGN could be able to continuously operate its reaction wheels for nearly one day before desaturation of the wheels and momentum dumping will be required (as indicated in Table 3). The RW-0.060-28 features a momentum of 0.06 Nm-sec

Table 2 Estimation of the worst case torque for disturbance rejection and the momentum storage in a reaction wheel (for an entire orbit)

Orbit altitude [km]	500	700	1,000
Torque from reaction wheel for disturbance rejection [Nm]	$7.75\ 10^{-7}$	$6.94\ 10^{-7}$	$6.11\ 10^{-7}$
Momentum storage in a reaction wheel [Nm-sec]	$4.40\ 10^{-3}$	$4.11\ 10^{-3}$	$3.85\ 10^{-3}$

Table 3 Estimation of the operating time without desaturation of the reaction wheel (using the Example of the RW-0.060-28)

Orbit altitude [km]	500	700	1,000
Number of orbits per day [–]	15.22	14.58	13.7
RW operating time (at maximum speed) [number of orbits]	13.63	14.59	15.57
RW operating time (at 80% speed) [number of orbits]	10.91	11.68	12.46
RW operating time (at maximum speed) [h]	21.5	24.03	27.29
RW operating time (at 80% speed) [h]	17.2	19.22	21.83

at 6,500 rpm, a power consumption of 2.0 W at full torque (0.5 W at 5,000 rpm steady-state) and a mass of 225 g [2].

Both attitude determination and control will be processed by one of two designated ACS computers. The second one will be passive until a watchdog switches the pair as a result of an error detection. The at the time active ACS computer will calculate the star tracking functionality for precise rotation speed determination between two attitude measurements of the star tracker as well. Since the on-board computer of BeeSat was designed for the picosatellite's attitude determination and control tasks, has a power consumption of less than 0.5 W and a mass of about 50 g it is suitable for NanoSiGN and will be applied as its ACS computer.

The on-board position determination will be realized by an orbit model. It will be updated via Two-Line Elements (TLE) from the ground station and during payload operation additionally via the navigational solution of the GNSS receiver.

2.3 Electrical Power Supply

To maintain the 3-axes stabilization and to operate the payload including the transmission of the payload data to the receiving ground station an average power of 11.6 W will be required. During the payload data transmission to the ground station a peak power consumption of up to 18.7 W could arise (reducible to about 15 W through operational constraints such as no measurements of the star tracker at the same time).

Due to the payload the surface area available for solar cells is limited in particular in zenith direction. In the baseline design (as shown in Figs. 1, 2 and 4 on the left side) the electrical power system utilizes mainly the spacecraft surfaces in flight direction and both cross-flight directions for body-mounted solar arrays. The solar arrays will consist of about 26 triple junction solar cells by Azurspace each with an efficiency of 28%. In numerical simulations alternative configurations with one-way deployable solar arrays on the cross-flight directions (towards the zenith direction as indicated in Fig. 4, no change in spacecraft dimensions) were also investigated. The conducted power simulations are based on the calculation of the irradiation incident on oriented surfaces of a spacecraft over the spacecraft trajectory by the Illumination Model of the Space Environment Information System (SPENVIS [3]) which is provided by the European Space Agency (ESA). A brief summary of some Results of

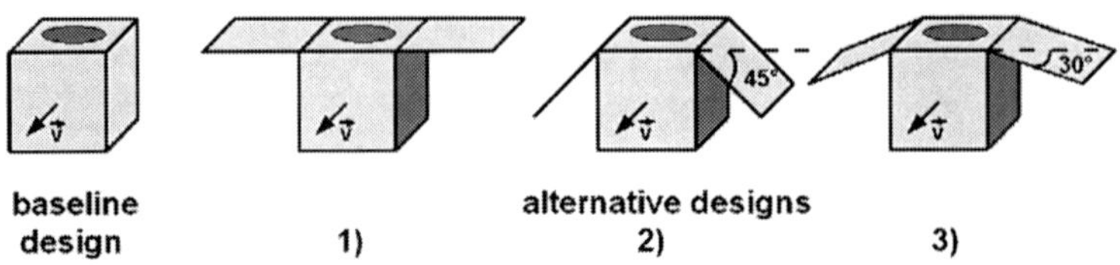

Fig. 4 Investigated design concepts for NanoSiGN's solar arrays

Table 4 Average power generated in circular sun-synchronous orbits with local time of ascending node of 6:00 and 12:00 a.m.

Spacecraft configuration (see Fig. 4)	Average power [W] generated in 6:00 a.m. SS-orbits	Average power [W] generated in 12:00 a.m. SS-orbits
Baseline design	32.2	9.1
Alternative design (1)	12.5	29.3
Alternative design (2)	25.1	23.2
Alternative design (3)	19.5	26.4

these power simulations for two extreme cases of orbits concerning illumination conditions are presented in Table 4.

Either of the baseline and the alternative design number one has its preferable illumination condition in one of the two sun-synchronous orbit cases and generates more power than the other configurations. But non sun-synchronous orbits are drifting and the undesirable condition occurs over a period of time. Hence the alternative designs number two and three which will generating a power of 24 and 22 W on average are more suited for non sun-synchronous (near) polar orbits whereas the baseline design is suited for a specific range of sun-synchronous orbits. The one-way deployable solar arrays could be realized as a mechanism with an adjustable stop position which will be set before the launch. Thus the more complex alternative design offers more flexibility concerning the orbit selection (in particular for a piggy-back launch strategy) compared with the baseline design.

For Energy storage and power supply during eclipse rechargeable batteries will be used. The 6S1P MPS176065 Li-ion module for nanosatellites by Saft is under consideration. This Li-ion module has a mass of 1.25 kg, an energy storage capacity of 120 Wh (at begin of life, 5.5 Ah) and contains its own thermal control sensors and heaters [4].

2.4 Communications

The communication system of the satellite will be divided into two separate systems. An half-duplex UHF/VHF system will be used for uploading telecommands from and downloading housekeeping information to the ground station of the TU Berlin located at the ILR. The configuration of the two orthogonal deployable UHF antennas (see Fig. 1) ensures contact under every given spacecraft orientation in particular during acquisition and safemode. For payload data downloading to the receiving near polar ground station the HISPICO system (highly integrated S-band transmitter for picosatellites) will be used. It consists of a digital S-band transmitter and a patch antenna on-board the satellite and a receiver at the ground station which contains the decoder and can broadcast the received data into the network. The HISPICO S-band transmitter was developed by IQ wireless and the TU Berlin.

It has a bit rate of 1 MBit/s, a maximum power consumption of 5 W and a mass of 110 g.

Due to the fact that the near polar receiving ground station will have contact in almost every revolution of NanoSiGN's (near) polar orbit the housekeeping informations will also be included into the transmission of the payload data and forwarded to the mission control center.

2.5 Thermal

The thermal design relies mainly on passive thermal control techniques to keep the internal temperature in the required range by payload and the bus systems. The batteries will have to be Boxed with additional thermal isolation and if necessary heaters to prevent them from freezing. An integrated solution provides the 6S1P MPS176065 Li-ion module for nanosatellites by Saft [4].

2.6 Payload

The primary payload will consist of a dual-frequency GNSS receiver and a GNSS antenna system (300 mm diameter in zenith direction). For on orbit verification of the GNSS based POD additional payloads will be carried by NanoSiGN including a passive Laser Retro-Reflector (LRR) prism (for satellite laser ranging, SLR) and a VLBI transmitter system (S- and X-Band). The LRR prism and VLBI antennas are located on the satellites nadir face as shown on the left of Fig. 1 and in Fig. 2.

The payload data collected by the GNSS receiver will be stored in the dedicated memory of the payload control unit (PLCU), complemented with attitude informations from the ACS and then transferred to the receiving ground station through the HISPICO system. During contact to the receiving ground station the PLCU will be able to forward the payload data straight to the S-band Downlink system as well.

3 Conclusion

The continuing revolution in the know-how from the longtime experiences with the TUBSAT series and the further development of the BeeSat picosatellite technologies by the TU Berlin had led to a robust nanosatellite design which is capable of an in its satellite class unprecedented permanent 3-axes stabilization for a sophisticated payload operation.

The internal software of the dual-frequency GNSS receiver which will enable the scientific interpretation of the GNSS signals is still under development. Therefore a continuative feasibility study for a microsatellite which will utilize a flight-proven American GNSS receiver is currently funded by the Technologiestiftung Berlin (TSB) and GFZ and carried out by a cooperation of the TU Berlin, GFZ and partners of the Raumfahrtinitiative Berlin-Brandenburg (RiBB).

References

1. A. Helm, O. Montenbruck, J. Ashjaee, S. Yudanov, G. Beyerle, R. Stosius, M. Rothacher. GORS – A GNSS occultation, reflectometry and scatterometry space receiver. Proceedings ION GNSS 2007, Fort Worth, Texas, September 25–28 (2007).
2. S. Doug. Sinclair Interplanetary, 268 Claremont Street, Toronto, Ontario, Canada, Standard Products: Nanosatellite Reaction Wheels, 60-mNm-sec datasheet, 60mNm-secwheel2008c.pdf (2008), Web: http://www.sinclairinterplanetary.com
3. European Space Agency and Belgian Institute for Space Aeronomy, Space Environment Information System – SPENVIS (2008/2009), Web: http://www.spenvis.oma.be/
4. D. Reulier. Key Account Manager, Saft – Defence & Space Division, Poitiers, France, Presentation, 6S1P MPS176065 Li-ion module for nanosatellites (2008).

Part V
Instruments and Technologies

The Vegetation Instrument for the PROBA-V Mission

L. de Vos, W. Moelans, J. Versluys, V. Moreau, J.F. Jamoye, Jan Vermeiren, L. Maresi, and M. Taccola

Abstract In this chapter recent progress in the development of the Vegetation Instrument (VI) for the PROBA-V mission is presented. PROBA-V is an earth observation mission, which will ensure continuity of the actual SPOT/ VEGETATION mission until Sentinel-3 will become operational.

The consortium will consist of VITO as prime investigator, VES as mission prime, OIP as prime contractor for the payload and Xenics and AMOS as most important subcontractors for the payload.

The payload is a multispectral spectrometer with 4 spectral bands. To guarantee daily coverage of the earth, a very large swath of 2,250 km is envisaged. The ground resolution of the instrument will be at least 1,000 m.

The payload will consist of 3 SIs (Spectral Imagers), each containing a very compact TMA (Three Mirrors Anastigmat) telescope, and a large VNIR and SWIR FPA (Focal Plane Array). The three SIs will be mounted on a common optical bench, and aligned to each other to cover the full swath.

Feasibility studies have revealed two major criticalities in the payload, being the TMA manufacturing and alignment, and the SWIR FPA manufacturing. To reduce potential risks during the very short development time of the payload, two pilot projects were already started. This chapter presents the first achievements of both development projects.

1 Introduction

The major challenge in designing the payload is to make it compatible with the resources available on a small satellite like PROBA and at the same time accommodate the large swath. The selected solution is to divide the FOV (Field Of View) in three smaller parts and to use compact reflective optics (TMA telescopes) for each part. Each TMA is equipped with large VNIR and SWIR FPAs to cover the large swath.

L. de Vos (✉)
OIP, Westerring 21, 9700 Oudenaarde, Belgium
e-mail: ldv@oip.be

R. Sandau et al. (eds.), *Small Satellite Missions for Earth Observation*,
DOI 10.1007/978-3-642-03501-2_28, © Springer-Verlag Berlin Heidelberg 2010

The TMAs and the SWIR FPA are not commercially off the shelf available and have to be developed for the Vegetation Instrument. These development projects are further elaborated in this chapter.

2 General Instrument Concept Description

The concept design of the Vegetation Instrument is shown in Fig. 1. The instrument contains three Spectral Imagers (SI), each providing multispectral images in three VNIR (Visible) and in one SWIR (Short Wave Infrared) band. Together with the two star trackers, the three SIs will be mounted on a common optical bench.

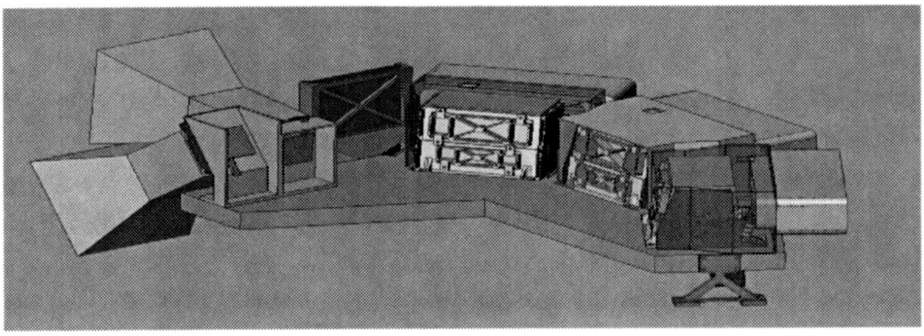

Fig. 1 Concept design of the vegetation instrument

Each SI contains one telescope, a beam splitter to separate the VNIR from the SWIR spectral bands, spectral bandpass filters to select the spectral bands, and the VNIR and SWIR focal plane arrays.

The optical design concept of the TMA is shown in Table 1 and in Fig. 2. The TMA has a FOV of 34°, a focal length of 110 mm and f-number 6, and contains one spherical and two aspherical mirrors. More details are provided in the chapter that is dedicated to the TMA development.

Table 1 Optical system characteristics

Parameter	Value
Focal length (mm)	109.6
Aperture (mm)	18.6
f/#	6
FOV	34.6° × 5.5°
Length x width x height (mm)	90 × 110 × 140

The spectral bands will be realized by spectral bandpass filters centered on 460, 658, 834 and 1,610 nm, with bandwidths of respectively 42, 82, 121 and 80 nm. The filters will be applied on the detector windows.

The optical axis of the central telescope will point to nadir and the two outer telescopes will point 34° from Nadir. Together the three TMAs will cover a complete

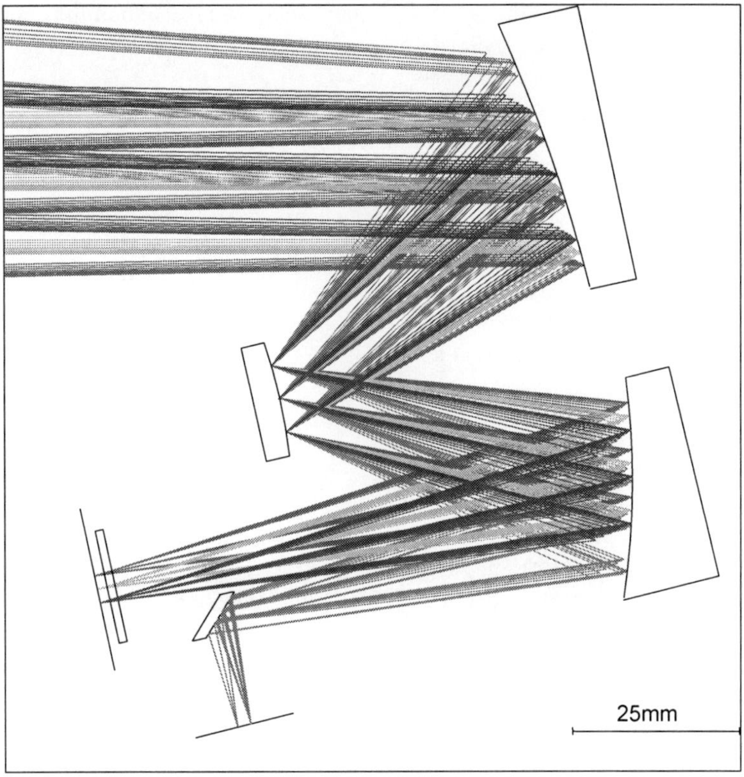

Fig. 2 Optical design concept of the TMA

FOV of 102°. The optical system is telecentric, and the aperture is located at the position of the second (spherical) mirror.

3 The TMA Telescope Development

The use of TMA telescopes for optical imaging instrumentation of Earth Observation missions has become more attractive, as the progress of optical manufacturing and integration technologies allowed expressing the full potential of those configurations in terms of broad spectral band, and un-obscured and wide field image quality.

The Vegetation instrument of the PROBA-V mission will make use of a set of three such telescopes, identical to each other. The purpose of the related GSTP development is to demonstrate the feasibility of one item of the set with respect to its required optical quality, and to secure the instrument development.

Taking into account the mission constraints and objectives, including the innovative features of the instrument, a full-aluminium design was selected [1]. This choice allows taking benefit from the last developments in ultra-precision milling and turning techniques, as well as in optical aluminium production [2]. Furthermore, this leads to a homothetic telescope behaviour.

The optical performance requirement of the telescope is expressed by a dedicated Modulation Transfer Function (MTF) table, inherited from the overall system MTF (Table 2).

Table 2 MTF required specifications

Channel	Nominal MTF (%)	2σ MTF (%)	Max. frequency (lp/mm)
Blue	68.1	53	38.5
Red	68.5	54	38.5
NIR	68	53.7	38.5
SWIR	71	62.4	20.0

The telescope MTF, shaped by the telescope aperture, is directly affected by the mirror residual surface form errors (SFE) and by the accuracy of the mirror alignment. The mirror surface tolerance is analysed by Zemax ray-tracing thanks to the TEZI operand. The algorithm used for the sensitivity analysis considers that the Zernike coefficients have an identical weight in the mirrors SFE.

To have a more realistic Zernike coefficients distribution, the budget has been validated making use of SFE measurements on similar mirrors, realized in the past. The measured surface grids amplitudes are sized to fit the calculated budget (30 nm RMS SFE) and introduced in the optical design. This budget is established to have the highest possible M2 SFE budget, because this mirror is the more critical one. The analysis of the MTF sensitivity with respect to the mirrors surface errors allows estimating the mirrors SFE budget for the manufacturing process. The adopted budget is 35 nm RMS, 20 nm RMS and 35 nm RMS for M1, M2 and M3 mirrors, respectively.

In order to allocate the corresponding manufacturing and alignment budgets to the subsystems, a preliminary analysis led to the tolerances as shown in Table 3

Table 3 Alignment tolerances

	Required tolerances			Manufacturing tolerances		
Tolerance	M1	M2	M3	M1	M2	M3
Dx (μm)	±15	±25	±25	7	7	7
Dy (μm)	±15	±25	±25	7	7	7
Dz (μm)	±60	±25	±50	4	4	4
Tx	0.75′	1.5′	1′	4″	4″	5″
Ty	0.5′	1′	0.75′	6″	4″	8”
Tz	2′	–	3′	12″	–	15″

The primary (M1) and tertiary mirrors (M3) of the configuration are steep off-axis aspheres that are directly generated by diamond turning during the last step of the manufacturing process. Their surface shapes are controlled by a null-interferometric tests involving the use of computer-generated holograms. The secondary mirror (M2) is a convex sphere, which hosts the telescope aperture stop.

The mirror manufacturing process entails several operations of milling and heat treatment, dedicated to provide the required shape and stability to the part. Presently, the telescope's structure and the M2 mirror have already been made (see Fig. 3), and the mirrors M1 and M3 are nearly completed.

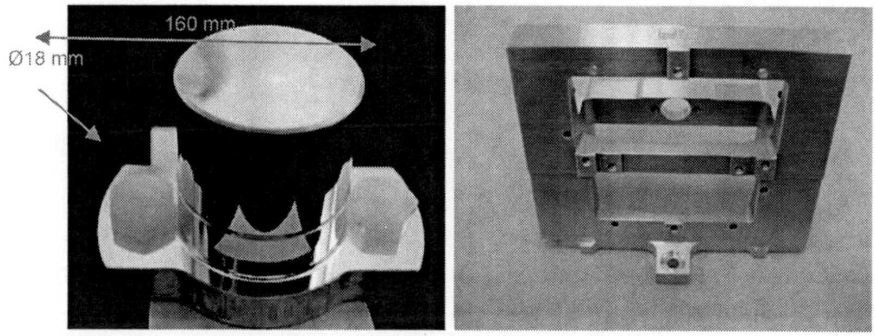

Fig. 3 PROBA-V GSTP M2 mirror and telescope's structure

The obtained Results are better than specified in terms of roughness, form and mechanical interfaces position. 3D measurements were performed on the structure after machining and were mainly related to the position and the flatness of the diamond turned mirror interfaces on the structure. The flatness of each interface is below 0.5 μm. The positions of these interfaces are listed below (Table 4).

Table 4 3D measurement of mirror interfaces (I/F) on structure

	Measured	Nominal	Error	Tolerance
Angle between M1 and M2 I/F [°]	0.9646	0.9612	0.0034	±0.024
Angle between M3 and M2 I/F [°]	0.4244	0.4262	–0.0018	±0.020
Distance between M1 and M2 I/F [mm]	73.418	73.4208	0.0028	±0.011
Distance between M3 and M2 I/F [mm]	75.193	75.1955	0.0025	±0.0115

The Surface Form error (SFE) of M2 mirror is tested by interferometry with an adapted spherical reference surface (f/1.5, λ/60) on WYKO RTI 4100 interferometer. The roughness was also measured by interferometric means. The requirement on the SFE is 20 nm RMS and was measured on the useful aperture at 8.41 nm RMS (see Fig. 4). Each roughness measurement is below 3 nm RMS, which is also better than the required 6 nm RMS.

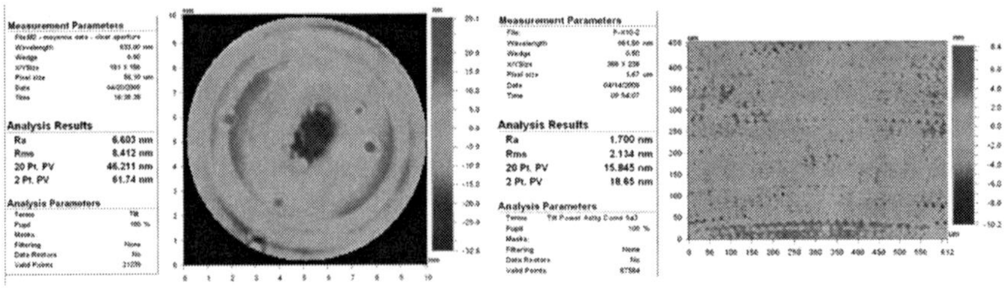

Fig. 4 PROBA-V GSTP M2 mirror SFE on clear aperture and roughness test

4 The SWIR Detector Development

This development concerns the large format SWIR focal plane array containing at least 2,704 pixels with 25 μm pitch. Several alternatives were considered:

- A monolithic approach with one long sensor line of 67.6 mm. This is too ambitious in view of the present status of the technology where detectors are manufactured on 3″ wafers
- Optical butting. This is feasible, but is a relatively expensive approach,
- Mechanical butting, with two further options:
  - Either end to end stitching of photodiode modules
  - Or an approach, where photodiode modules overlap.

Finally it was decided to select the last solution, with 3 overlapping detectors of 1,024 pixels and approximately 80 pixels in the overlap area (see Fig. 5).

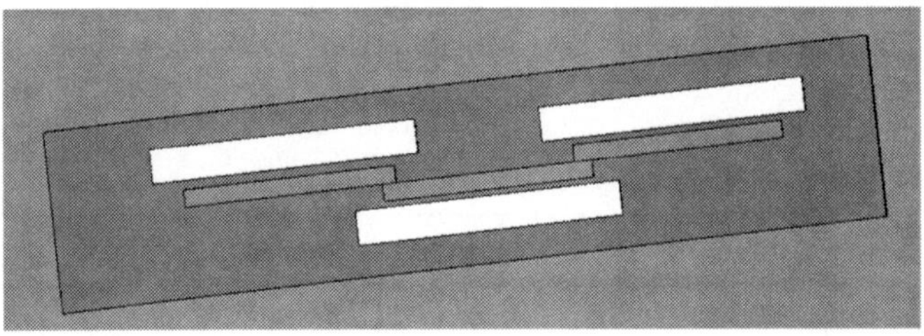

Fig. 5 Concept drawing of the mechanically butted array. In *light grey* the linear detector arrays are shown, while in *white* the ROIC are represented

Taking into account a maximum separation of 2 mm (or 80 sample lines) between the sensor lines, there will be always an overlap between imaged ground pixels, which allows for a proper reconstruction of the image. The target separation is 1.5 mm or 60 sample lines. By selecting 2,704 active pixels, the SWIR FPA will have the same physical length as the VNIR detector.

Due to the large overlap of the pixels, the requirements on in-plane alignment (X and Y axis) can be strongly reduced. By definition the alignment error in X (parallel to the 3 linear detector arrays) is always smaller than 12.5 μm or half a pixel pitch.

In the Y direction the error shall be smaller than 25 μm. The baseline is to use a reference jig for positioning the outer sensor lines in conjunction with accurate dicing of the arrays. Then the middle array will be pushed against the fixed outer arrays.

The bigger challenge is the height control of the sensor line (perpendicular to the sensor plane or Z-alignment). This requires accurate polishing of the individual photodiode arrays with a precision better than 40 μm. Recent experiments indicate that a precision better than 20 μm can be routinely met. The die attachment process will introduce an additional height error smaller than 50 μm, which Results in a Z-axis precision better than 100 μm, well within the depth of focus of the telescopes.

Further experiments are ongoing, where the reference jig that is designed for the Y-alignment is also used as a tool foor the height adjustment. The detector arrays are then placed on a thicker layer of low outgassing epoxy and pushed down in the epoxy till they match the height of the reference jig.

The ROIC shall have 1,024 inputs in a single sided configuration. Due to the short development time, it was decided to extend an existing design with 512 pixels with an additional 512 inputs. Due to this upscaling the maximum readout rate will drop from 25 to 12.5 Msamples/sec. This frequency is still much higher than the minimum requested data rate of 105 ksamples/sec for the 30 msec nominal integration time. The read-out strategy is still under Discussion as two alternatives are possible:

- The slow readout as mentioned above, where slow AD converters can be used. The disadvantage is, however, that the power dissipation of the output drivers can only be adjusted in a rather narrow range, which Results in a higher than necessary output power dissipation.
- A fast readout rate, where the pixels are read at 5–10 Msamples/sec and where the output amplifiers are switched off during 95–98% of the line time, resulting in a fifty-fold reduction of the output power. The disadvantage is however that fast, low-power and high-accuracy space qualified AD converters are necessary.

The analog path of this precursor device is shown in Fig. 6. It is a high performance circuit capable of generating >3,000:1 dynamic range at low sensitivity. Due to the very high detector and input capacitances it was absolute mandatory to design an on-board correlated double sampling stage, which eliminates the kTC noise of resetting the detector and integration capacitor. In order to maintain the SNR in the

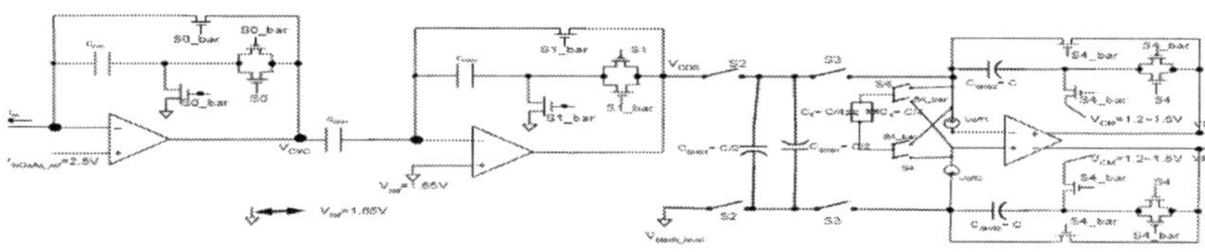

Fig. 6 Analog path of the ROICL. *Left*: the CTIA with variable gain and reset; *Middle*: AC-coupled CDS stage; *Right*: balanced Sample&Hold stage

output stage and to keep the non-uniformity low, the internal Sample&Hold is built around an offset compensated amplifier.

The precursor device is set up for a broad variety of detector geometries and applications. Consequently, the 512 inputs have a selectable conversion capacity ranging between 5 and 2,130 fF. This is accomplished by selecting or deselecting a 25, 100, 500 and 1,500 fF capacitor in addition to a fixed 5 fF value. This Results in a sensitivity ranging between 32 and 75 nV/e^-. The full well capacity ranges between 53,000 e^- and 22 Mio e^-.

For the new extended design the largest capacitor will be divided in 3: 2* 200 and 1,068 fF, so that more values can be achieved round the optimal value of 600 fF for the vegetation mission. The baseline combination is: 5, 25, 100, 200 and 500 fF, which allows optimizations around 500, 600, 700 and 800 fF.

A noise curve, expressed in ADU_{rms}, for a 12 bit AD converter is given in Fig. 7 for the different feedback capacitors. The noise increase for lower capacitance values is not yet explained and further investigations are ongoing.

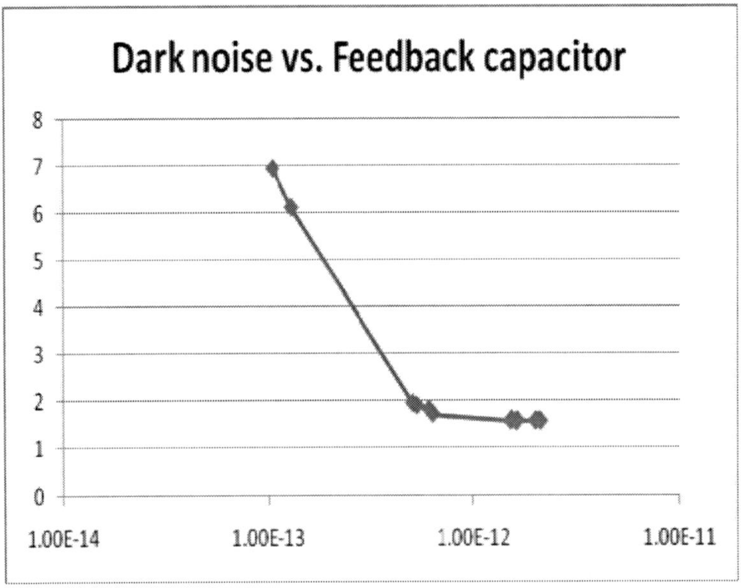

Fig. 7 Noise (dark conditions) as a function of feedback capacitor for the 512 input ROIC

For the new ROIC circuit with 1,024 inputs, noise estimates are made based on the noise Results of the precursor device with 512 inputs. The Results of the performance analyses are summarized in Table 5. The SNR is defined as the ratio between the number of signal electrons for a given signal level and the overall noise for that signal level, and the dynamic range is the ratio between the full well capacity and the overall noise at a given signal level.

In order to make use of the full performance of the array a signal conversion on 12 or preferably 14 bits shall be used. As a part of the signal swing of the SWIR

Table 5 Performance analysis for the new ROIC and associated FPA

Input values					
Light level	L1	L2	L3	L4	Unit
Photon flux	3.55E+06	1.18E+08	2.25E+08	3.43E+08	Phot/sec/pix
SNR		333			
Performance estimates					
Tint	0.03				sec
Photon budget	106,500	3.54E+06	6.75E+06	1.03E+07	phot/pix
QE	75%				%
Signal charge	79,875	2,655,000	5,062,500	7,717,500	e^-/pix
Max output	2				V
Safety margin	1.5				–
Signal swing			1.33		V
CFB			6.08E-13		F
% Full well	1%	35%	67%	102%	%
Shot noise	283	1,629	2,250	2,778	e-rms
Circuit noise	3,476				e-rms
Dark noise(40°C)	644				e-rms
Overall noise	3,546	3,893	4,190	4,496	e-rms
SNR	23	682	1,208	1,716	
Dynamic range	2,141	1,951	1,812	1,689	

image sensor is used for fixed pattern noise and photo response non-uniformity, the effective bit depth can be reduced to 10, max 11, bits after 2 point non-uniformity correction.

In order to realize the optimal capacitance value of 600 fF, the largest feedback capacitor is split in 3 parts of 200, 200 and 1,070 fF, respectively, which are mask programmable. The nominal value used is 200 fF, which allows to vary the integration capacitor from 100 to 800 fF in steps of 100 fF (except 400 fF).

References

1. S. Kirschstein, A. Koch, J. Schöneich, F. Döngi. Proceedings of SPIE, Vol. 5962, 59621 M, 2005.
2. G.P.H. Gubbels, B.W.H. van Venrooy, A.J. Bosch, R. Senden. Rapidly solidified aluminium for optical applications, Proceedings of SPIE, Vol. 7018, 70183A, 2008.

Software Defined LFM CW SAR Receiver for Microsatellites

Naveed Ahmed and Craig I. Underwood

Abstract There is an increasing demand for a low cost, day-night, all weather spaceborne imaging capability using synthetic aperture radar (SAR) on small satellites. Traditional pulsed SAR payloads have been too expensive and too power demanding to be employed on low-cost microsatellite platforms. Recent developments based on Continuous Wave (CW) techniques have proved successful in minimizing the cost, mass and power of SARs for small airborne platforms (UAVs). However, when considering the use of CW techniques for space based SARs, other considerations come into place: A major one being need of high isolation between transmit and receive antennas, in short, a bi-static configuration. Conventional receiver designs based on analogue demodulation techniques remain inflexible and are complicated to change to suit changing operational requirements in a dynamic bi-static satellite formation. This chapter focuses on the research being carried out in the Surrey Space Centre (SSC) on the design and development of a software defined linear frequency modulated (LFM) CW SAR receiver that can be used on bi-/multi-static microsatellites for remote sensing.

1 Introduction

Satellite remote sensing carried out with SARs is unaffected by environmental conditions such as clouds and low light conditions, but these payloads are very expensive, heavier in weight (compared to optical payloads) and consume high powers. As a result, SAR payloads have generally been limited to large and expensive space platforms.

Surrey's Disaster Monitoring Constellation (DMC) has proven the utility of remote sensing microsatellites employing optical sensors. A SAR based DMC is the next logical step, however, this can only be affordable if the cost of the platform and payload can be made sufficiently low, circa the $10 M of a typical optical microsatellite mission.

N. Ahmed (✉)
Surrey Space Centre (SSC), University of Surrey, GU2 7XH, Guildford, UK
e-mail: N.Ahmed@surrey.ac.uk

R. Sandau et al. (eds.), *Small Satellite Missions for Earth Observation*,
DOI 10.1007/978-3-642-03501-2_29, © Springer-Verlag Berlin Heidelberg 2010

Since the inception of SAR, the focus of research, design and development has primarily been on pulsed technology [1]. Recently, new technology developed for small aerial platforms (unmanned air-vehicles – UAVs) has proved successful in minimizing the cost, mass and power of SARs. In these payloads, chirped signals are transmitted continuously (continuous wave – CW) instead of in the form of high power narrow pulses [2]. The average power remains the same in both the CW and pulsed cases, however, the ease of generating and handling low continuous average power, rather than producing short duration high power pulses, lends itself to power-limited systems such as UAVs and indeed microsatellites. Such systems also have other significant advantages including reduced circuit complexity, efficient duty cycles, greater waveform versatility, cost effectiveness and low weight and volume [3].

CW technology has not yet been employed in space due to obvious difficulties associated with the high isolation needed between the transmitter and receiver, thereby necessarily resulting in a bi-static solution [4]. New designs for space application based on this technique have to address the challenges raised by this configuration.

2 Problems of Pulsed SARs

Pulsed SAR technology is well proven in airborne as well as spaceborne platforms. Transmission and reception of pulses in different time slots avoid the suppression of an echo signal by the high peak power transmission pulse (typically 1–2 kW) [5]. This makes it possible to employ a single platform to form a mono-static radar. However, the radar design is complicated [6]. Generating narrow pulses (of a few μs) at a high peak power (a kW or higher) is a complex process involving special hardware. On a spacecraft, this high radio-frequency (RF) power demand is met by deploying large solar panels to provide the necessary electrical power. As an Example, TecSAR (a 300 kg satellite) bus has to deploy solar panel to provide 1.6 kW during imaging [7]. Pulsed SARs with phased array front ends incorporating large numbers of transmit/receive (T/R) modules and RF cables to switch narrow, precise and high powered pulses are both bulky and expensive [8].

If a satellite based pulsed SAR using 1 kW of peak power is employed, and if it is expected to operate for only 10 min out of a 90 min orbital period, then during the entire day, it would need 40 million pulses. If the bus is designed to operate at 30 V and if the power amplifier rise time is 0.5 μs, then the power system needs to provide instantaneous rate of change of current (dI/dt) = ~33 A in 0.5 ms = 66 MA/s – and this process is required 40 million times per day! Now besides other associated complications, the satellite bus power system has to be designed to withstand this requirement. Space batteries have limited depth of discharge and number of charge and discharge cycles. Operating such a payload directly from the battery would seriously reduce its lifetime. Thus, this demand is generally met by employing especial types of capacitors that have compatible charging and discharging characteristics. All this adds up to additional weight that give rise to a very high price for the entire system. As this system is operating at high peak powers, the amount of overall power loss in the form of heat would also be significant. All of which bodes ill for a microsatellite implementation.

3 Linear Frequency Modulated CW (LFM CW) SAR

This technique is based on transmitting a linear frequency modulated (LFM) CW signal through a dedicated antenna. A copy of the transmitted signal is also fed to the demodulator. The echo signal is received through a separate antenna. Both the received signal and the copy of the transmitted signal are mixed together to extract the baseband, containing the target information. Because of the ramped nature of the transmit frequency, the frequency offset between the transmitted and received signal corresponds to the time delay and therefore the range to the target. Thus, analysis in frequency allows (cross-track) range information to be extracted. Similarly, as the transmitter and receiver are moving with respect to the target (in both the airborne and spaceborne cases) there exists a Doppler shift in the signal which can be used to give (along-track) azimuth information. The received signal is therefore digitized for further processing to extract these parameters. Direct demodulation is possible due to a short travelling time at aircraft altitudes and the transmission and reception cycle can be completed in the same pulse period [2]. However, in the spaceborne case we have an additional complication in that the time-of-flight of the signal from the spacecraft to the ground is much longer and several cycles are in-train at any one time.

An attraction of a CW system for space use is the higher processing gain (Time-Bandwidth Product) when compared with a pulsed system [1, 5]. For Example the signal-to-noise ratio (SNR) that was achieved for a spaceborne pulsed SAR (Seasat) with transmit RF power $P_t = 1$ kW and pulse width $\tau = 33$ μs and a pulse repetition frequency (PRF) of 1,630 Hz was 16 dB for a normalised target radar cross section $\sigma^0 = -14$ dB, whereas the average RF power was only 54 W. The same 16 dB SNR may be achieved by increasing the chirp period (making it CW) to 0.613 ms (reciprocal of the PRF) and keeping all other parameters the same with 54 W power. Generating this low power in continuous form is much easier then generating narrow and high powered pulses through amplifier technology. Now in this case the bus may not need any additional capacitor network as (assuming a 30% DC-RF efficiency) the 6A required (at 30 V) can be either drawn directly from the solar panels or through batteries. Thus operating the system at low instantaneous powers remains a significant advantage in reducing the weight and cost of the entire system.

3.1 SAR Antenna

Our studies have shown that for a disaster monitoring role, an L-band SAR operating with a 30 m resolution over a 30 km swath would be sufficient. Keeping the requirements of low cost, light weight and low power consumption in mind, a simple deployable stressed parabolic dish antenna design is proposed, with a diameter of 3.66 m (12′). The simulation Results for this antenna with a 2.7 m focal length feed are shown in Fig. 1. The radiation pattern in the direction of 90° and 270° with respect to the bore-sight is below –12 dBi, whereas the gain in the bore-sight direction is +32 dBi. The first side lobes are at ~0 dBi. Based on these Results the parameters for LFM CW SAR have been worked out and summarized in the table of Fig. 1.

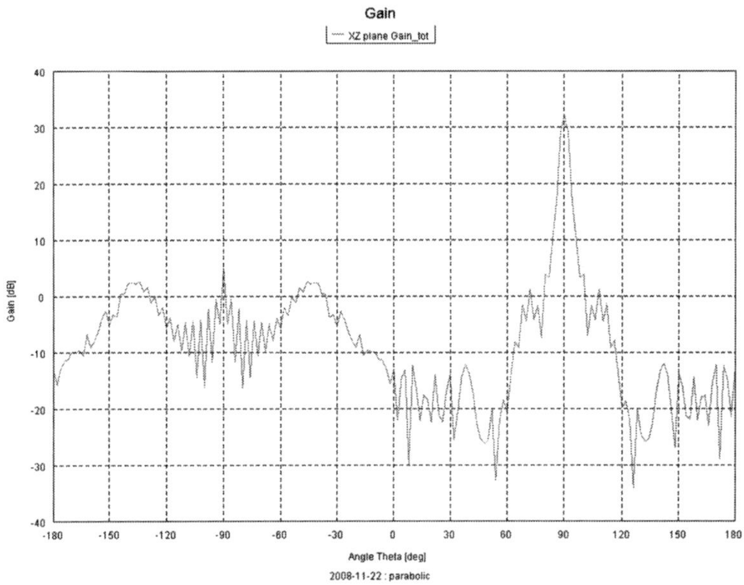

Fig. 1 Radiation pattern of 3.66 m parabolic dish antenna with 2.7 m focal length

3.2 Challenges for Spaceborne LFM CW SAR

3.2.1 Antenna Isolation

In a CW system the transmitter and receiver work simultaneously. High transmitted power will suppress the receiver if it is co-located. Isolation can be achieved through a duplexer (circulator) with a common antenna, but commercially available circulators can only provide isolation of the order of 40 dB or so. We need more isolation than this and therefore we choose a bi-static configuration, where the satellites are separated along the orbit by a small baseline (Fig. 2). In order to meet the "far field" conditions for the radar equations, the satellites need to be more than 2 km apart. However, if the satellites are too far apart, problems arise with relative position control and with maintaining the antennas sharing a common footprint. The

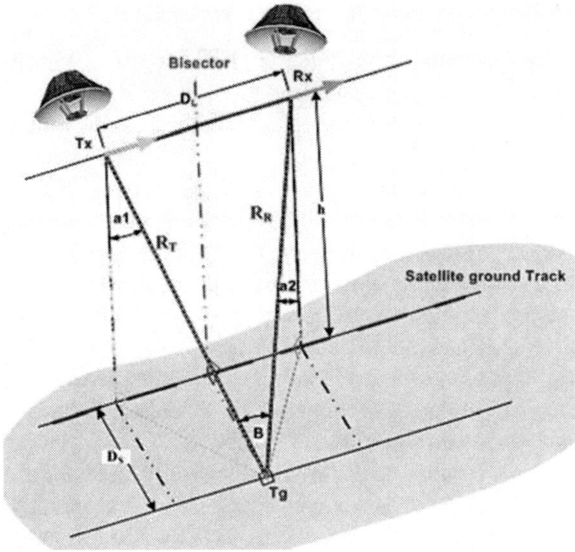

Fig. 2 Bi-static LFM CW SAR configuration (Source: T. Wanwiwake, C. Underwood, "A Bi/Multi-Static Microsatellite Constellation" IAA-B7-0212P, 2009)

maximum angular separation along the baseline which allows a pseudo-monostatic approximation is ~5°–80 km at most.

Our radar, with 60 W transmit power, operating at a slant range of 708 km at 23.5 cm wavelength, with a ~30 dB gain antenna, gives –124 dBW received power for a –10 dB cross-section. We desire any interference from the transmitter to be well below this level. The antenna pattern shows that the gain in the 90 and 180° directions (with respect to the bore-sight) is below –12 dBi, which gives –142 dBW of interference power at 30 km baseline separation – a signal-to-interference noise ratio of 18 dB in this case. Physical separation of the antennas therefore seems to offer adequate isolation. The situation can be further improved by increasing the inter-satellite distance up to the 5° limit.

3.2.2 Effect of Nadir Echo Due to First Side Lobe

The nadir echo from the first side-lobe will arrive at the receiver at an earlier time than the echo from the target in the main lobe. In the case of a pulsed system its effect is overcome by adjusting the PRF so that the receiver is not "listening" when it arrives. However in the CW case, adjusting the PRF will not work due to the continuous nature of the transmission. By considering the simulation Results shown in Fig. 1, the power from the side lobe (12° off nadir, $G = 0$ dBi) will be –186 dBW – assuming the ground return cross-section is unchanged. In practice the cross-section is likely to be larger – however there still appears to be plenty of margin, given this antenna pattern.

3.2.3 Use of a Software Defined Receiver Architecture

There are two major advantages achieved in operating the LFM CW SAR systems from aerial platforms: one is operating the entire system at low instantaneous power levels and the second one is that the separate antennas can be mounted on a single platform, allowing the use of homodyne receivers. Unfortunately, in the case of a spaceborne system, the long slant ranges (~1,500 km) mean long propagation delays (~5 ms) and very weak signal returns, which necessitates the use of separate transmit and receive platforms – i.e. a bi-static pair. We would therefore adopt a heterodyne scheme in which the incoming signal is mixed down through intermediate stages. Such a system could be implemented using an analogue architecture, however we propose the use of a software defined architecture, to give flexibility over parameters such as pulse-repetition frequency (PRF), bandwidth, look angle, inter-satellite distance, operating frequency (switching over to a different frequency band), etc., if these are required during the mission.

Recent developments in microelectronics, especially in analogue-to-digital conversion (ADC) technology, have given birth to the software defined approach. Commercial-off-the-shelf (COTS) systems are now available comprising the necessary ultra-wide-band ADCs, digital signal processors (DSPs), memory devices and interfaces, combined with the software tools needed to configure (and potentially re-configure on orbit) a flexible payload system.

The preliminary design is shown in Fig. 3. The received echo signal is amplified by low noise amplifiers (LNAs), passed through a band-pass filter (BPF) to select only the desired signal band, and then this echo signal is sampled by the ADC. An inter-satellite link will be used to provide a reference signal, in conjunction

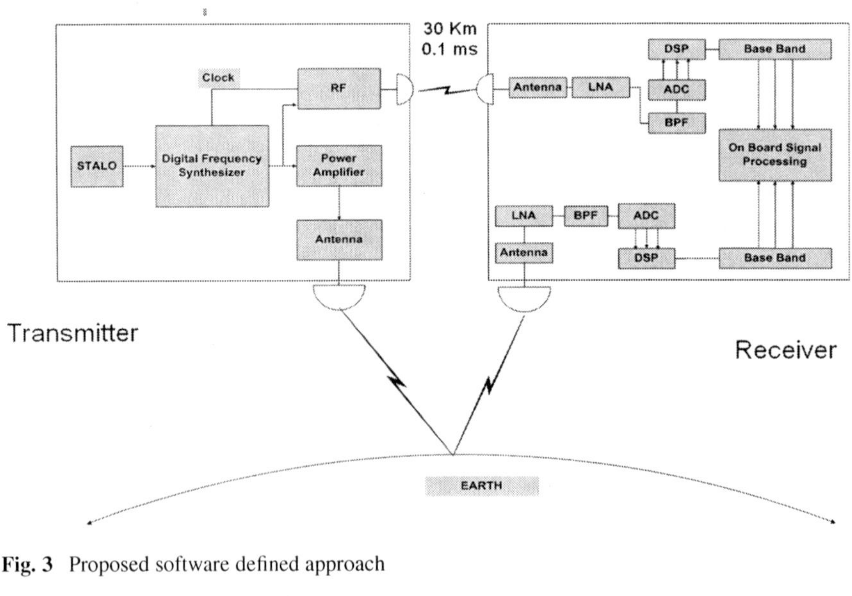

Fig. 3 Proposed software defined approach

with a differential GPS navigation system. Synchronisation issues are still being investigated.

Two sampling approaches are possible. In Pass Band Sampling, the entire incoming signal is sampled at a rate at least twice the highest frequency component present in the signal. If the carrier frequency is 1.28 GHz, and the baseband bandwidth is 10.65 MHz to meet the radar's resolution requirements, then to meet the Nyquist criteria a sampling frequency of 2 × 1.29 GHz = 2.58 GHz would be required. Although there are COTS ADCs able to sample at this frequency, they consume high power during the process. It will also need significant processing power that can be achieved through using DSPs in parallel. Once the incoming signal is sampled, the separation of baseband from the carrier is achieved through onboard software, and only the digitized baseband containing the target information is transmitted to the ground station for further processing into an image.

The second techniques is band pass sampling. This is comparatively new, and is based on the intentional Introduction of aliasing, caused by under sampling the incoming signal band. The choice of sampling frequency is critical and remains confined within a restricted region defined by [9]:

$$\frac{2f_u}{n} \leq f_s \leq \frac{2f_l}{n-1} \tag{1}$$

Where f_l and f_u denote lower and upper bounds of the signal respectively and n is an integer given by:

$$1 \leq n \leq \left|\frac{f_u}{B}\right| \tag{2}$$

By incorporating the previously mentioned values, n is given by:

$$1 \leq n \leq \left|\frac{1.29065\ \text{GHz}}{10.65\ \text{MHz}}\right| \Rightarrow 1 \leq n \leq 121$$

Hence the sampling frequency should be in the range:

$$21.33\,\text{MHz} \leq f_s \leq 21.60\,\text{MHz}$$

If f_s is selected in this region, then the baseband can be extracted directly using lower powered components.

4 Initial Simulation Results

Matlab code for the Pulsed Stripmap Range Doppler Algorithm (RDA), as given by [10], has been simplified and modified to perform this CW Simulation.

For the time being, it is assumed that both the transmitter and receiver are co-located and are following a straight trajectory with constant velocity during the imaging period. This pseudo-monostatic approximation is justified due to the short baseline separation distance (30 km) compared to the slant range (708 km). Also, we assume a flat Earth. At this stage Range Cell Migration Correction has not been incorporated into the simulator. The simulator assumes the baseband converted signal is sampled at the Nyquist rate. The following input parameters are used during the simulation:

Azimuth Resolution = 7.5 m–	Range Resolution = 40 m
PRF= 4,200 Hz–	Slant Range = 707 km
Operating frequency=1.28 GHz–	Number of point scatterers = 5

Figure 4 shows the SAR image after "fast time" matched filtering i.e. range compression. The red colour shows the location of the targets in respective range bins.

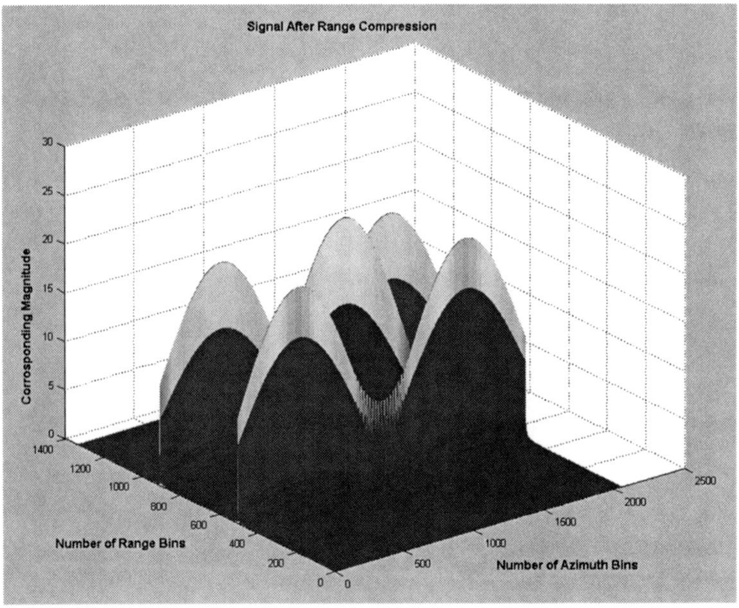

Fig. 4 Stripmap SAR signal after range compression

Figure 5 shows the SAR image after azimuth compression. The horizontal axis shows the range bins in terms of delay time for each target and vertical axis represents the length of the synthetic aperture centred at the bore sight.

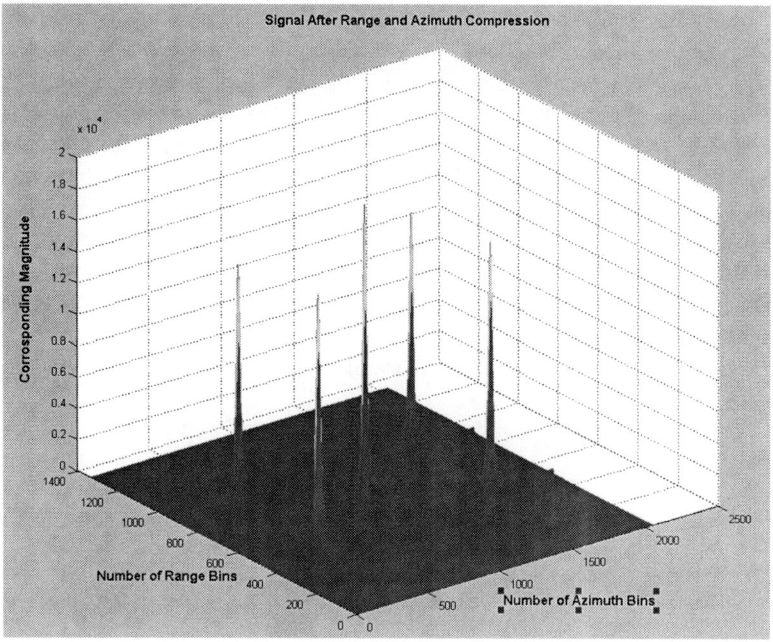

Fig. 5 Fully processed stripmap SAR signal after range and Azimuth compression

5 Future Work

There is still much work to do to refine the concept, however this initial work shows promise. Future work includes:

- Incorporating Range Cell Migration Correction into our own RDA.
- Making the model more realistic by incorporating accurate orbital models.
- Simulating the Software Defined Design Architecture and then amending the signal processing algorithms according to Results.
- Investigating issues of synchronisation, orbit control and attitude stability.

6 Conclusions

Current SAR satellites typically have complex, large volume, large mass and high power payloads incompatible with true microsatellite (~100 kg) platforms. We believe that the use of LFM CW techniques, similar to those now used on aerial platforms, could provide the breakthrough needed to implement SAR on a microsatellite. An L-Band system with modest resolution (30 m) and modest swath (30 km) is proposed, which seems both feasible and useful. Our study has shown that it would be possible, using a simple parabolic mesh antenna design and bi-static configuration, to achieve the isolation required between transmitter and receiver. This is critical in a spaceborne system, given the much larger slant ranges (and thus weaker return signals) associated with spacecraft compared to aircraft. We further propose that a software defined approach, making use of COTS digital technology, would be useful in enabling system flexibility – potentially enabling in-orbit reconfigurability enabling a dynamic bi-static constellation to be contemplated. Initial simulations are promising, but much work remains to be done, particularly with regard to synchronisation, orbit control and attitude stability requirements.

References

1. B.R. Mahafza MATLAB Simulations for Radar Systems Design, ISBN 1-58488-392-398.
2. E.C. Zaugg, D.L. Hudson, D.G. Long. The BYU μSAR: A Small, Student-Built SAR for UAV Operation, ASPRS Annual Conference Proceedings, Denver, Colorado, 23 May, 2004.
3. K. Morrison. Increasing the Effective Bandwidth of an SF-CW SAR using Frequency Agility, University of Cranfield, RMCS, 2005.
4. R.L. Smith. Micro Synthetic Aperture Radar USING FM/CW Technology, Msc Theses, BYU.
5. C. Elachi. Spaceborne Radar Remote Sensing: Applications and Techniques, ISBN 0-87942-241-6.
6. A. Currie. Synthetic aperture radar. Electronics and Communication Engineering Journal, 3, 159–170, August 1991.
7. U. Naftaly, R. Levy-Nathansohn. Overview of the TECSAR Satellite – Hardware and Mosaic Mode. IEEE Geoscience and Remote Sensing Letters, 5(3), July 2008.
8. W. Wiesbeck. SDRS: Software-Defined Radar Sensors, University of Karlsruhe, Germany, 2001.
9. J. Bae. An Algorithm for Down converting Signals Using Complex Band pass Sampling in SDR System, The 14th IEEE 2003 International Symposium.
10. M. Soumekh. Synthetic Aperture Radar Signal Processing with Matlab Algorithms, Wiley Inter science, ISBN 0-471-29706-2.

PhytoMapper – Compact Hyperspectral Wide Field of View Instrument

L. Maresi, M. Taccola, M. Kohling, and S. Lievens

Abstract Recent advancement on detectors, optics fabrication, metrology and interference filters are the basis of a new and compact hyperspectral instrument to map vegetation and soil in applications requiring fast revisit time and medium spatial resolution.

The instrument stems from the design of the Proba-V, a small satellite developed to ensure continuation of the Spot-Vegetation products. The instrument presented in this paper has been optimized to accommodate a detector with a slightly different size that provides the necessary format for a hyperspectral instrument.

With a mass of only 5 kg, a power consumption of 5 W, an overall size of 300 × 220 × 160 mm (W × L × H), and with a staggering 2,400 spectral channels covering the VNIR spectral range, the instrument provides unprecedented ratio between mass and resolution, opening the possibility to perform hyperspectral imaging to both small satellites for Earth Observation and small interplanetary probes.

The paper presents the design, performance of this instrument, and the development status of the critical technologies, namely optics manufacturing, development of the linear variable spectral filter, and of the detector.

1 Introduction

The instrument presented in this paper benefit form a number of recent developments, namely:

- a fully reflective telecentric telescope developed for Proba-V, a small satellite that will ensure continuation of the Spot-Vegetation products;
- a large format CMOS detector developed in the frame of the Medusa Project;

L. Maresi (✉)
ESA/ESTEC, Keplerlaan 1, 2201 AG, Noordwijk, The Netherlands
e-mail: Luca.Maresi@esa.int

R. Sandau et al. (eds.), *Small Satellite Missions for Earth Observation*,
DOI 10.1007/978-3-642-03501-2_30, © Springer-Verlag Berlin Heidelberg 2010

- a linear variable filter, result of a number of technology development activities sponsored by the Technology Research Programme of the European Space Agency.

The conceptual design presented in this paper resulted in an instrument with 2,400 spectral channels covering the VNIR spectral range, with unprecedented ratio between mass and resolution.

The paper presents the design, performance of this instrument, and the development status of the critical technologies, namely optics manufacturing, development of the linear variable spectral filter, and of the detector. Furthermore the paper provide a short overview of the possible application both for Earth remote sensing instruments and for interplanetary probes.

2 Instrument Design

The instrument is composed of a TMA telescope, with a proper baffling system to prevent straylight, a double CMOS detector with a spectral optical window coated with a linear variable interference filter.

2.1 Optics

The optical system is an all reflective TMA telescope with a field of view of 34°. The primary and tertiary mirrors are very complex aspheric shape, manufactured with a single point diamond turning and interferometrically tested with the aid of computer generated holograms. The mirrors and structure holding the mirrors are both manufactured in aluminium providing an athermal design. The telescope is telecentric: the variation of the angle of incidence of the chief rays on the focal plane is lower than 1° over the entire field of view, ensuring a negligible shift of the peak transmission over the interference filter. The simplicity of the design is achieved by using state-of-art Single Point Diamond Turning and 3D metrology system. The combination of these technologies allows manufacturing innovative optical system which complexity is beyond the manufacturability by using classical methods. A sketch of the optical layout is reported in Fig. 1. Figure 2 shows the opto-mechanical layout based on the Proba-V concept.

2.2 VNIR Detector

The very large FOV of the telescope is fully exploited with a large format detector. The detector baselined for the PhytoMapper is the very large CMOS sensor developed by Cypress for Medusa, an airborne instrument developed under ESA contract. The whole detector is composed of two separate detectors mounted on

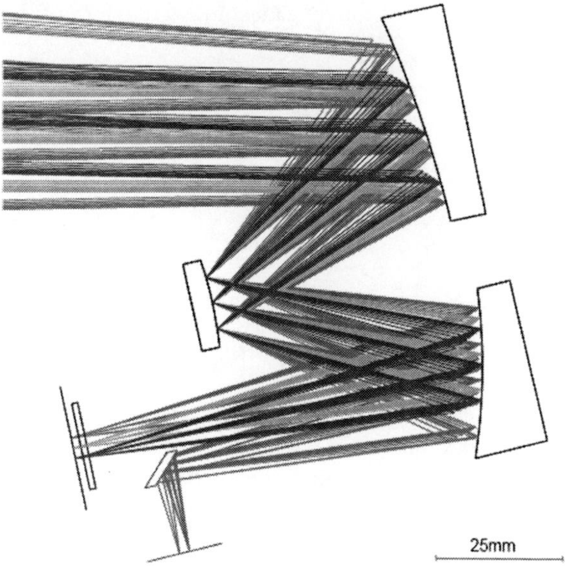

Fig. 1 Telescope optical layout

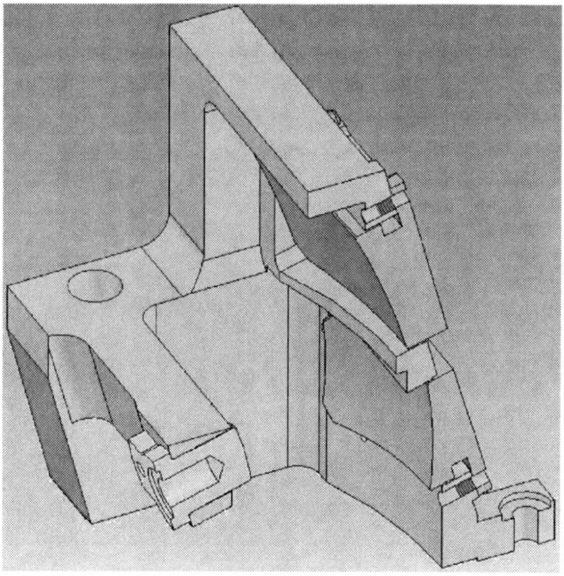

Fig. 2 Opto-mechanical layout

the same die. The main characteristics of the sensor are reported in Table 1, the quantum efficiency and the diagram of the detector are shown in Figs. 3 and 4, respectively.

Table 1 Detector characteristics

Parameter	Specifications
Pixel architecture	6 transistor pixel
Pixel size	5.5 × 5.5 μm
Format	10,000 × 1,200 (each detector)
FWC	>30,000 e^-
Dark Current	600 e^-/s @ 20°C
Operational Temperature	–70°C ÷ 70°C

2.3 Linearly Variable Filter

The linear variable filter (LVF) is glass coated with an interference filter of increasing thickness along one direction. The peak of the transmission curve varies with the thickness of the deposition. One limitation of the linear variable filters is the presence of *second order light* when the spectral range is large. An interference filter designed for a wavelength λ_0 also transmits peaks at $2\lambda_0$, $3\lambda_0$, etc. This limitation is overcome using two detectors placed next to each others, one detector is dedicated to observe the spectrum from 400 nm up to 700 nm and the second from 700 nm up to 1,000 nm.

The characteristics of a linear variable filter manufactured by Selex Galileo under an ESA technology demonstration programme are reported in Table 2 and Fig. 5. The LVF provides a continuum change in wavelength spanning from 440 nm up to 900 nm. By matching the deposition of the LVF with the dimension of the detector it is possible to sample the spectrum is a step of 0.25 nm and a bandwidth of 10 nm. The PhytoMapper uses two separate filters optimized to cover

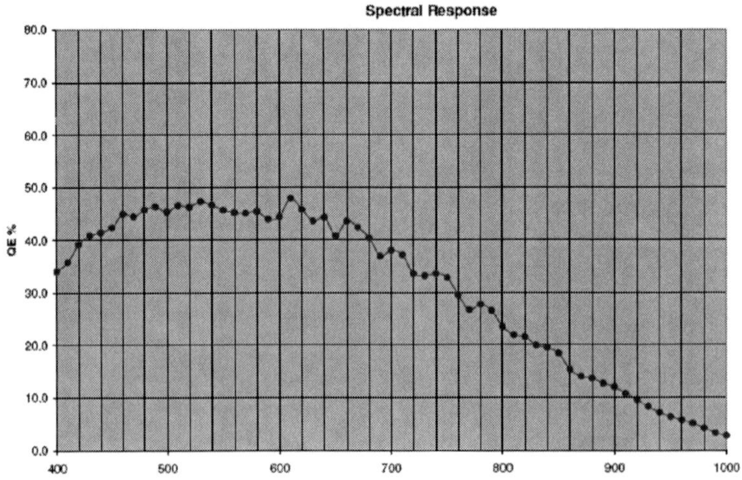

Fig. 3 VNIR detector quantum efficiency

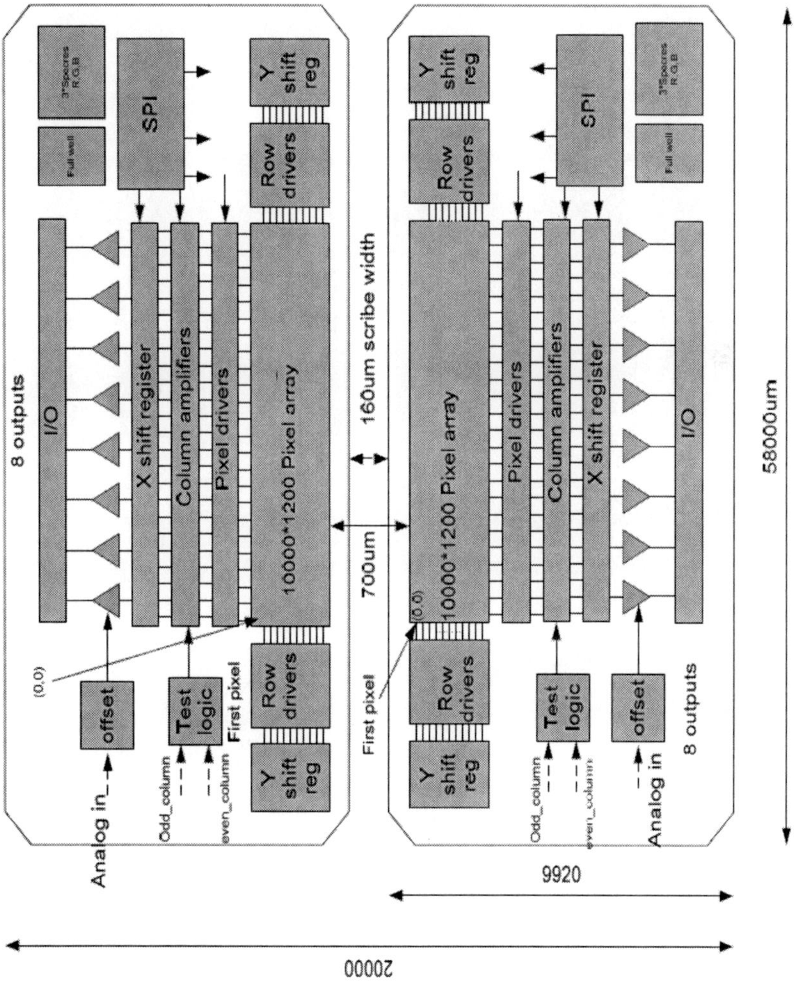

Fig. 4 VNIR detector layout

Table 2 Performance of linear variable filter

Centre wavelength (nm)	Max transmission (%)	FWHM (nm)
443	50.31	10
490	60.83	10
560	65.84	11
665	65.58	11
705	65.26	11
740	64.28	11
775	64.08	11
865	57.94	12
940	53.13	12
Mean values	60.81	11

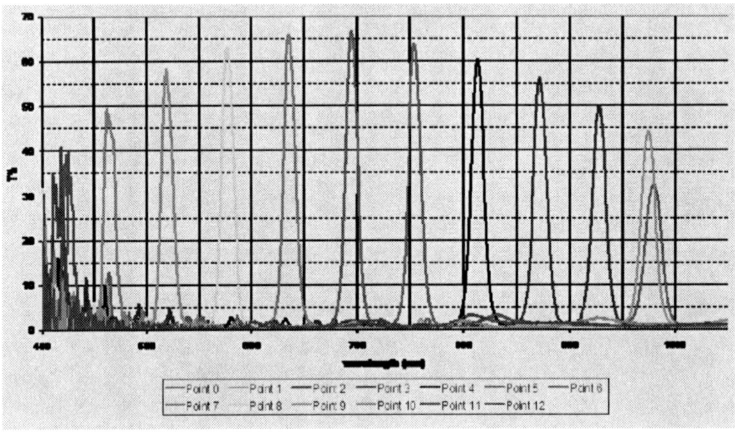

Fig. 5 Transmission curve of the linear variable filter

the spectral range between 400 and 700 nm and between 700 and 1,000 nm. The requirements of the filter of the PhytoMapper are simpler than the performance obtained with the LVF manufactured as technology demonstrator. It will be possible to improve performances in terms of peak transmission, out of band rejection, and bandwidth.

The detector baselined for this instrument has a format of 2 times 1,200 rows by 10,000 pixels, providing a staggering 2,400 spectral bands spaced of 0.25 nm and with 10 nm bandwidth. The resolution and span of spectral bandwidth will allow to investigate almost any composition of the atmosphere and soil.

3 Applications

Hyperspectral Instruments have been already designed and built for several applications, but the combination of using the technology of Single Point Diamond

Turning with large format CMOS detector and with the Liner Variable Filter gives the possibility to realize a very compact hyperspectral instrument that will open up possibilities to small satellites for applications traditionally done by larger Earth Observation satellites. The following paragraphs just provide an Overview of possible applications. A detail analysis of the physical properties that can be retrieved by using the PhytoMapper on-board on a small satellite goes beyond the scope of this paper, but is surely a topic of interest of the Authors.

3.1 Extending the Analysis of Crops

Multispectral Imagers have been extensively used to monitor the health of vegetation. These types of instruments have a limited number of bands centred on specific wavelengths. By using the hyperspectral capabilities, the wide field of view, and the ground sampling distance of only 50 m, it will be possible to extend the possibility to discriminate between varieties of different plants. Furthermore, the hyperspectral capabilities will enhance the compensation of soil and atmospheric effects.

Forestry studies can especially benefit from hyperspectral data, since this data allows identification and mapping of specific species, as well as a way to monitor the impact of pollution and other environmental factors. By using bands with carefully chosen wavelength responses, the chlorophyll concentrations can be studied in detail and linked to individual forest canopies.

Studies of inland and coastal waters as well as oceans are also very good study objects for a hyperspectral imager. Well chosen spectral bands, which in this case mean detailed narrow bands in the blue spectral region, allow very accurate determination of the “ocean colour”. This can reveal biophysical properties of the oceans upper layers: detection of phytoplankton, dissolved organic material, suspended matter, plankton, etc.... Multi-temporal imagery of these can in turn aid understanding of the role of the oceans in the climate system and improve the ability to forecast change. The spectrum of vegetation is characterized by the absorption of specific substances as shown in Table 3. Figure 6 shows the spectrum of different plants.

3.2 Planetary Observations

The compactness and the low mass of PhytoMapper are very appealing also for planetary missions. The optomechanical concept is based on an all reflective a-thermal design providing unchanged performances also when the telescope is exposed to very low or high temperatures, typically seen by low mass interplanetary body.

Table 3 Absorption features of some key plant materials (Vegindex, 1997)

Material	Wavelength (nm)
A-carotene	420–500
Arabinogalactan	990, 1,210, 1,450, 1,550, 1,740, 1,930, 2,100, 2,280, 2,320, 2,500
B-glucan (a hemicullulose)	1,450, 1,770, 1,930, 2,090, 2,330, 2,500
Carnauba wax	930, 1,040, 1,210, 1,390, 1,410, 1,540, 1,730, 1,820, 1,930, 2,010, 2,050, 2,140, 2,310, 2,350, 2,390, 2,430
Cellulose	1,480, 1,930, 2,100, 2,280, 2,340, 2,480
Chlorophyll-a	380–450, 675
Chlorophyll-b	410–470, 610
D-ribulose 1,5-diphosphate carBoxylase	1,500, 1,680, 1,740, 1,940, 2,050, 2,170, 2,290, 2,470
Humic Acid	400–700, 1,920, 2,300, 2,340
Lignin	1,450, 1,680, 1,930, 2,040–2,140, 2,270, 2,330, 2,380, 2,500
Lutein (a xanthophyll)	400–500
Pectin (apple)	1,440, 1,720, 1,920, 2,090, 2,240, 2,360, 2,480
Pectin (citrus)	980, 1,190, 1,440, 1,560, 1,680, 1,730, 1,780, 1,930, 2,080, 2,250, 2,320, 2,360, 2,480
Protochlorophyll	410–470, 580
Starch	990, 1,220, 1,450, 1,560, 1,700, 1,770, 1,930, 2,100, 2,320, 2,480
Tannic Acid	990, 1,120, 1,460, 1,660, 1,930, 2,130, 2,260, 2,320, 2,500
Xylan (a hemicellulose)	1,210, 1,450, 1,720, 1,790, 1,930, 2,090, 2,260, 2,330, 2,500

A possible application of Phytomapper could be the mapping the soil composition of a planet, or of a small body, as an asteroid or a comet. An Example of the different spectra of iron Martian rocks is shown in Fig. 7.

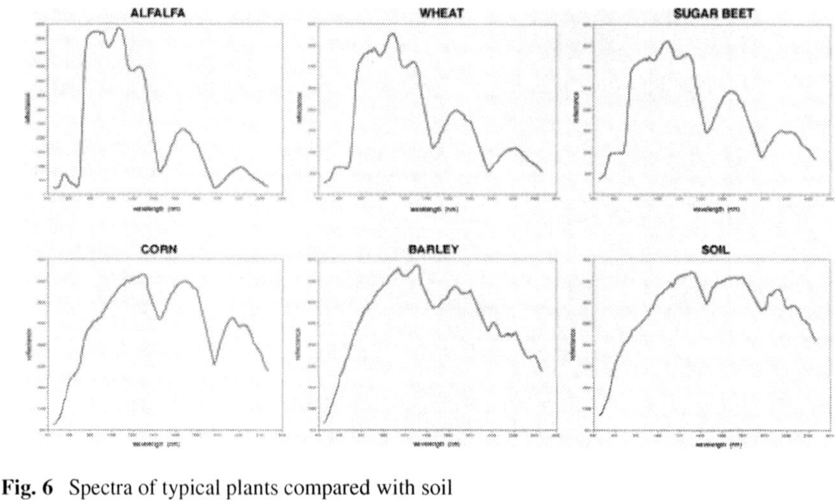

Fig. 6 Spectra of typical plants compared with soil

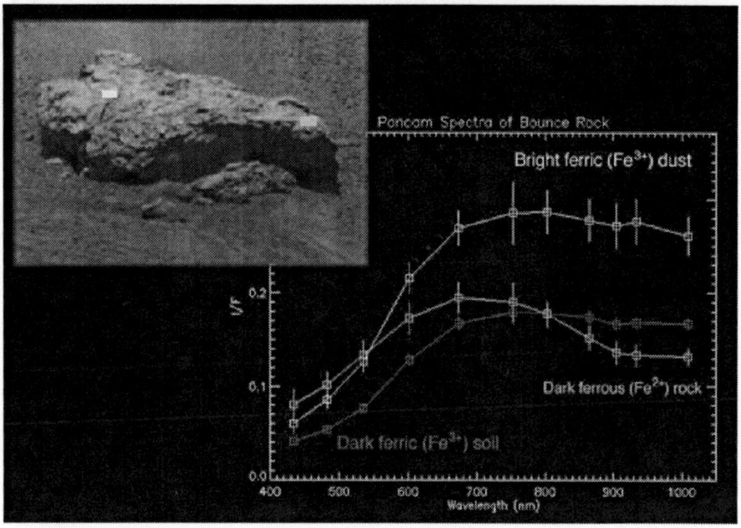

Fig. 7 Spectral signature of the different forms of iron from the surface of Mars. Image credit: NASA/JPL/Cornell

4 Instrument PERFORMANCES

The performances of the PhytoMapper are summarized in the Table 4 and compared with the Multispectral Imager developed for the Proba V satellite. Figure 8 shows the MTF of the optics. The PhytoMapper offer a reasonably good SNR: more then 50 in the spectral bands where the detector has a poor quantum efficiency and a definitely good and smooth optics MTF up to the detector cutoff frequency equal to 91 lp/mm.

Table 4 Main characteristics of Phytomapper compared with Proba V instrument

Parameter	PhytoMapper	Proba-V spectral imager
Focal length	90 mm	110 mm
F#	5	6
Field of View	34° × 5°	34° × 5°
Ground Sampling Distance at Nadir	48 m	100 m
Spectral range (VNIR)	400 nm ÷ 1,000 nm	450 nm 640 nm 850 nm
Spectral range (SWIR)	1.65 μm	1.65 μm
Number of bands	Up to 2,400	4
Spectral resolution	10 nm FWHM	50 nm FWHM
SNR	50	200
System MTF	0.15 @ 91 lp/mm	0.2 @ 38.5 lp/mm

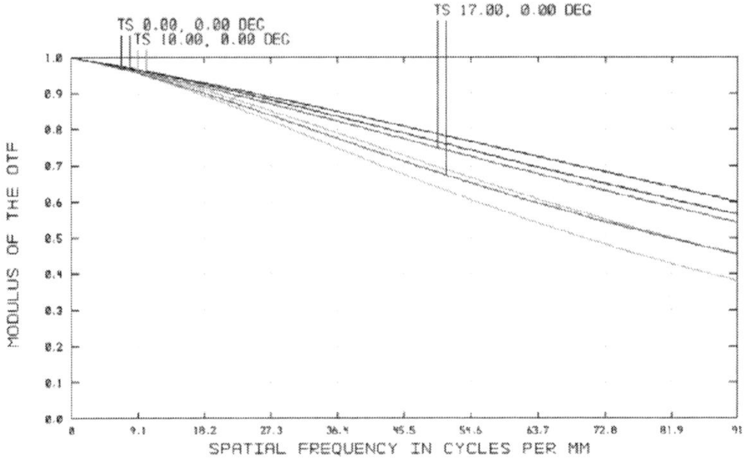

Fig. 8 Optics MTF

Development of Spaceborne Small Hyperspectral sensor HSC-III for Micro Satellite

Yoshihide Aoyanagi, Shin Satori, Tsuyoshi Totani, Toshihiko Yasunaka, Akihiro Nakamura, and Yusuke Takeuchi

Abstract The earth observation micro-satellite "TAIKI" is 50 kg satellite which has low-cost and small bus-subsystem. TAIKI is characterized by a low-cost spaceborne small hyperspectral sensor "HSC-III". HSC-III is targeted at the performances of 30 m ground sampling distance, visible and near infrared wavelength range, 10 nm spectral resolutions, 61 spectral bands and 10 kg weight. HSC-III consists of the telescope, the imaging spectrometer, the electrical system, the on-orbital calibration equipment. The telescope has a pupil diameter of 0.2 m, and has two mirror configuration of Ritchey-Chretien type. The spectrometer has the transmitting grating with the slit and relay lens unit, and array sensor using back-illumination type CMOS image sensor. As a SNR model of HSC-III, we did some calculations and concluded that SNR is approximately 340. Last year, we succeeded to develop the breadboard model of HSC-III optics instrument, and we obtained result of more requirement specification. Also, we have developed the on-orbit spectral calibration equipment. It achieved 0.02 nm of spectral calibration accuracy.

1 Introduction

The Space-Science Industries Program has a goal which is building some businesses on space. The program was kicked off in 2003 by the volunteer group that consists of students, researchers and engineers. In the program, we are planning to launch Micro-satellite as the demonstration for the space industries models. The earth observation micro-satellite "TAIKI" is 50 kg satellite which has low-cost and small bus-subsystem for remote sensing. The bus-subsystem will be developed as manufactured products and keeps a lid on development cost to within 1 million dollars. In 2006, we also have successfully launched pico-satellite "HIT-SAT" which has mission of demonstrated bus-subsystem using COTS (Commercial-Off the Shelf)

Y. Aoyanagi (✉)
Hokkaido Institute of Technology, Maeda 7-15, Teine-ku, Sapporo, Hokkaido, Japan
e-mail: q07601@hit.ac.jp

R. Sandau et al. (eds.), *Small Satellite Missions for Earth Observation*,
DOI 10.1007/978-3-642-03501-2_31, © Springer-Verlag Berlin Heidelberg 2010

components for TAIKI. This satellite is characterized by a low-cost spaceborne small hyperspectral sensor "HSC-III" [1]. The hyperspectral sensor is a combination of a spectral radiometer and an imaging sensor. The hyperspectral image acquires more spectral information from objects with a high spectral resolution compared with conventional multispectral sensors. A typical image generated by hyperspectral sensor is shown in Fig. 2. The hyperspectral sensor enables to distinguish a targeted object with a high accuracy, and give us lots of important information such as water quality measurement, classification of vegetation distribution, biomass and vegetation indexes for agricultural application. The mission objectives of TAIKI are summarized in the following: (1) To provide hyperspectral image for agricultural remote sensing, (2) To acquire visualization of the effect of climate change on plant distribution. This chapter describes actual statuses in the development of HSC-III.

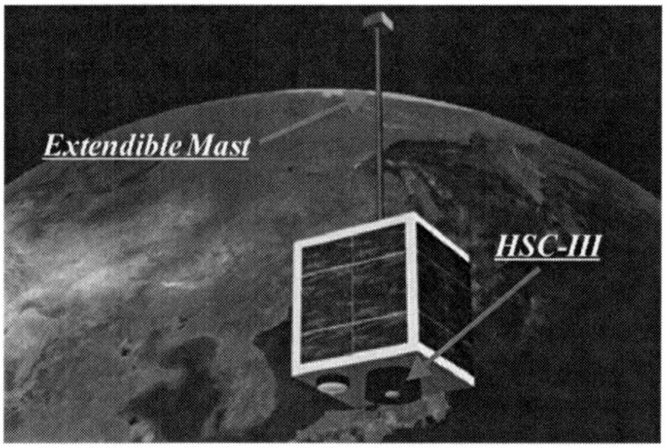

Fig. 1 Artist's view of the TAIKI

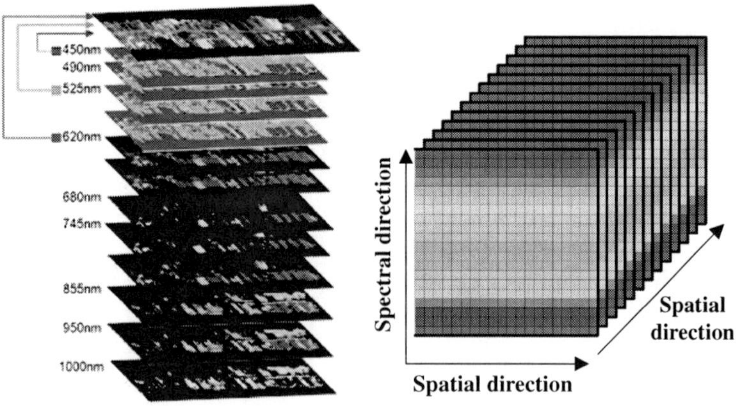

Fig. 2 A typical image generated by hyperspectral sensor

2 Hyperspectral Sensor Outline

The development of the HSC-III was started in January 2008 based on the optical design of the HSC-1700. In 2005, we have succeeded to develop a hyperspectral camera as the spin-off product named HSC-1700. After the release of the spin-off product HSC1700 to the general market, many big companies have been keenly interested in and purchased the products. The key requirements of the HSC-III instrument call for a GSD (Ground Sample Distance) of 30 m, a spectral range of 400–1,000 nm containing 61 bands of 10 nm resolution, SNR of >300, and an instrument mass of ≤10 kg. The imaging method employs push-bloom technique. The overall objectives are to demonstrate the technology in orbit and to use the imagery in support of practical applications such as agricultural monitoring. HSC-III consists of the telescope, the imaging spectrometer, the electrical system, the on-orbital calibration equipment and inertial reference unit. The telescope has a pupil diameter of 0.2 m, and has two mirror configuration of Ritchey-Chretien type. The spectrometer has the transmitting grating with the slit and relay-lens unit, and array sensor using back-illumination type CMOS image sensor. As a SNR model of HSC-III which is targeted at albedo 30%, we did some calculations and concluded that SNR is approximately 340. HSC-III optical design is presented in Fig. 3. The requirement specifications are summarized in Table 1. The data presented in Fig. 4 are results of calculated SNR model which is targeted at soils and grasses.

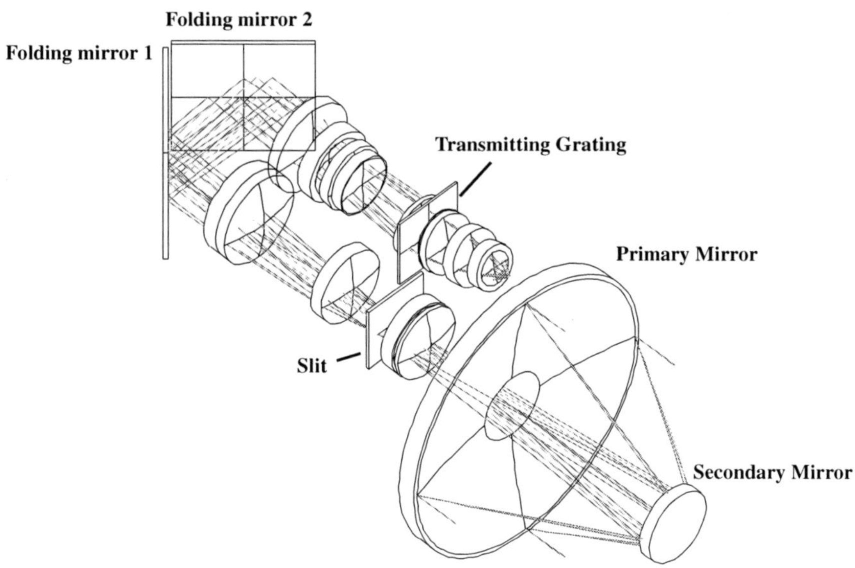

Fig. 3 HSC-III optics instruments

Table 1 Requirement specifications of HSC-III

Item	Requirement
Size	$78 \times 24 \times 24$ cm^3
Weight	10 kg
Power consumption	10 W
Telescope aperture	0.20 m
Ground sampling distance	30 m
Swath width	20.6 km
Wavelength range	400–1,000 nm(61 band)
Spectral resolution	10 nm
Signal to noise ration	>300@620 nm
Digitization	10 bit

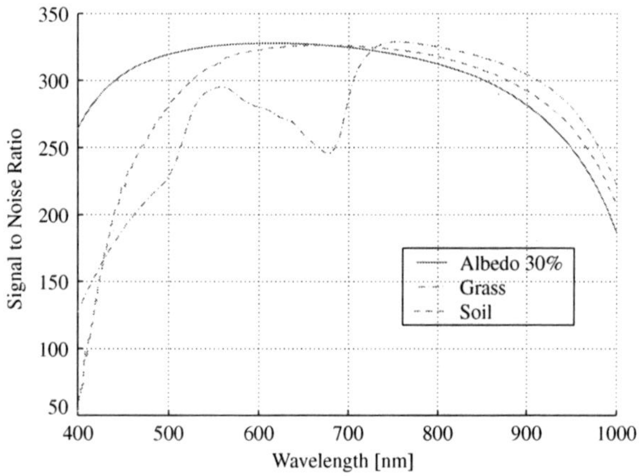

Fig. 4 SNR model

3 Breadboard Model of HSC-III Optics Instrument

A BBM (Breadboard Model) of the HSC-III optics instrument was developed in 2008. The BBM features a spectral range of 400–1,000 nm, 61 spectral bands, a radiometric resolution of 10 bits, and a sampling rate of 200 frames/s. The BBM optics instrument is shown in Fig. 5. The BBM optics instrument installs both the imaging spectrometer unit and a scanning mechanism using digital servo motor inside for the image mapping. The spectrometer includes the entrance slit, the transmitting grating and relay-lens unit. The entrance light from the slit is dispersed by the grating onto the detector which is an area array CCD image sensor. The slit is 15 mm long by 15 μm wide. The transmitting grating has an advantage which enables dispersion of entrance light into a spectrum of constant angle, so a spectral band width will become constant. The grating is 25 mm long by 25 mm

Fig. 5 Photo of the BBM optics instrument

wide, 300 lines/mm grating frequency and approximately 75% maximum efficiency, and the grating made of B270 glass. The BBM optics instrument is evaluated the spectral and the radiometric performances. The radiometric performance evaluations have measuring of radiometric accuracy and SNR. As result shown in Fig. 6, the calculated SNR model was experimentally verified. The spectral characteristics are summarized in Table 2 and Fig. 7. As Results, the spectral band width was approximately 10 nm, the central error was less than 1.0 nm, and out of band power is less than 20%. It has sufficient value compared with requirement.

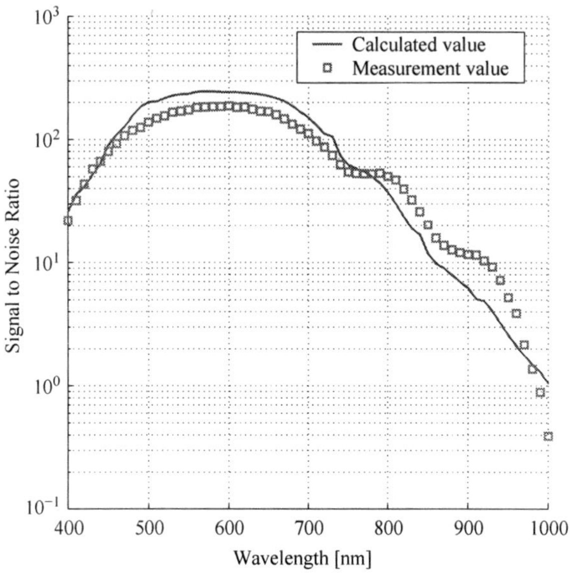

Fig. 6 Signal to noise ratio

Table 2 Spectral characteristics

Wavelength (nm)	Band width (nm)	Central error (nm)	Out of band power (%)
400	7.30	+0.85	7.38
500	9.40	+0.40	11.66
600	9.10	+0.15	11.37
700	9.25	−0.38	12.34
800	8.90	+0.75	9.98
900	12.40	−0.40	28.0

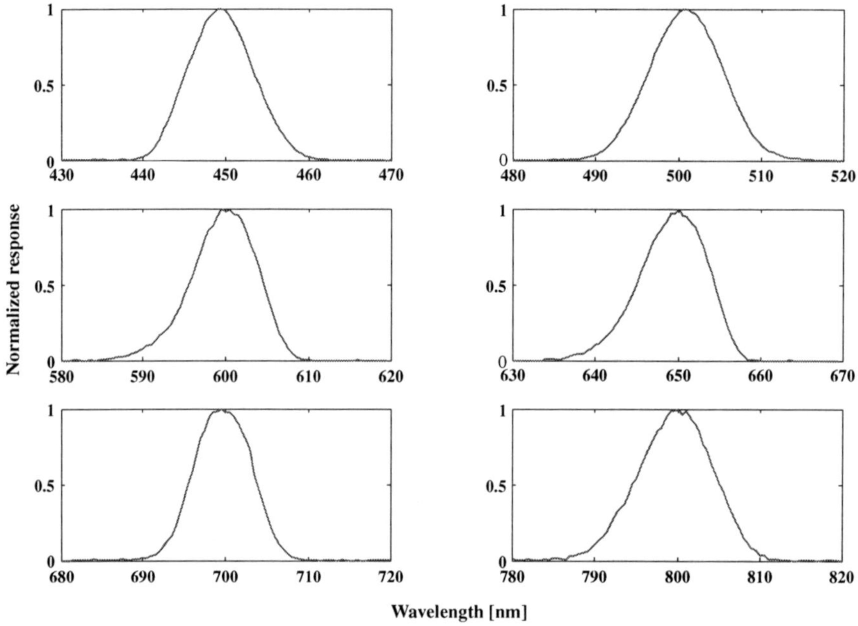

Fig. 7 Normalized response of band 6, 11, 21, 26, 31 and 41

4 Mission Data Handling Unit

The block diagram of Mission Data Handling Unit (MDHU) is shown in Fig. 8. The frame rate needs constant speed to get the constant GSD if hyperspectral imaging method employs push-bloom technique. HSC-III has 30 m GSD, so it requires 235 frame rates. And data volume is 384 Kbytes per 1 image, so required bit rate is 90 Mbytes/s. Thus, the communication interface between detector and MDHU employs the Camera Link Standard. The Camera Link Standard uses 28 bits to represent up to 24 bits of pixel data and 3 bits for video synchronous signals. The data transmission rates of the Camera Link up to 2.38 Gbits/s support the required

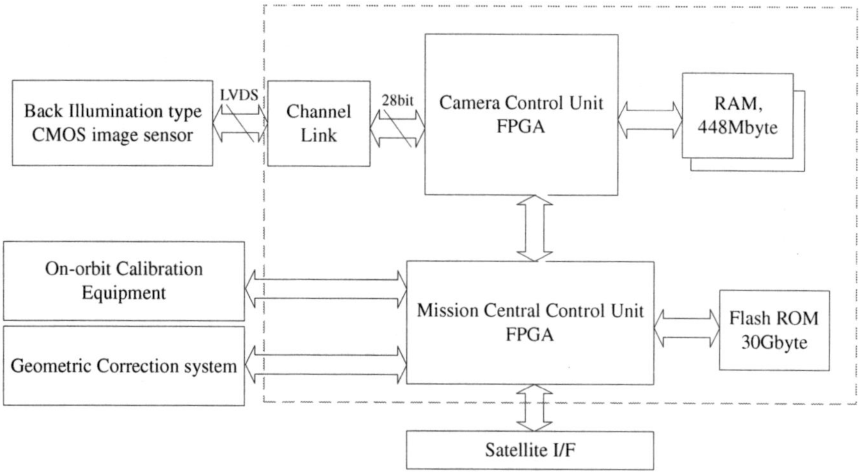

Fig. 8 Block diagram of MDHU

transfer speeds of hyperspectral data. The FPGA (Field Programmable Gate Array) is used for data logger system between Camera Link channel IC and SDRAM.

5 On-Orbit Calibration Equipment

In general, the earth observation satellites have to provide the observational data which has high accuracy assurance. Additionally, the hyperspectral sensor has to have the high spectral accuracy of each band because its sensor has more spectral bands. But, the conditions such as contamination, radiation, high temperature, aged deterioration and vibration environment affect the accuracy which was managed before flight. The hyperspectral data that have wavelength errors provide erroneous atmospheric correction with potential for erroneous classification of objects. Therefore, we need on-orbit calibration technique being valid. This chapter examined the effect of erroneous atmospheric correction on radiometric accuracy. Figure 9 shows simulation result of radiometric error under the conditions of wavelength bias error in case of vegetation as observational target. Figure 8 strikingly revealed that the radiometric error of H_2O and O_2 bands is influenced by the wavelength error. Also, we examined the effects of erroneous atmospheric correction on erroneous classification of the objects which consists of 3 kinds of pasture plants (Alfalfa, Orchard grass and Timothy). As a result shown in Fig. 10 and Table 3, we decided 0.25 nm of spectral calibration accuracy.

The demonstrated spaceborne hyperspectral sensor to date was Hyperion from NASA (National Aeronautics and Space Administration) and CHRIS from ESA (European Space Agency). These sensors were calibrated by solar spectrum and atmosphere absorption bands [2]. On the other hand, these calibration processes which used O_2 and H_2O absorptions are not enough accuracy that is required for

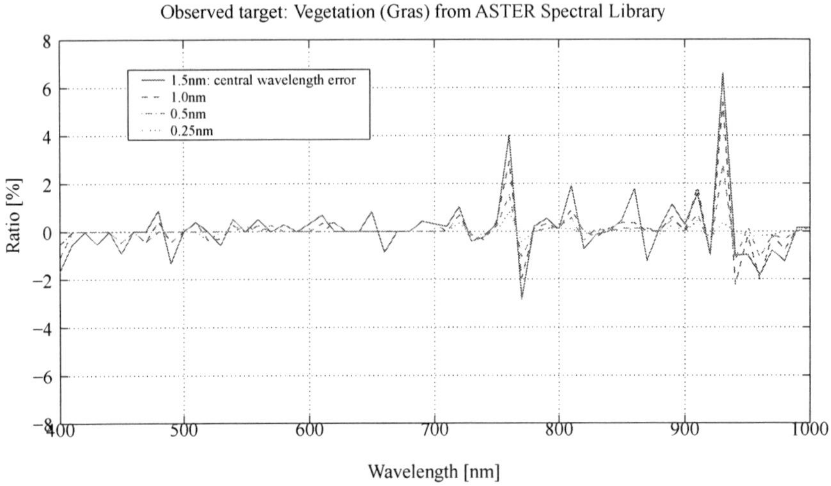

Fig. 9 Radiometric error under the conditions of wavelength error

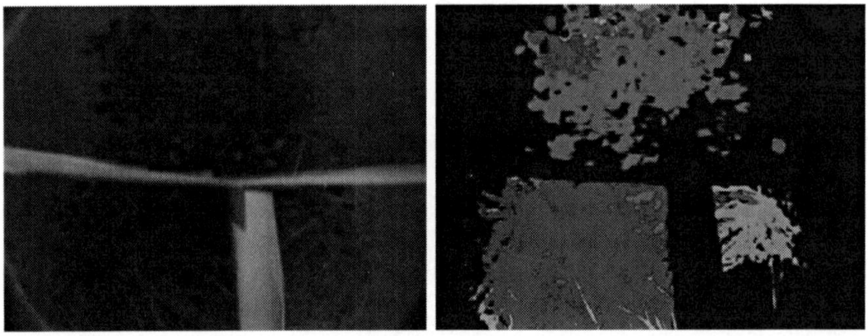

Fig. 10 Classification result of the pasture plants (*left*: sample picture)

hyperspectral sensor. Thus, we proposed LED (Light Emitting Diode) as the inner calibration radiance source. The advantages of LED light have high energy efficiency, long life, and small size and design flexibility. We developed the On-orbit Calibration Equipment (OCE). OCE consists of 6 high-intensity visible LEDs, an infrared LED, diffuse calibration panel and electrical circuit. It can switch between

Table 3 Overall accuracy of the classification

	Wavelength error [nm]				
Objects	0.00	0.25	0.50	1.00	1.50
Alfalfa	81.3%	83.1%	73.5%	74.9%	93.3%
Orchard grass	86.9%	88.5%	89.7%	90.4%	90.4%
Timothy	84.8%	83.1%	68.5%	65.4%	60.9%

observation mode and calibration mode by changeover mechanism which used stepping motor. The diffuse panel made by Spectralon. We evaluated the performance of OCE to compare with conventional method using spectral line of mercury. The performance of OCE was excellent so that 0.02 nm spectral calibration accuracy was achieved

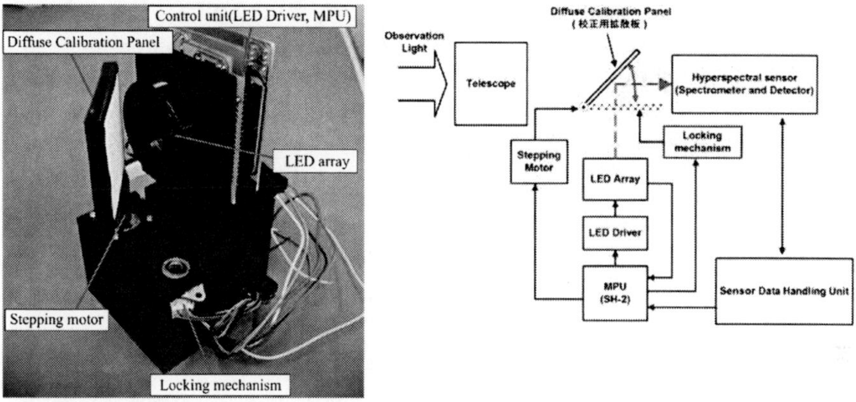

Fig. 11 OCE Overview

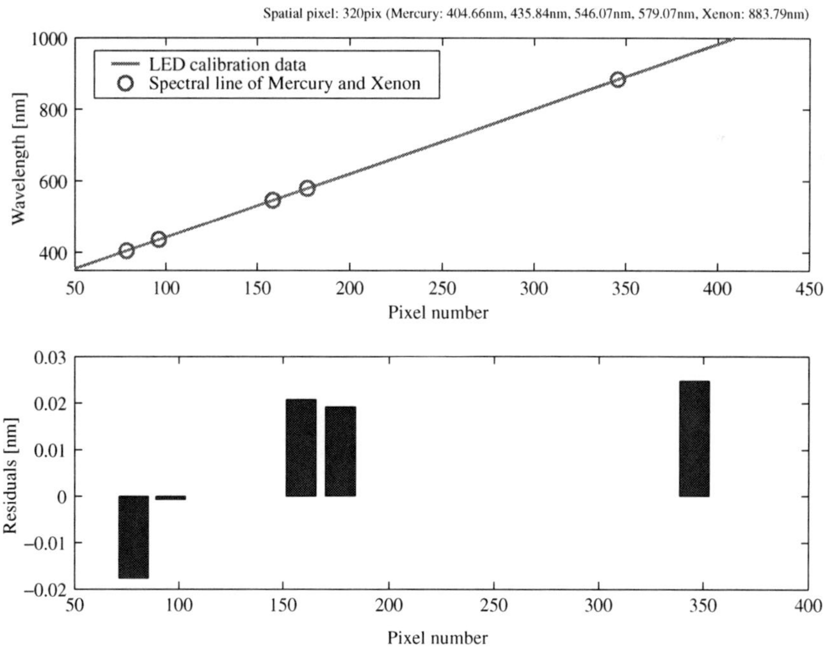

Fig. 12 Result of spectral calibration accuracy

6 Conclusion

The Space-Science Industries Program is performing investigations for the micro-satellite and hyperspectral remote sensing missions. The spaceborne small hyper-spectral sensor HSC-III is targeted at 30 m GSD, VNIR wavelength range and 300 SNR. We are planning to utilize for practical application of agricultural remote sensing. We also have successfully developed breadboard model of HSC-III optics instrument, mission data handling unit and on-orbit calibration equipment. As the experimental Results, we obtained more achievements than requirement performances. Additionally, we proposed LED as the inner calibration radiance source. We developed the OCE which consists of LED array. OCE achieved better 0.02 nm of spectral calibration accuracy compared with mercury spectral lamp.

References

1. S. Satori, Yoshihide Aoyanagi, S. Nishizato, T. Totani, U. Hara, T. Yasunaka, A. Nakamura, Y. Takeuchi. "Development of Hyperspectral sensor for Hokkaido Satellite 'TAIKI'," Proceedings of the 59th IAC (International Astronautical Congress), Glasgow, Scotland, UK, Sept. 29 to Oct. 3, 2008.
2. B. Pamela, S., John, S., Carol. "Hyperion on-orbit validation of spectral calibration using atmospheric lines and an on-board system", Proceedings of the SPIE, Vol. 4480, pp. 231–235, 2001.

Towards a Miniaturized Photon Counting Laser Altimeter and Stereoscopic Camera Instrument Suite for Microsatellites

S.G. Moon, S. Hannemann, M. Collon, K. Wielinga, E. Kroesbergen, J. Harris, E. Gill, and D. Maessen

Abstract In the following we review the optimization for microsatellite deployment of a highly integrated payload suite comprising a high resolution camera, an additional camera for stereoscopic imaging, and a single photon counting laser altimeter. This payload suite, the "Stereo Imaging Laser Altimeter" SILAT has been designated for deployment aboard the FAST microsatellite formation mission for Earth observation. This instrument suite has been designed for a Jupiter mission, but has been redesigned and optimized for an Earth observation mission. This chapter reviews the simulated Earth observation performance, the design modifications made for the mission and the optimization of the design for microsatellite use. Mass and power budgets are used to demonstrate the changes and the performance analysis is represented trough the simulation results. It is expected that the optimization will reduce the mass of the instrument by approximately 20% without compromising the performance of the instrument. In addition, results from breadboarding experiments of individual instrument components will be presented to show the progress from design optimization towards the FAST flight model.

1 Introduction

Understanding global climate change is a task that could be well suited to microsatellites, which can be efficiently designed to uniquely address key issues and specific scientific questions. A collaborative mission consisting of two spacecraft has been proposed jointly by TU Delft (The Netherlands) and the University of Tsinghua (China) for launch in 2011. This mission will be dedicated to the investigation of climate change by examining, among other objectives, atmospheric aerosol content and surface topography. Both spacecraft will carry a spectropolarimeter, which measures the intensity and polarisation of reflected and scattered radiation as a function of wavelength. In addition, each will also carry a topographic imager: one imager will be a radar, and the other a laser altimeter.

S.G. Moon (✉)
Cosine Research B.V., Niels Bohrweg 11, 2333 CA Leiden, The Netherlands
e-mail: s.moon@cosine.nl

R. Sandau et al. (eds.), *Small Satellite Missions for Earth Observation*,
DOI 10.1007/978-3-642-03501-2_32, © Springer-Verlag Berlin Heidelberg 2010

Discussed here is the design and development of SILAT, a Stereo Imaging Laser AlTimeter, which could satisfy the requirements of such a mission. This instrument combines a high resolution nadir-pointing camera, a low resolution offset stereo camera, and a single photon counting laser altimeter. The laser altimeter is characterized by a low power, high repetition rate design. SILAT features a highly integrated optical design and shared data processing unit, enabling a compact and low-mass package. In addition, the single photon counting mode of operation uses considerably less power than traditional laser altimeters. In addition to the instrument itself, factors involved with accommodating SILAT on a microsatellite (referred to as the HIPS microsatellite in the following) will also be addressed. This includes configuration of the instrument, integration with the microsatellite bus, and the resource requirements imposed on the satellite by SILAT. The microsatellite bus has been designed on the basis of a Europe-oriented Earth observation mission.

2 Mission Overview

2.1 The FAST Mission

In early 2008 the University of Technology Delft in the Netherlands and the University of Tsinghua in China proposed a collaborative formation flying mission consisting of two microsatellites. These spacecraft would each be equipped with two main payloads. The first is a spectropolarimeter, which is used to measure the intensity and polarisation of reflected and scattered radiation in the Earth's atmosphere as a function of wavelength. The second payload is a topographic imager; and, in the interest of collecting as much distinct data for a target location, each microsatellite will be carrying a distinct type of topographic imager. One microsatellite will carry a radar instrument and the other will carry a laser altimeter. In the case of the FAST-D satellite, SILAT will fulfil the laser altimeter role. SILAT is an ideal payload as it can provide not only a photon counting laser altimeter, but also a high resolution visible spectrum camera and a single band stereo camera. This instrument suite is heavily integrated and optimized, which reduces the mass, volume and power requirements, making it ideal for deployment on a microsatellite in the 25–50 kg range. The SPEX instrument will fly as a second payload on the FAST mission to collect data on aerosol particles in the atmosphere. In addition to providing scientific data for the FAST mission, the use of the SILAT payload would act as an in situ validation of the single photon counting laser altimeter concept.

3 Scientific Objectives

SILAT is capable of delivering varied scientific data, owing to its modular design and removable instruments. The mission parameters were formulated based upon a mission to provide regular high resolution maps of Europe in general and the Netherlands in particular. In addition, stereoscopic images and height information for incorporation into digital elevation models (DEM)s will be provided by SILAT. This aspect of the mission will be able to provide high resolution height and spectroscopic detail of the Dutch dykes and dunes with relatively high frequency. In conjunction, SPEX can provide information about atmospheric particles, which in combination can result in valuable data for monitoring and predicting the threats to sensitive areas of the Netherlands (Court, 2007).

The FAST mission, which will feature two microsatellites, has science objectives that include, but are not limited to the above. The two satellites will monitor fixed and liquid particles present in the atmosphere, such as soot particles. The two satellites can also generate stereoscopic images, and altimetry measurements of snow and ice levels on earth. The goal of the formation flying is to provide new, valuable scientific data in the fields of air quality and climate change while utilizing novel satellite concepts; namely, microsatellite platforms and formation flying. Doing so explores the benefits of alternatives to large, single satellite Earth observation missions.

4 Description of SILAT Integrated Payload Suite

SILAT is an highly integrated payload that is composed of three separate instruments. SILAT is the product of design studies conducted at cosine for the purpose of developing the concept of the use of highly integrated payload suites for future planetary exploration missions (Collon, 2005). These studies have evaluated the SILAT design at both Mercury (Collon, 2005) and Europa (Bentley, 2007). This section will describe each component of SILAT and detail the science objectives achievable, as well as the requirements, for each component (Fig. 1).

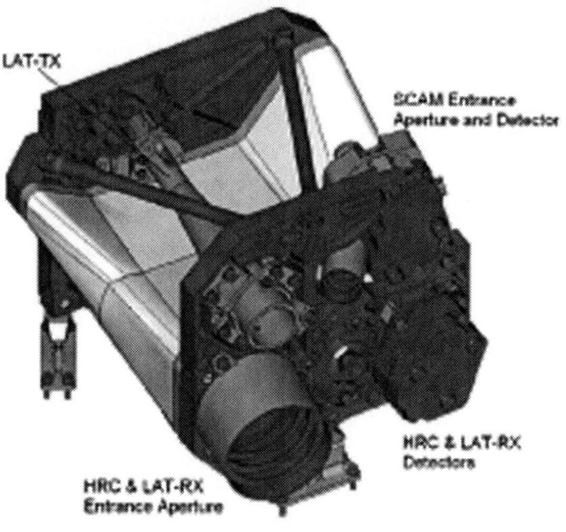

Fig. 1 SILAT integrated payload suite with sub-instruments identified

4.1 SILAT Features

SILAT utilizes a number of unique features that allow it to be used in a variety of planetary theatres. Thermally, SILAT is as isolated as possible from the spacecraft temperature gradient: radiatively through the use of multi-layer insulation (MLI) on the exterior of the instrument suite, and conductively via the mounting legs, which act as a thermal choke. Furthermore, the body of SILAT is, as much as possible, composed of silicon carbide (SiC). This material has a thermal conductivity of 180 W/(m·K) (on par with that of 6061-T aluminium) but its Coefficient of Thermal Expansion (CTE) an order of magnitude lower (2.2×10^{-6} K^{-1} against 23.9×10^{-6} K^{-1}). This promotes the elimination of thermal gradients in the instrument suite, thus inducing uniform expansion. This renders SILAT resistant to misalignment from thermal expansion.

SILAT also features custom fasteners for the optical components. These fasteners allow the optical components to be minutely adjusted, to the micrometer level. They are made from SiC as well, keeping in line with the SILAT design philosophy.

Another facet of SILAT is that it employs passive elements as much as possible. While the LAT is an obvious exception, SILAT contains no moving parts, and all cooling devices are passive; only cold fingers and radiators are used. This increases the reliability of the payload suite, which is essential for deep space missions.

4.2 Earth Orbit Optimization – Radiation Shielding

The materials, and material thicknesses, specified in the original SILAT design were based on an evaluation of the radiation environment in an orbit around Europa at Jupiter. A Low Earth Orbit (LEO), even one close to the inner Van Allen radiation belt, experiences a significantly less harsh flux of harmful radiation. As such, the radiation shielding of the instrument suite has to be re-evaluated. This includes both the materials used, their thickness and their orientation if there is a directional dependency.

Densimet, a tungsten alloy, is an excellent radiation shield, but is very dense. Its use on the Jupiter configuration as a shield for the electronics was driven by the desire to keep the envelope of SILAT as small as possible. Given the reduced radiation environment, aluminium can provide the same respective shielding while maintaining the volume profile of the instrument. As such, the use of Densimet is not necessary for the FAST configuration, and is replaced by aluminium.

Since the radiation environment is less harsh, the thickness of the SiC baseplate can be reduced as well. Based on the evaluation of the radiation environment, the thickness can be reduced by approximately 50% saving approximately 1 kg of mass.

The new structural dimensions of the redesigned SILAT will be presented at the 7th IAA Conference on Small Satellites.

5 Performance of SILAT for Earth Observation

5.1 Orbital Parameters

Based on the scientific goals described in §3, a sun synchronous orbit at 700 km has been chosen for the microsatellite. The details of the orbital parameters are summarized in Table 1.

Table 1 Parameters of sun synchronous orbit

Type	SSO
Inclination	98.7°
Altitude	700 km
Velocity	7.51 km/s
Orbital period	98.6 min
Sunlit time	61.1 min
Eclipse time	37.5 min
Sunlit fraction	0.62

This orbit has been chosen based on a trade-off between the resolution, swath width and revisit of the SILAT payload with respect to the mapping of Europe. 700 km established a reasonable spatial resolution of the HRC of 21.9 m (Table 2) while allowing 100% coverage of Europe to be achieved in 100 days, and the Netherlands in 52 days. The sunlit fraction of the orbit is a parameter used to gauge the amount of power available from solar irradiance, and drives the solar array design.

Table 2 Orbit dependant HRC characteristics

Ground velocity	6.77 km/s
Focal distance	320 mm
F#	4.9
IFOV	31.5 μrad
Single pixel footprint	21.9 m
Integration time	1.62 ms
FOV	3.67°
Swath width	44.8 km

This orbit is also beneficial in that it is launched to relatively frequently. This makes it a convenient selection for a piggy-back launch strategy, which is the preferred for a microsatellite launch.

5.2 Instrument Performance

In order to quantify the performance of SILAT for Earth observation, simulations have been carried out at cosine. For this purpose, MathCAD sheets have been developed to calculate performance parameters for the HRC and LAT. The program uses orbital parameters described in §5.1 as the mission input. It calculates the radiometric level of the HRC and LAT detectors, based on parameters such as the Sun-Earth distance, the Earth-satellite distance, the albedo of different surfaces, the atmospheric transmittance, the aperture size and overall transmission efficiency of the imaging optics and filters. From the detector specifications (dark count noise, readout noise, quantum efficiency, pixel size and dead time) the simulations calculates the SNR.

In addition the program takes the basic characteristics of the laser into account. Thus, the performance calculations of LAT are based on parameters such as repetition rate, pulse energy, wavelength and pulse duration. For both the HRC and LAT the simulation has additional variables to reflect various illumination conditions such as cloud, clear sky, day, night.

5.3 HRC Performance

Based on the orbital parameters given in Table 1 and the CCD detector characteristics described in Moon et al. [1] the performance parameters for the HRC, presented in Table 2, are calculated. These parameters are fed into an Earth observation simulation code (or in the case of non-Earth planetary exploration, to the HIPSIM simulator), and the SNR is obtained.

The integration time is defined as the time the spacecraft requires to travel across half a pixel footprint. Although the resolution of the HRC does not compete with larger, more powerful optical payloads, it will be sufficient to carry out the environmental surveys described by the HIPS microsatellite mission and the FAST science objectives.

From the radiometric level and the performance characteristics on the HRC detector the signal to noise ratio was calculated for all three wavelength bands corresponding to the filter attached to the HRC detector. The result is shown in Fig. 2,

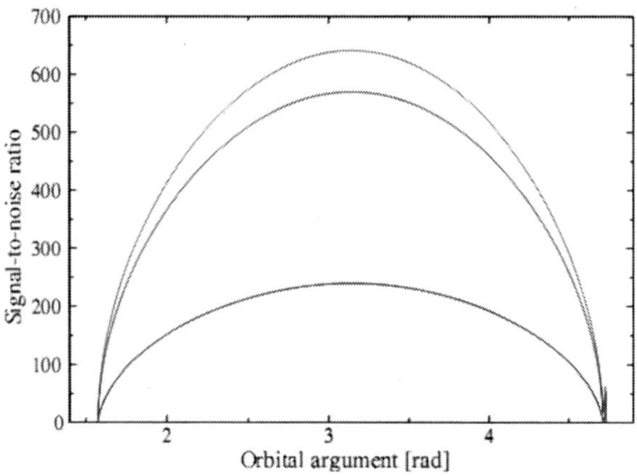

Fig. 2 Plot of Signal-to-Noise vs. orbital argument for an SSO SILAT. Note that the argument at which the SNR falls to zero are the entry/exit points from eclipse

where the SNR is depicted as a function of the orbital argument. The maximum SNR for one orbit exceeds 200 in all three wavelength bands. The green band even reaches values of 500. However, for historic reasons, atmospheric effects are not fully taken into account yet. While transmission losses through the atmosphere have been taken into account, the simulation that generates the SNR does currently not reflect the increased background light levels caused by atmospheric Rayleigh scattering. Work has begun on including this in the simulation model, however the additional module is not complete and results have not yet been generated.

5.4 SCAM Performance

Due to the limitations of the poster session, the performance of the SCAM instrument has been omitted. The Earth Observation performance of this instrument on SILAT is covered in Moon et al. [1].

5.5 LAT Performance

The performance of the LAT can be characterized at different levels. Since the instrument is based on time-of-flight measurement, the ultimate limit of the achievable height resolution is correlated with the pulse duration of the laser in use. Light travels 30 cm in 1 ns. Since a photon emitted by the LAT-TX will travel the distance from the instrument to the Earth's surface twice during its flight, a height difference of 15 cm will result in a path length difference of 30 cm, which will be detectable by the detector if it operates at a speed of 1 ns. Thus the ultimate height resolution is about 15 cm. With good statistics, that value can be improved by the familiar $1/\sqrt{N_{sh}}$ factor, where N_{sh} represents the number of shots taken. Parameters for the LAT performance are shown in Table 3.

Table 3 Performance parameters of the laser altimeter

Ground velocity	6.77 km/s
Detector size	20 μm
IFOV-RX	63 μrad
Footprint size (RX)	43.8 m
Footprint size (TX)	17 m
Repetition rate	10 kHz
#-pulses/footprint	25
Time-of-flight	4.67 ms

The next important set of parameters consists of the ground speed of the spacecraft, the repetition rate of the laser, and the footprint sizes of the transmitter and receiver optics. These parameters determine the statistics with which the ground track will be scanned. Within the time the spacecraft passes a given distance a number of shots will be emitted where mutual overlap of the individual footprints of the laser pulses will provide the redundancy for improving the statistics. With the current parameters (10 kHz repetition rate, 700 km altitude) about 25 pulses will be emitted during the time the spacecraft passes the distance equivalent to one LAT-TX footprint size.

From the height of the spacecraft, the atmospheric transmission, the transmitter and receiver properties and the laser characteristics it is possible to calculate the number of detected photons per second (Altimeter equation, [Degnan, 2002]). Preliminary results indicate that the rate is approximately 25 detected photons per second. More detailed investigations on how that rate can be increased on the HIPS microsatellite orbit are still underway.

As stated earlier, the CRR converts the returned pulses to a resolution of the height based on statistical distributions of the TOF of the returned photons, thus this factor contributes directly to the spatial resolution achievable by the LAT.

It is also of crucial importance that the orbital height corresponds to a TOF of the LAT signal photons of slightly less than 5 ms. Within that time, the spacecraft pointing must be stable within a fraction of the instantaneous field-of-view (IFOV) of the LAT-RX. Otherwise the LAT will lose signal and fail to register any of the returning photons.

6 Earth Orbit Optimization – Breadboarding Results

Cosine has recently presented a successful demonstration of the photon-counting laser altimeter breadboard. This breadboard is capable of controlling a pulse emitted from a prototype microchip laser, detecting said pulse and then sorting and analyzing the resulting data to form a distance profile. The demonstration so far have been limited to distance measurements of objects in various environments within labs, indoors and outdoors in fair weather, with an effective range of 100 m. Preliminary demonstrations for external parties have been given, but the components have recently undergone fine tuning and additional development in order to increase the performance of the breadboard.

The current status of the laser altimeter is within the integration stage. The full breadboard including emitter, receiver, control electronics and data-processing FPGA will be assembled during the first week of May, with the full field testing conducted during May in Leiden, Netherlands. The instrument is expected to exit the breadboard stage during May and enter full prototype status on an engineering model in the summer.

The current goal of this project is to produce two pre-flight models, one of SILAT, and one of the MPS, for a micro- and nano-satellite launch opportunity respectively, in partnership with the FAST mission with TU Delft and the Delfi-NexT mission, also with TU Delft.

In addition to the breadboarding, SILAT is also being optimized for the FAST project by optimizing the design for a LEO mission. This includes a detailed look at the radiation environment in the proposed FAST orbit, and a reduction of the radiation shielding materials and a change in the shielding strategy, while maintaining the desired performance. A parametric study, which employs a comparative analysis of previous SILAT operations for Jupiter against the LEO radiation environment, has resulted in a new mass of 5.2 kg and the replace of very dense, very radiation hard shielding materials with aluminium or stainless steel. Previous studies have shown that the thermal and control electronics subsystems will be appropriate for a LEO application of SILAT. With this knowledge, and the knowledge of the behaviour of the optical systems with respect to Earth's atmosphere, SILAT is ready to move into a full, integrated engineering model.

In the context of the FAST mission, there is approximately 1 year to develop and test an engineering model of SILAT, followed by a 6 month period to produce the FAST-ready flight model. cosine and the University of Technology in Delft are working closely to keep the satellite and payload development on schedule.

7 Other Micropayloads

In addition to developing the laser altimeter, cosine has developed two other micropayloads using the same development methodology as SILAT. These two micropayloads are the Highly Integrated Broadband HyperSpectral Imager and Spectrometer (HIBRIS) and the Multifunctional Particle Spectrometer. As these are represented on the symposium poster, they are described briefly below.

7.1 HIBRIS

HIBRIS has been developed in parallel with SILAT, using the same approach to development. Instruments have been integrated and miniaturized to provide maximum scientific value within a small volume, with low mass and power requirements. HIBRIS consists of a hyper-spectral Near Infrared (NIR) spectrometer combined with a low resolution Thermal Infrared (TIR) imager. The instrument provides wide area thermal imagers with detailed spectroscopy over the NIR wavelength range. Like SILAT, HIBRIS is meant as a payload module; it is connected as a fully independent unit to a small or large satellite. HIBRIS is thermally and vibrationally isolated, and requires less than 4.5 W of input power from the spacecraft. HIBRIS has a mass of approximately 7 kg. HIBRIS is currently in the breadboard stage. A fully functional engineering model is planned for 2010.

7.2 MPS

The Multifunctional Particle Spectrometer is a variable role energetic particle detector. The MPS is only $6 \times 6 \times 8$ cm large module, occupying a very small volume. The instrument has multiple configurations however, allowing it to fulfil a variety of roles from simple particle monitoring collecting counts of different particle types, to full-fledged particle analysis. Its full capabilities include determining the type, energy and angle of incidence of incoming electrons, protons, gamma rays and ionized particles. The MPS can operate at a rate on the order of kHz.

The MPS is currently in the prototype phase. The control electronics breadboards have been completed, and the optical and semi-conductor are being prototyped. The short term goal is to complete an engineering model before Autumn 2009. The MPS currently has a launch opportunity on the TU Delft nano-satellite Delfi-NexT.

8 Summary and Outlook

SILAT is a highly integrated payload suite that contains three separate visible spectrum remote sensing instruments: a high resolution camera; a reduced resolution stereo camera and a photon-counting laser altimeter. SILAT has been designed for use in exploration missions to Mercury and Europa, and this chapter has evaluated the feasibility of using such a payload for Earth observation such as the upcoming FAST mission at the Technical University of Delft. Contained within this chapter has been an evaluation of the performance of the SILAT integrated payload suite for an Earth observation mission, and a preliminary investigation into the microsatellite support platform that SILAT would require. Performance calculations have been done by cosine, and the microsatellite design by ISIS. The MathCAD simulations indicate that the HRC and SCAM will have high SNR ratios. Though the full impact of atmospheric effects must still be modelled, it is not expected that this will appreciably reduce the utility of SILAT in an Earth observation role. The resolution of these two instruments will also be increased if used in the upcoming FAST mission, as FAST is proposed to fly 50 km lower than the orbit indicated in this chapter.

References

1. S. Moon et al. A miniaturized laser altimeter and stereo camera for a microsatellite formation mission. In ESA *Small Satellites Systems and Services – The 4S Symposium*. Rhodes, Greece, 26–30 May 2008. ESA, The Netherlands.
2. J.J. Degnan. Photon-counting multikilohertz altimeters for airborne and spaceborne topographic measurements. *Journal of Geodynamics*, 35, 503–549, 2002.
3. M. Bentley, F. Varlet, S. Kraft. *Consolidation of HIPS FM Preliminary Design*. cosine internal communication [CR-HDD-TN-01, Issue 02]. June 06, 2007.
4. M. Collon et al. *HIPS Breadboard Engineering Plan*. Cosine internal document [CR-HIPS-PL-EP, Issue 01]. September 8, 2005.
5. A.J. Court et al.(TNO). 2007. *TNO-SPEX (Spectropolarimeter for Planetary Exploration)*. Available at http://www.tno.nl , accessed May 1, 2009.

A Plan of Spaceborne ISAR Satellite Imaging System Aiming at Space Objects

Guodong Xu, Xinghui Cao, and Fulin Su

Abstract Propose a plan of spaceborne inverse synthetic aperture radar (SISAR) imaging to space object based on present techniques and equipments, and provide practical approach from on-board-computer (OBC) to payload to carry out mission requirements. Space objects are classified into three categories through the analyses of motion characters, three-axis-stabilized satellite, spin-stabilized satellite and space debris. Then simulation models are established respectively to achieve imaging investigations. The result shows that objects belonged to three-axis-stabilized and spin-stabilized satellite can be reconstructed clearly if proper parameters are set, and space debris can be detected and estimated correctly. The primary results verify the feasibility of SISAR system.

1 Introduction

Spaceborne synthetic aperture radar (SSAR) has been well-investigated in recent two decades in both civil and military services, focusing on global terrain mapping, ocean survey, digital elevation mapping (DEM), ground moving target indication (GMTI) and other pertinent requirements [1, 2]. Correspondingly, spaceborne inverse synthetic aperture radar (SISAR) is empty of in-depth investigation whereas its value is equivalent significant on account of its potential abilities of high resolution imaging of space object [3].

At present the observation to space object mainly depends on ground-based equipments, using large aperture antenna to get high space resolution as well as obtaining far observable range by vast transmitting power. But the measurement with ground-based equipments will be effected by atmosphere attenuation leading to restrict further surveillance ability. Naturally, spaceborne vehicles surveillance concept is proposed and then investigated to obtain a broader and clearer figure of space object, some has carried through [4–6]. But those systems all based on visible light sensors could only operated in a restricted time, space scope and exploring

X. Cao (✉)
Research Center of Satellite Technology, Harbin Institute of Technology, Harbin, 150001, China
e-mail: caoxinghui1982@163.com

R. Sandau et al. (eds.), *Small Satellite Missions for Earth Observation*,
DOI 10.1007/978-3-642-03501-2_33, © Springer-Verlag Berlin Heidelberg 2010

range owing to light illumination constraint. As ISAR imaging don't require visible light illumination and that its resolution is independent with detected-range between radar and target, the advantages of spaceborne ISAR technique make it provide a more attractive approach to observe space object in future missions. With high 2-D resolution, it will be a challenge of considerable importance to multiple application areas such as space traffic control, satellite defense, space debris elusion via wide area scanning, tracking and identification.

Spaceborne ISAR imaging is a novel exploration revealing attractive research perspective. Nevertheless, object in space is rather different with object in the air or on the sea. Two main difficulties should be involved: (a) objects distribute in the so vast space that to meet an object is not frequent, which requires the platform take the ability to observe all kinds of objects in certain area; (b) The state of motion will be very complex on account of space perturbation and controlling modes of man-made satellites. In order to deal with these problems, this chapter proposes a material plan of space object imaging via spaceborne ISAR observation. The investigation concentrates on general concept of the whole project programming including task orientation, spatial movement analyse and relationship of spatial relative motion.

2 Mission Concept and Requirements

SISAR imaging is almost a wholly original concept to observe space objects, trying to transform into practical project by succeeding to present spaceborne SAR (SSAR) and small satellite vehicle techniques. Linked with distribution of space objects in the extra space of the earth, the starting point of the mission requirements is based on low orbit and ground segment cooperation. The preliminary system requirements have been derived, displayed in Table 1.

Table 1 Mission and system requirements

Mission requirements	Preliminary system requirements
Observation Distance 50 ~ 500 km	Carrier frequency f_0: 35 GHz
Objects size: imaging: 1 ~ 50 m	Bandwidth: 1 GHz
Detection: 0.05 ~ 50 m	
Observation belt: whole round belt	Incident angle: –25° ~ 25°
Observation intersection angle:	Pulse represent frequency (PRF): 1 ~ 200 Hz (depending on object orbital height)
Mode 1: –6° ~ 6°	
Mode 2: –4° ~ 4°	
Mode 3: –3° ~ 3°	
Objects type: artificial objects three-axial stabilization spin stabilization uncontrolled state	Pulse width: 50~1000 μs (depending on object orbital height)
Range resolution: 0.15 m	Downlink capacity: 10 MHz/s
Azimuth resolution: 0.15 m	Data compression: 8:3/8:4

SISAR imaging system takes ability of two main missions: one is to observe and reconstruct image of space station, conventional satellite and small satellite, which are commonly considered more then 1 m, then form into database; the other is to detect and parse space debris, which can be as small as 5 cm, position orbit then form into database.

Operational SISAR imaging will perform based on three kinds of modes, including inward, upward and on orbit, where inward mode is that antenna beam scans the scene around the earth core, sketched in Fig. 1(a), here scalar velocity of targets is faster than the speed of ISAR satellite; while upward mode denotes the wave beam back to the earth core, sketched in Fig. 1(b), here scalar velocity of targets is slower; and on orbit mode means both targets and ISAR satellite on the almost same orbit with almost same speed, shown in Fig. 1(c). All three modes are designed to achieve –25 ~ 25° swing angle via coordination control between satellite platform and phased array antenna, as shown in Fig. 1(d), here the mission requires –3 ~ 3° orbital intersection angle in inward mode, –6 ~ 6° in upward mode, –4 ~ 4° in the rest one, respectively. As pulse repetition frequency (PRF) and pulse width are varied by the distance and relative velocity and orbital intersection angle between objects and satellite, model database is upbuilt to directly transfer radar parameters for various cases. Table 2 presents partial data of calculated results list, where V_{rev} is the relative velocity between satellite and objects, T/s synthesize aperture time, $\Delta\alpha$ is the change of geocentric angle in the time of T, Δf_{d1} is Doppler frequency variety by movement, Δf_d is Doppler frequency variety by rotation.

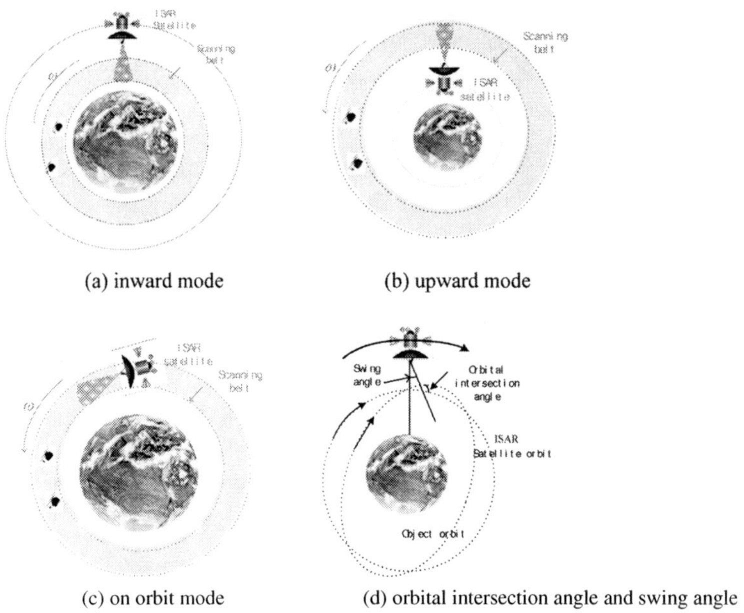

(a) inward mode (b) upward mode

(c) on orbit mode (d) orbital intersection angle and swing angle

Fig. 1 Sketch of space geometry relationship of SISAR imaging

Table 2 Partial data of calculated results

Orbital height/km	Velocity /m/s	V_{rev}/m/s	T/s	$\Delta\alpha$/°
$H_0 = 500$	7617	–	–	–
$H_0 = 400$	7673	56	94	6
$H_0 = 300$	7730	113	93	6
$H_1 = 200$	7788	171	92	6
Intersection angle/$H_0 = 500$ m	V_{rev}/m/s	T/s	Δf_{dl}/Hz	Δf_d/Hz
1°	145.0	36.2	507	0.97
3°	405.2	13	1417	2.7
4°	538.0	9.8	1883	3.6
6°	804.3	6.5	2814	5.4

2.1 D Radar Parameter Selection

In order to avoid the influence of the earth and atmosphere background and try to miniaturize payload, the design selects 40 GHz millimeter wave to be carrier frequency aiming at space objects imaging, so it can escape electromagnetic wave atmosphere window and reduce a great deal of the volume of radar antenna. Generally speaking, the size of spatial artificial objects is small, thereby the bandwidth of the transmitting signal is determined to be 1 GHz for high resolution imaging. At the same time, relative motion of spatial orbit in similar direction is low, and it should take a longer time to get azimuth accumulated angle, so it is consider to enhance the pulse width as well as to reduce PRF to depress the apparatus' pressure from peak transmitting value and sampling rate. Moreover, STRETCH method, namely twice mixing to echoes, is used to reduce A/D transfer sampling rate and data quantity for storage and calculation.

3 Imaging Simulation

Based on mission truss of Sect. 2 and space objects classification of Sect. 3, in this section, design flow and imaging results are given. The process of SISAR imaging can be reduced to the following 6 steps: (1) constitute simulation model; (2) setup simulation parameters and record echoes data; (3) estimate orbital motion parameters of objects; (4) coarsely compensate to echoes data using orbital parameters; (5) accurately compensate, involving range alignment and phase correction; (6) image reconstruction.

Set up a satellite object model consisting of 179 scattering points, shown as Fig. 2, Satellite object located in the orbit of 300 km height, while SISAR platform is 500 km height, and the task is to reform images to three kind of objects discussed above, respectively. Firstly, three-axis-stablized satellite is considered, presume the intersection angle between SISAR platform and object orbit is zero degree, and the orbital motion of the object include the effect of various perturbations, such

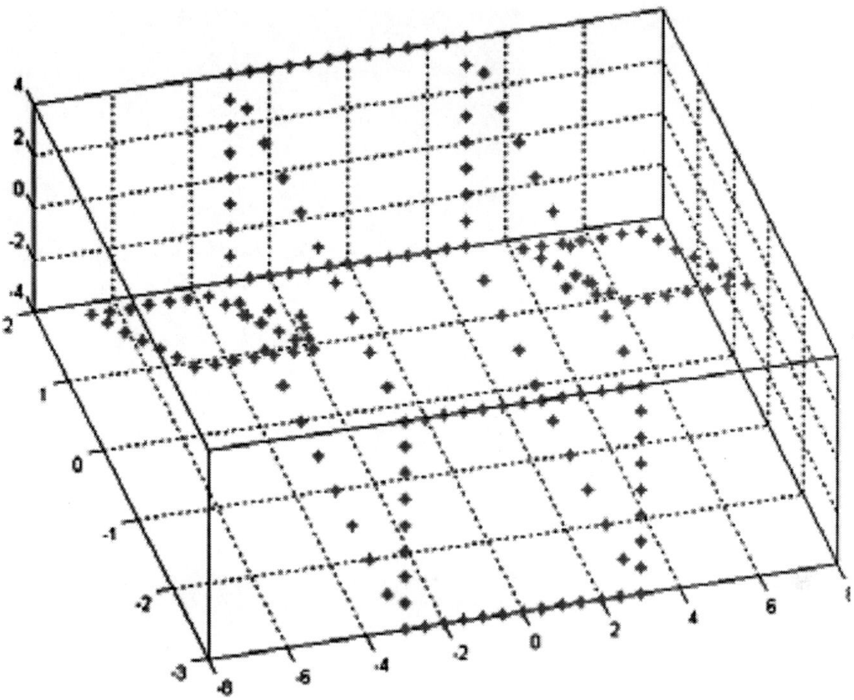

Fig. 2 An original satellite object model

as gravitation of three bodies, perturbation of anomaly shape of the earth, here the data of orbital motion is produced by STK, and the rest radar parameters refer to Sect. 2.

Based on the method mentioned in reference [6], we record echoes data and then estimate orbital motion parameters utilizing low resolution scanning radar to draw up the trace, following that is to complete coarse compensation, one of which aim is to reduce echoes data, the other is to take the pressure from further compensation. Figure 3 provides the result of coarse compensation after range compression. It can be seen from Fig. 3, by the effect of perturbations, echoes can be reformed directly as they are not orderliness in range direction. The next step is accuracy compensation, which contains range alignment and phase correction, while envelope minimum entropy is used on range alignment and rank one is used on phase correction. The result of accuracy compensation is shown in Fig. 4. After this process, azimuth compress can be carried through to get the 2D reconstructed image, Fig. 5 gives the ultimately result.

Next, the second kind of object is considered, spin-stablilized satellite, here the case of single spin-stablilized is investigated. Presume angle velocity of spin is 0.025 rad/s, the accumulated time is 1s, others radar parameters are the same as above. The results of range compression and azimuth compression are shown as Figs. 6 and 7, respectively.

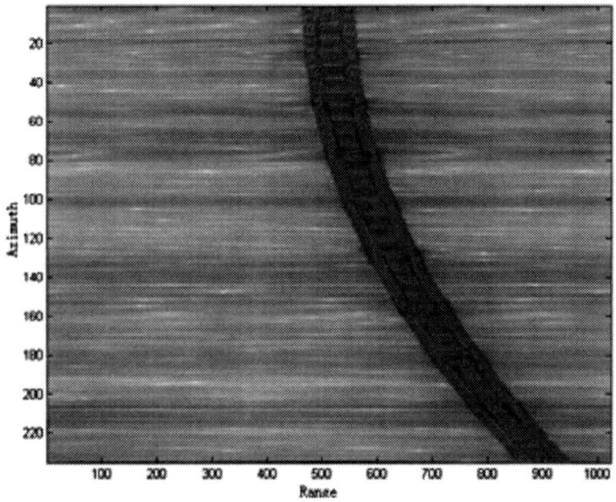

Fig. 3 Coarse compensation and range compression

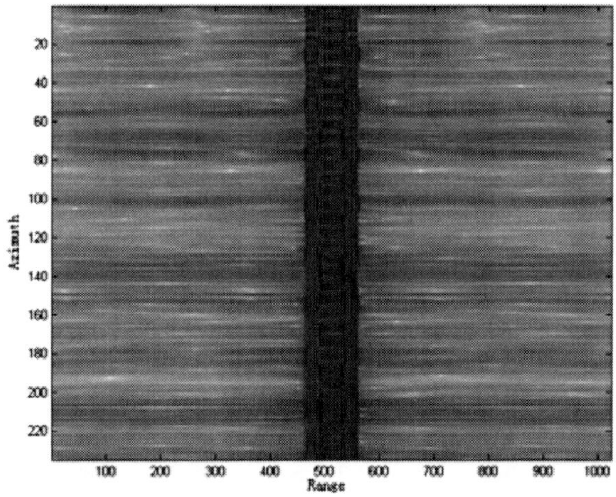

Fig. 4 Accuracy compensation

For the third kind of object, space debris, whose motion is not controlled in free space, partly can get ISAR image depending on smooth motion, but most can't reconstruct clear image because of motion scrambling. So radar scanning and estimating abilities are paid more attention to adapt practical requirements of space debris information. Set up simulation experiment of space debris, while the orbit and attitude motion are reduced by STK, 26 dB Gaussian noise is added to receiving signals, and 50 groups echoes data are received. Range compression result is displayed in Figs. 8 and 9. It can be seen signals are almost merged in noise because of its weakness, so here we use relative accumulation method to strengthen signals' amplitude. The result of relative alignment and accumulation of 50 groups data is shown in Fig. 10. From Fig. 10, the signal of the object can be easily detected, and the width of the echoes signal is the object's size in the direction of range.

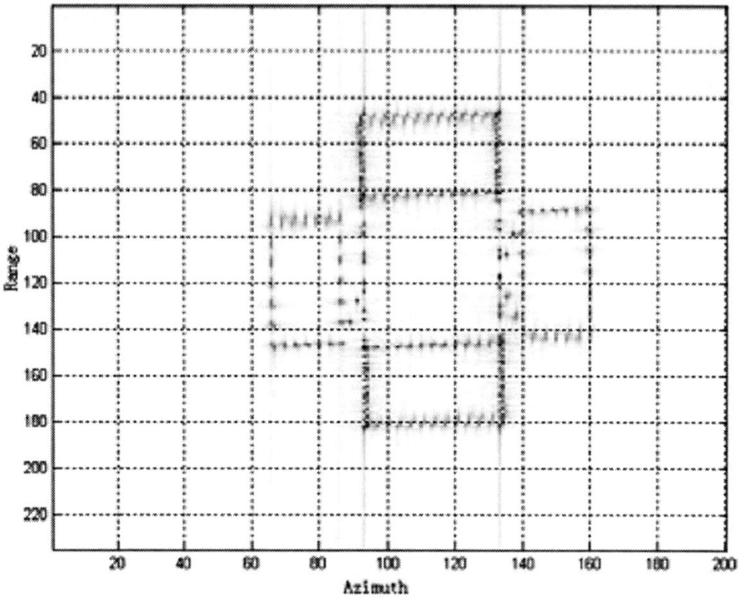

Fig. 5 SISAR reconstructed image of class one

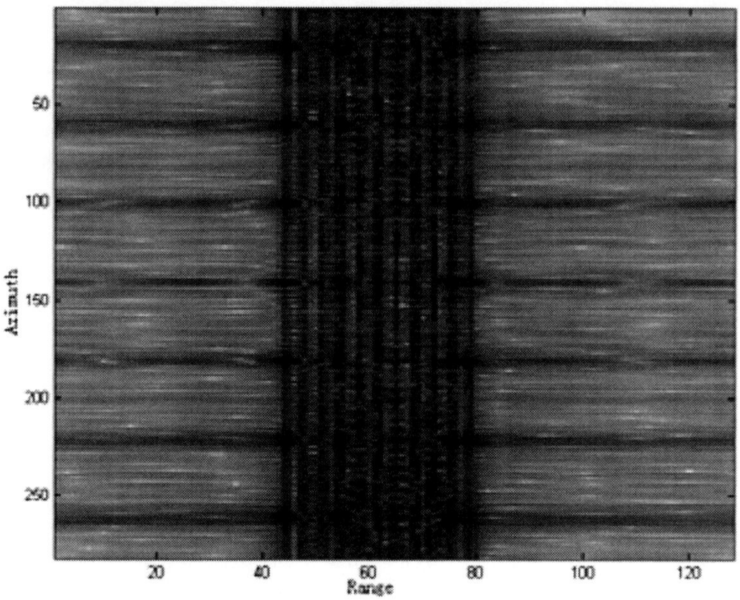

Fig. 6 The result of range compression of spinning satellite

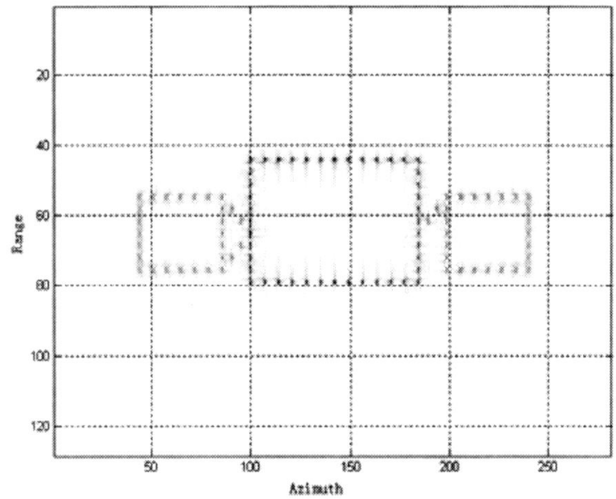

Fig. 7 2D reconstructed ISAR image of spinning satellite

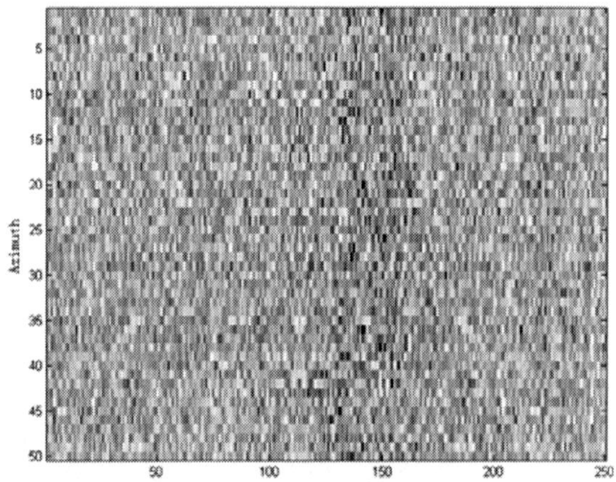

Fig. 8 Range compression result of space debris (S/N:-26 dB)

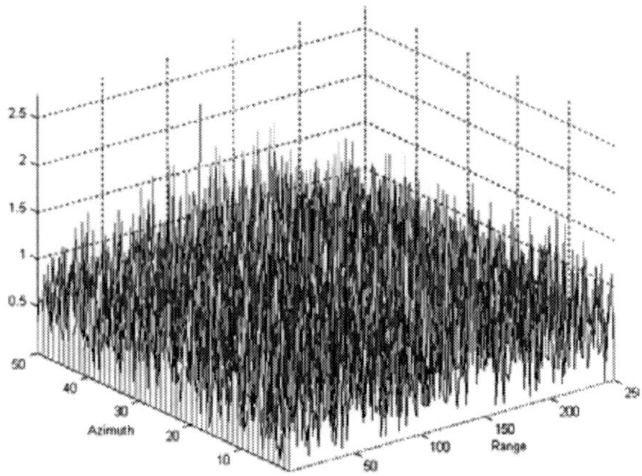

Fig. 9 3D amplitude display of range compression result of space debris (S/N:-26 dB)

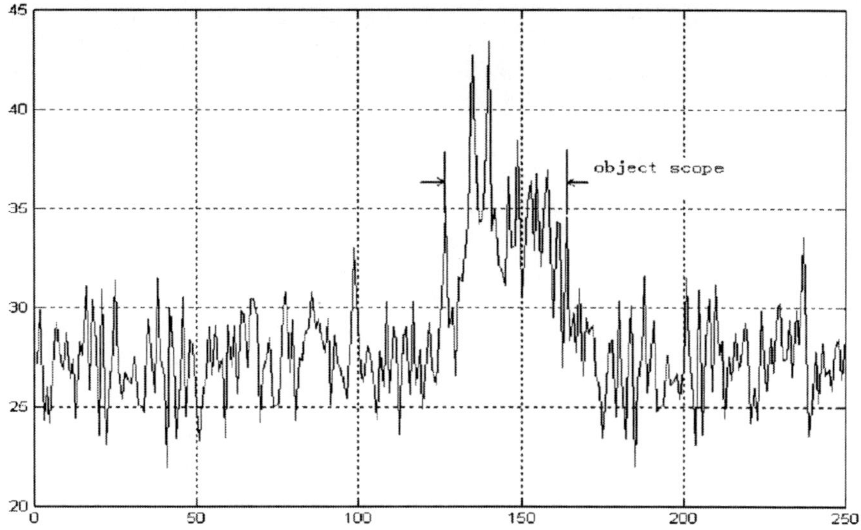

Fig. 10 Accumulation of 50 groups data

4 Conclusion

To observe space object via space-based radar antenna is a novel notion in recent years. In this chapter, we propose a whole task of SISAR plan in both assignment level and processing level, demonstrate SISAR imaging to space object can be carried out based on present techniques and platform situation. As far as three classes of space objects are concerned, models to three-axis-stablized satellite, spin-stablilized satellite, space debris are set up respectively, and echoes data are obtained according to radar and system parameters design as well as the perturbations simulation through STK. SISAR reconstructed images results to different class of space object obtained shows (1) space-to-space observation can be achieved through reasonable parameters setup; (2) motion compensation is one of the most important process to reform radar image.

References

1. J.L. Kretsch. Shuttle radar topography mission overview [J]. Applied Imagery Pattern Recognition Workshop, USA, Proceedings 29th: 276–282, 2000.
2. D. D'Aria, A.M. Guarnieri, F. Rocca. Focusing bistatic synthetic aperture radar using dip move out. IEEE Transactions on GRS, 42(7), 1362–1376, 2004.
3. X. Cao, F. Su, H. Sun, G. Xu. Space Debris Observation via Space Borne ISAR Imaging. The 5th International Conference on Microwave and Millimeter Wave Technology, Guilin, 1406–1410, 2007.
4. H. Breit, W. Knopfle, N. Adam, M. Eineder. SRTM X-SAR DEM of Europe-Results and algorithmic improvements. Geoscience and Remote Sensing Symposium, 2002. IGARSS ′02. 2002 IEEE International. German Aerospace Center, Wessling, Germany, 155–157, 2002.

5. G.D. Arndt, B.A. Bourgeois, J.R. Carl et al. Signal processing requirements for a spaceborne radar for detecting orbit debris. Proceedings of the SPIE, Vol. 2214, 102–110.
6. X. Cao, F. Su, Y. Jiang, G. Xu. Interferometric ISAR 3D Imaging of Target Satellite in Low Earth Orbit. The 7th International Symposium on Test and Measurement (ISTM), Beijing, 1037–1040, 2007.

Increasing the Data Volume Returned from Small Satellites

A. da Silva Curiel, A. Haslehurst, P. Garner, M. Pointer, and A. Cawthorne

Abstract An X-band Antenna Pointing Mechanism has been developed in order to improve the power efficieny of data downlinking on small, low cost Science and Earth Observation satellite missions. This permits the accompanying payload chain to support high speed data storages, processing, and downlinking in X-band. The system is scalable, and can be implemented in redundant or single string configuration. The unique aspect of the downlink chain is the implementation of an antenna pointing mechanism. Such systems have not been used before on small satellites before, but it permit the system designer to either maximise the downlink data throughput, or minimise the power resources on the spacecraft. As such, the small antenna pointing mechanism is an enabling technology in commercialisation of small EO spacecraft missions.

1 Introduction

SSTL develops Earth Observation missions for a wide range of applications, and aims to develop cost effective imaging missions that are suitable for carrying out commercial missions. The SSTL-300 mission is designed to meet demanding high-duty cycle imaging application, whilst maintaining the benefits of a small satellite design which can be delivered on timescales of 18–24 months, on limited budgets up to US$30, and could be batch launched in constellation to provide higher temporal coverage.

In 2006, NASRDA, the Nigerian National Space Research and Development Agency commissioned its first operational high-performance Earth Observation system based on the SSTL-300 platform. The system is a turnkey solution, comprising space segment, ground segment and image processing facilities, together with an extensive training programme.

A. Silva Curiel (✉)
Surrey Satellite Technology Ltd (SSTL), Tycho House, Surrey Space Centre, Surrey Research Park, Guildford, Surrey, GU2 7YE, UK
e-mail: A.da-Silva-Curiel@sstl.co.uk

R. Sandau et al. (eds.), *Small Satellite Missions for Earth Observation*,
DOI 10.1007/978-3-642-03501-2_34, © Springer-Verlag Berlin Heidelberg 2010

NigeriaSat-2 is designed to achieve 2.5 m imagery in a panchromatic waveband along with 5 and 32 m imagery in four mutli-spectral channels. The spacecraft will deliver high data throughput on an agile platform, whilst still maintaining high levels of pointing accuracy. Figure 1, below, shows an illustration of the spacecraft. The spacecraft is designed for 7.25 year lifetime, and is scheduled for launch in 2009.

Fig. 1 The NigeriaSat-2 spacecraft

As illustrated in Fig. 1, the spacecraft is designed around the high resolution instrument, providing a large solar array area, whilst maintaining a stiff structure to support rapid slewing.

The key requirements for the mission were to maximise the data capacity, whilst maintaining a very modest total system cost. The majority of the elements of the spacecraft, including the instruments and payload support chain are developed by SSTL, so that the cost and schedule can be carefully controlled. Early design focused on making the system suitable for commercial operations, and consequently it was necessary to develop ways of maximising the downlink data throughput without significantly affecting the mission cost.

2 Communication System Trades

Small satellites are typically limited in the amount of power that can be devoted to the payload. This becomes a severe limitation in small Earth Observation missions, where a large volume of data is generated, which must then be transmitted to the ground. Many simpler spacecraft systems are limited in their ability to image and downlink simultaneously, to retrieve real-time or stored data during imaging. Simple spacecraft and downlink configurations tend to limit either the data rates achievable, or the timeliness with which imagery in the groundstation vicinity can be returned. SSTL carried out a number of system trades in the SSTL-300 platform design in order to maximise the data throughput. The various configurations considered and choices are discussed further in this section.

The requirement for the SSTL-300 platform is illustrated in Fig. 2. There is a need to image, whilst at the same time downlinking stored or real-time image data. In the worst case, the angle between image target and groundstation could be up to ~100°.

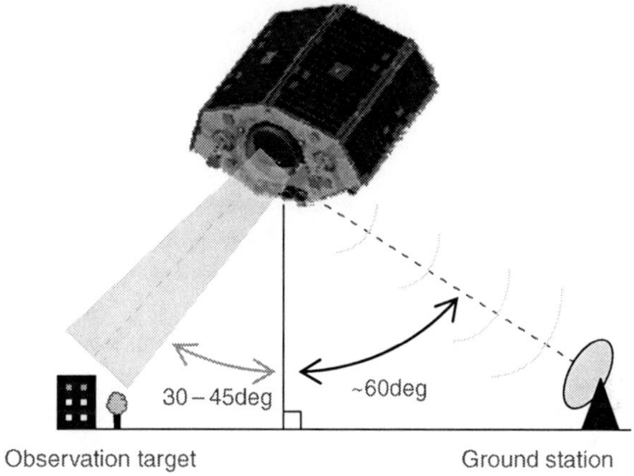

Fig. 2 Image and downlink configuration

An omnidirectional spacecraft antenna is the simplest and cheapest way to support the downlink. It copes well with the needs for an agile mission, but it provides poor gain and drives up the size of the groundstation dish and the cost of the ground segment. In this case, the link budget would be sized for the range with the spacecraft near the local groundstation horizon, but the resulting design will be significantly overdesigned for the case with the spacecraft at zenith.

An isoflux antenna can be used to counteract the change of free-space loss throughout an orbital pass to maintain a constant link margin. The SSTL Beijing-1 and RapidEye platforms includes such isoflux antennas. An Isoflux antenna provides equal power flux density on the ground. Due to its beam shaping, an Isoflux antenna can have gain of up to 2–3 dBi above that of an omni-directional antenna on the horizon. This configuration is well suited to mapping missions which mostly image at nadir, however when the satellite platform is required to downlink and off-point simultaneously, for instance to image areas in the same coverage circle of the groundstation, the isoflux antenna solution is no longer efficient other than under some very limited circumstances.

There are various means of improving the link budget, and these include data compression, more efficient modulation schemes, or improvements in power efficiencies of the on-board communications system. Although each of these provide small factors of improvement, none of these can provide the orders of magnitude of improvement demanded in the SSTL-300 system design. On-board data compression seems attractive, but also requires consideration of efficient use of the limited available power. Improvement in the groundstation dish size can provide significant improvement, but this also rapidly increases the life-cycle costs of the groundstation. To gain larger improvement factors, it is necessary to consider the downlink antenna further.

The spacecraft antenna could be configured as a high gain antenna, to improve the link budget. In this case the RF beam is narrower, and the spacecraft must be slewed to track the groundstation during transit. Again, this configuration can not support simultaneous downlinking and imaging in the groundstation coverage circle. In this case, imagery can only be stored on-board, and retrieved on a subsequent pass, unless a second groundstation is included downrange in the system design.

This situation could be improved by switching between multiple high antennas, but large numbers of antennas would be required in order to make significant improvements to the link budget. It is also possible to equip the instrument with a steering mirror, but this is also difficult and costly to implement in practice.

In considering this dilemma SSTL studied how these constraints have been solved on larger spacecraft systems, and has developed a small satellite Antenna Pointing Mechanism (APM) for use on-board its highly agile satellites. This permits SSTL to steer a high gain downlink antenna to focus the available on-board energy in a tight beam to the groundstation, whilst leaving the spacecraft free to point a larger Earth Observation instrument. The system provides a greater than 10-fold improvement in the communications link, which can be used to increase the date rate, or reduce the amount of power required for downlinking. The cost of using the mechanism is more than offset by the savings that can be made, and has been carefully considered in the trade with an electrically steered antenna. Although a mechanical system requires careful design in order to avoid mechanical disturbances, this is not unlike the problem faced with the on-board reaction wheels. SSTL can build on its mechanical and electrical system heritage with its line of reaction wheels, to minimise the development risks for such a mechanical system. A mechanical system also tends to have fewer implementation losses than an electrically steered system, allows different gain and polarisation antennas to be implemented for different missions, and thus provided a very attractive solution to improving the SSTL-300 downlink capacity.

A summary of the various trades are presented in Table 1, and these solutions are not very attractive for high performance missions. Consequently, steered antennas were considered in more detail. A summary of the various trades for steered antennas are presented in Table 2.

3 Downlink Chain Development

In the SSTL-300 downlink chain, the APM is complemented by a high capacity solid state data recorder, and high rate X-band Transmitter modulator, amplifier and

Table 1 Trades for fixed antenna

Solution	+ Positive points – Negative points
Iso-flux antenna	+ Relatively simple implementation + Constant power flux density – Difficult to implement for an agile mission – Careful beam shaping required –
Omni directional antennas	+ Simple implementation + Broad coverage – Very poor antenna gain – Inefficient for supporting simultaneous imaging and real-time
High gain antenna	+ High gain – Would need to slew spacecraft to track groundstation – Can not support simultaneous imaging and real-time
Key	Non-compliant Compliant Selected

Table 2 Trades for steered antennas

Solution	+ Positive points – Negative points
Switched antennas	+ Next simplest implementation + High gain – Fixed number of positions possible – Need many switched antenna elements before benefit is gained – Switch matrix can become complex
Electrically tracked antenna	+ Can have high gain – Highly complex/costly development – Limited cone of pointing – Poor Axial ratio
Mechanically tracked antenna	+ Can have high gain – Potential mechanical disturbance and wear-out issues
Steered mirror on instrument	+ Can have high gain – Potential mechanical disturbance and wear-out issues
Key	Non-compliant Compliant Selected

antenna to provide a cost effective yet highly capable downlink chain. As illustrated in Fig. 3, the payload is supported through a serial data interface, and can be complemented by a spacecraft GPS receiver to support time synchronisation, and allow timation and location data to be included in the image meta-data.

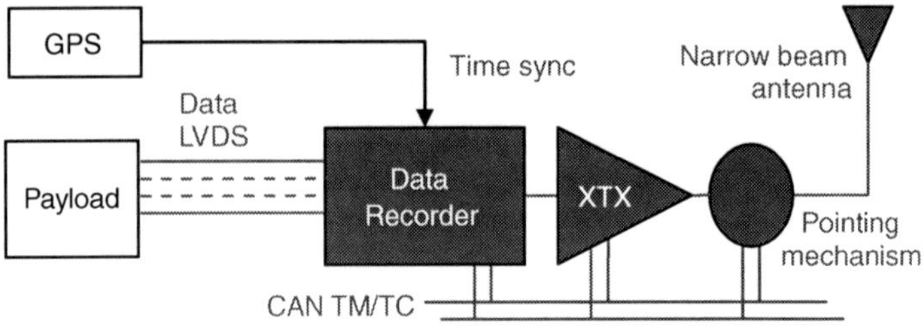

Fig. 3 Payload downlink solution

On larger satellites, where typically more power is available, a high power X-band downlink is often implemented using Travelling Wave Tubes leading to a high downlink power demand. By using the APM, the SSTL-300 can achieve similar downlink rates and data capacities on much smaller spacecraft.

SSTL has flown data recorders and X-band downlinks on several of its previous missions, but for the SSTL-300 the systems needed to be modified to cope with the higher data rates and storage requirements.

4 Data Recorder

The SSTL data recorder employs a PowerPC processor, and solid state memory. The unit was upgraded for the SSTL-300 to provide a more modular design, and a number of serial data inputs and outputs rated at up to 150 Mbps. Through its processor, the data recorder can be used in various onboard data processing tasks, for instance to provide on-board data analysis, compression or encryption. With a large number of parallel inputs and outputs, multiple units can be cross-linked, or multiple payloads can be supported simultaneously.

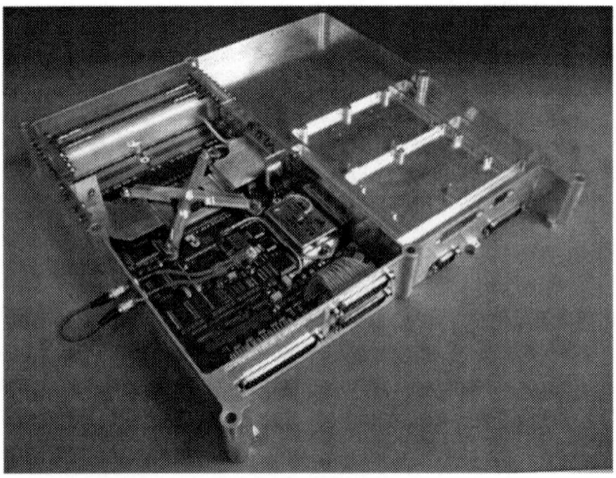

Fig. 4 High speed data recoder (HSDR)

Table 3 HSDR specifications

Parameter	Specification
Capacity	16 GByte
Inputs	20 LVDS @ 150 Mbps 5 SerDes @ 1 Gbps
Outputs	10 LVDS @ 150 Mbps
Mass	1 kg
Volume	300 × 150 × 30 mm
Power	15–50 V
Temperature	–20 to +50°
Vibration	>15 Grms all axis
Radiation Tolerance	10 kRad (>7 year LEO)

5 X-Band Downlink

The X-band unit has heritage from the SAC-C, GALEX, Beijing-1 missions as well as the RapidEye constellation of 5 spacecraft. The unit supports downlink rates up to 300 Mbps, and can be configured to switch between two preset rates.

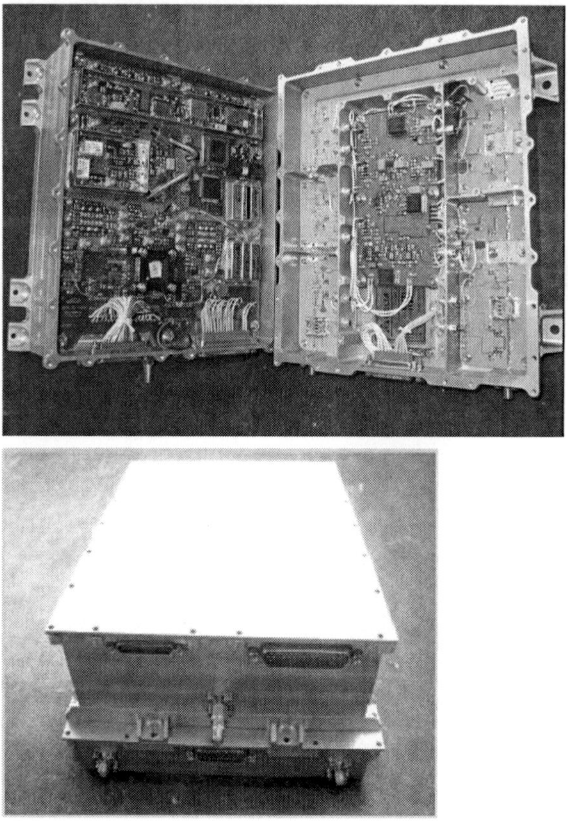

Fig. 5 X-band transmitter (XTX)

Table 4 X-band downlink specifications

Parameter	Specification
Centre frequency	8.0–8.5 GHz
RF power	6 W
Modulation	BPSK or QPSK
Mass	3.25 kg
Volume	200 × 191 × 129 mm
Power	15–50 V, 55 W
Temperature	–20 to +50°
Vibration	>15 Grms all axis
Radiation Tolerance	20 kRad

6 Antenna Pointing Mechanism

The Antenna Pointing Mechanism was developed using largely heritage components and technologies, but required testing to qualify it for a 7 year mission. Although antenna pointing mechanisms are used extensively on geostationary missions, for LEO missions such mechanisms are much rarer, and must cover a much greater angular range.

Several configurations of the two axis mechanism were considered and prototyped, but an X–Y mechanism was eventually selected and developed balancing compactness and simplicity in operation. Although flexible coax cables were tested, the final design employs RF and power connections internal to the two hinges. A small modular X-band horn antenna was designed, which can be configured for left-hand or right-hand circular polarisation.

A qualification unit was developed, which was then tested extensively. The unit was vibration tested using random and sine profiles, and thermally tested in a vacuum chamber. The unit was also life tested in a thermal vacuum chamber, simulating sufficient cycles for nearly 20 years of on-orbit operation. RF tests were performed before and after these tests to ensure there were no changes in performance.

In order to provide a simple interface, the driver electronics are integrated in the unit, and the TM/TC interface is via CAN. A microcontroller and memory are included, so that a predefined track, containing time, elevation and azimuth angles can be uploaded to the unit prior to use.

7 Conclusions

The design of a payload support solution for the SSTL-300 platform is described, which employs a unique element for a small Earth Observation satellite that enables improvements in the link budgets that can be sustained by an order of magnitude. It is shown how such a chain can provide a far superior solution in terms of mass, power and cost compared with a more traditional approach. Indeed the APM is an enabling technology for small satellites, which permits them to achieve downlink

Fig. 6 Antenna pointing mechanism (APM)

capacities and downlink rates previously only available on larger and more expensive systems. This translates to a greater data return for the same sized spacecraft, or operation through a significantly smaller ground station.

Table 5 Antenna pointing mechanism specifications

Parameter	Specification
Articulation	2-axis ±110° elevation ±270° Azimuth
Pointing	0.024° steps Up to 19°/s rate 0.72° accuracy
Gain	15 dBiC
Beamwidth	25°
Polarisation	LHCP or RHCP
Power	3.4 W
Mass	2.7 kg
Volume	240 × 196 × 185 mm

The design trades leading to the development of this gimballed antenna this have been described. The APM design and its qualification for its first flight are described, and the downlink chain will be flown on the N2 mission in 2009.

Further Reading

1. A. Cawthorne, M. Beard, A. Carrel, G. Richardson. Launching 2009: The NigeriaSat-2 mission – High-performance Earth observation with a small satellite.
2. SSC08-III-7, 22nd Annual AIAA-USU Small satellite conference, Logan, Ut, August 2008.

Integrated Design Based Plug-and-Play Small SAR Satellite Project

Zhang Jinxiu, Cao Xibin, Lan Shengchang, and Zhao Dan

Abstract RCST-HIT (Research Centre of Satellite Technology, Harbin Institute of Technology) is now in the process of designing a new Plug-and-Play (PnP) satellite, carrying X-band, HH polarization spaceborn Synthetic Aperture Radar (SAR), offering the 1 meter resolution image in ScanSAR mode, less than 300 km sun-synchronism orbit. The satellite contains a light weight high accuracy Paraboloid deployable reflector antenna. The SAR component electronic beam steering capability is achieved by using a feed array in the focal plane. This project will make the small SAR satellite advance the modular satellite with open standards and interfaces, self describing components, and an auto-configuring system. And the integration design based Plug-and play is simplified and testing tasks can be automated.

1 Introduction

It is becoming increasingly important that satellite be developed in direct support of responsive missions [1, 2], which indicated that satellite need to be rapidly constructed and tested from standard sub-systems based on varying mission needs, and/or integrated rapidly with payloads.

A new scientific space project consisting of a satellite carrying SAR (X-band, HH polarization) which adopted the PC-based concept of Plug-and-Play (PnP) [3–5], offering the 1 meter resolution image in ScanSAR mode, less than 300 km sun-synchronism orbit. This project will make the small SAR satellite advance the modular satellite with open standards and interfaces, self describing components, and an auto-configuring system. And the integration design based Plug-and play is simplified and testing tasks can be automated.

J. Zhang (✉)
Research Centre of Satellite Technology, Harbin Institute of Technology, Harbin, China
e-mail: jinxiu@hit.edu.cn

R. Sandau et al. (eds.), *Small Satellite Missions for Earth Observation*,
DOI 10.1007/978-3-642-03501-2_35, © Springer-Verlag Berlin Heidelberg 2010

The well integrated electronic system based on "system on chip" is used as the on-board Task Management and guidance, navigation and control (GNC) computer and TT&C (non-antenna). In order to improve the satellite integration level and rapidly test performance, the Plug and Play Payloads have two parts function, one is to achieve the original observation data form target areas, the other is to realize the X-band data downlink to the ground station, which is traditionally in the charge of satellite platform. The power sub-system consists of three gallium arsenide solar array and Li-ion battery. And the Integrated Design Based Plug-and-play small SAR satellite has the other advantage of Plug and Play LV Interfaces and Plug and Play Propulsion.

This chapter is organized into four main sections. In Sect. 2, the planning of small SAR satellite is introduced. The well integrated electronic system is presented in Sect. 3. The Plug and Play Payloads is described in Sect. 4.

2 Planning for Small SAR Satellite

The satellite features a low-mass design and low moments to provide a maximum of pointing agility. Combining high maneuverability with electronic antenna beam steering, satellite offers high-resolution coverage of large areas. The satellite is 3-axis stabilized, using a newly developed plug-and-play bus, with a launch mass of 250 kg (including a payload mass of 95 kg). The design keeps the bus and payload well separated so that changes due to growth potential in either one element will have a minimum effect on the other element. Electrical power of 550 W (EOL) is provided by three-junction GaInP2/GaAs/Ge solar panels.

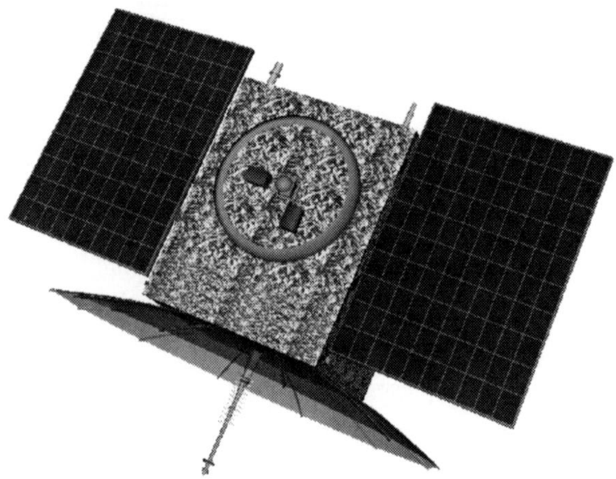

Fig. 1 Satellite configuration primary plan

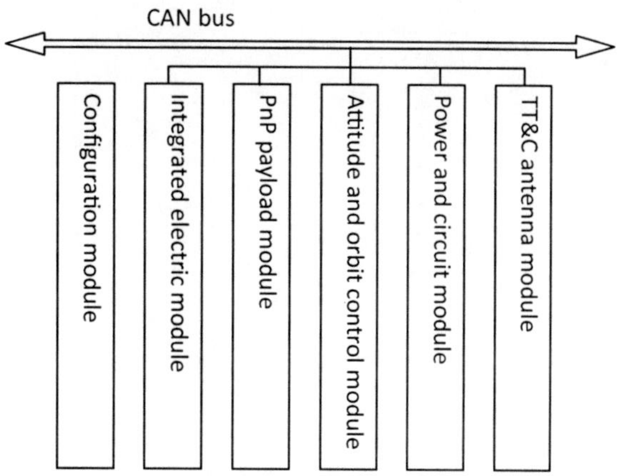

Fig. 2 Satellite module based on CAN bus

3 Integrated Electronic System

The well integrated electronic system based on "system on chip" is used as the on-board Task Management and guidance, navigation and control (GNC) computer and TT&C (non-antenna). The key function are as follows: (1) Determination and control of attitude and orbit; (2) Data collection and information processing; (3) Task management and real-time monitoring; (4) Comm-link for satellite-ground and inter-satellite.

It consists of processor unit, configuration unit and RF unit, in Fig. 3. It is possible to meet with the conversion among multi-requirements and hardware assembler of software module, based on processor unit with FPGA. The configuration unit is composed of micro-controller for FPGA monitor, reconfiguration and upgrade, configuration memory. According to various requirements, the configuration unit can locate the essential software module into processor unit, with fault monitor and handling. RF unit is used to achieve the satellite software upgrade for task management and control arithmetic on-board, and TT&C.

4 Plug and Play Payload

In order to improve the satellite integration level and rapidly test performance, the Plug and Play Payloads have two parts function, one is to achieve the original observation data from target areas, and the other is to realize the X-band data downlink to the ground station, which is traditionally in the charge of satellite platform. The antenna configuration principle is shown in Fig. 4.

The double-mode payload employs electronic beam steering, which can be operated in various observation modes.

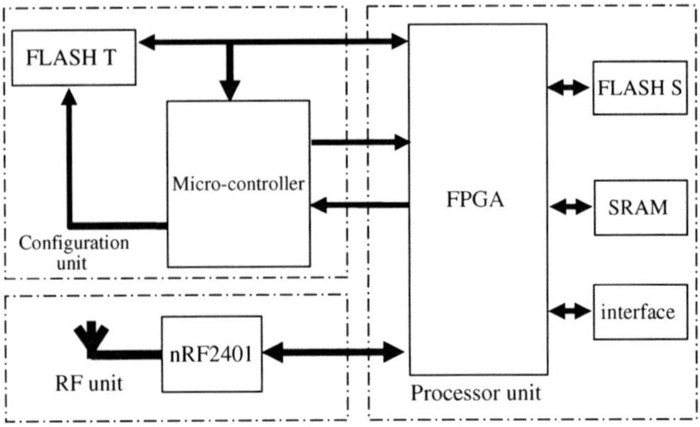

Fig. 3 Structure of integrated electronic system

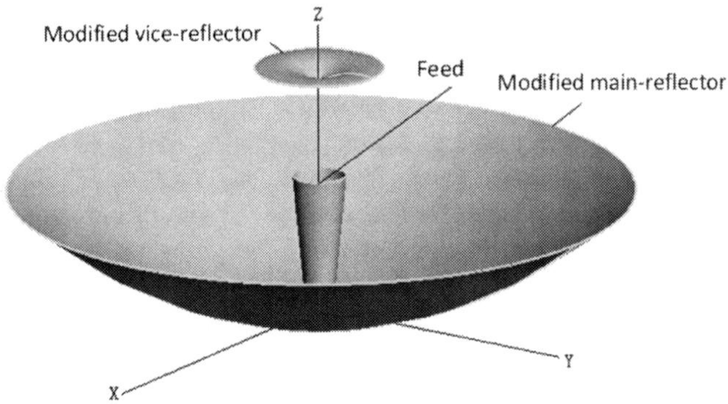

Fig. 4 Antenna configuration principle

Table 1 Parameters of ScanSAR mode

SAR mode	ScanSAR
Spatial resolution	1 × 1 m
Imaging breadth	7.5 km
Max imaging	200 km

- ◇ ScanSAR mode: The coverage of large strips is achieved by electronic beam steering. The nominal coverage mode has only one beam in the target area. In principle, the swath width may get to more than 15 km for some incidence angles. However, this reduces the ground resolution to about 2 m.
- ◇ Spotlight mode: This focuses on a specific, pre-assigned target. In spotlight, the satellite performs mechanical steering to halt the antenna footprint in

a specific target area. The longer integration time over the spot target area yields an improved azimuth resolution. The range resolution is achieved in adjusting the bandwidth to the incidence angle.

5 Conclusion

In a word, a new approach for implementation of X-band SAR satellite for responsive missions is described in this chapter. It is based on the use of plug-and-play and integrated design technology. Though further studies are needed, preliminary results on the concept are very promising, and this approach to plug-and-play and integrated design technology implementation is likely to find applications also in other small satellite for responsive missions.

References

1. J. Lyke, S. Cannon, D. Fronterhouse, D. Lanza, T. Byers. A Plug-and-play System for Spacecraft Components Based on the USB Standard, Proceedings of the 19th Annual AIAA/USU Conference on Small Satellites, Logan, UT, 8–11 August, 2005.
2. Col. (Ret.) Yoram Ilan Lipovsky, Tal Inbar. Responsive Tactical Space Using Micro Satellites and Ariel Launching: The Perspective of a Small Nation. RS4-2006-3002. 4th Responsive Space Conference, Los Angeles, CA, 24–27, April 2006.
3. Y. Sharay, U. Naftaly, U. TecSAR. Design considerations and programme status. IEE Proceedings: Radar, Sonar and Navigation, Vol. 153, no. 2, pp. 117–121, 2006.
4. U. Naftaly, R. Levy-Nathansohn. overview of the TECSAR satellite hardware and mosaic mode. IEEE Geoscience and Remote Sensing Letters, Vol. 5, no. 3, pp. 423–426, July 2008.
5. Don Fronterhouse. J. Lyke. Steve Achramowicz. Plug-and-play Satellite (PnPSat). AIAA 2007-2914. AIAA Infotech@Aerospace 2007 Conference and Exhibition, Rohnert Park, CA, 7–10, May 2007.

Part VI
Attitude Control Systems

Star Sensor Development Based on the TUBSAT Experience

M. Buhl and U. Renner

Abstract Already the first satellite of the Technical University Berlin, TUBSAT-A (launched in 1991) contained a University built CCD star sensor with a unique technique to reduce the sensor data by three orders of magnitude. In the following 15 years of operation this technique proved to be very successful. On the other hand, TUBSAT-A was one of the first satellites to discover and investigate the impact of proton radiation that is responsible for the hotspot phenomenon.

Based on this extensive experience a simple hot spot filter was conceived and successfully tested on MAROC-TUBSAT that was launched in 2001 and is still operational.

A star sensor of the third generation is scheduled to fly on the Indonesian missions LAPAN-A2 and AMSAT-ORARI in the beginning of 2010. It excels, based on the on board experience of the previous designs by its compact size (one unit of 700 g), high signal to noise ratio even in the presence of hotspots and robustness in star extraction and identification.

The chapter will present the results of the in-orbit experiments with TUBSAT-A and MAROC-TUBSAT and discuss the design principles and improvements that lead to the design of the latest model.

1 In-Orbit Experiments with TUBSAT-A and MAROC-TUBSAT

1.1 TUBSAT-A

TUBSAT-A (Fig. 1) was a pioneer in many areas of micro satellite technology including the operation of a CCD based star sensor [1]. Launched July 17, 1991 with an Ariane-4 rocket it was operated for 16 years (last contact was made in 2007). The primary mission was Store & Forward communication and many experiments,

M. Buhl (✉)
Institute of Aeronautics and Astronautics, Technical University of Berlin, Marchstraße 12, D-10587, Berlin, Germany
e-mail: matthias.buhl@ilr.tu-berlin.de

R. Sandau et al. (eds.), *Small Satellite Missions for Earth Observation*,
DOI 10.1007/978-3-642-03501-2_36, © Springer-Verlag Berlin Heidelberg 2010

including animal tracking and polar expedition support, have been carried out successfully. One of the technology demonstration payloads was the new CCD based star sensor.

Fig. 1 TUBSAT-A

The star sensor provides a Top-10 list containing the coordinates of the 10 brightest objects in the image. In Fig. 2 we present 15 successive Top-10 lists of the star sensor whereas the attitude is determined from the first one. Since the data is taken all 0.5 s one can also determine the angular velocities. In this case it is given by

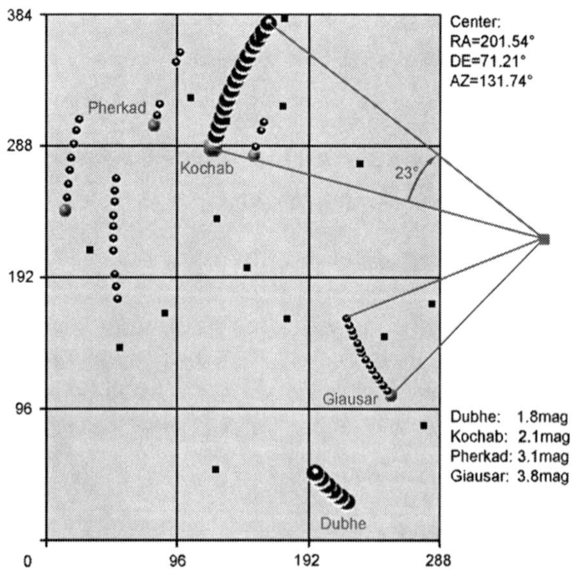

Fig. 2 Star sensor image TUBSAT-A (1991)

23°/7 s. One can also observe static objects which are easily identified as hotspots on the image sensor.

The algorithm to extract the star positions from the image is quite simple but strongly effective. From each sensor line only the brightest pixel is taken to create a map of possible star candidates. The most important fact is that the data is reduced by a factor of the number of rows, in this case 288. It is a very fast data reduction algorithm. The resulting map was named Panorama Image. That algorithm allows rapid update rates. On TUBSAT-A the Panorama Image is now sorted by brightness and the 10 brightest pixels are extracted. These are not necessarily stars since there are other bright objects on the sky (planets, moon) and on the image sensor itself (hotspots). All pixels are now cross-checked against a star catalogue to identify the stars. The attitude of the satellite is then calculated from the image position and the unit vector in the inertial frame of reference of the identified stars.

TUBSAT-A proved the working of the CCD star sensor and its Panorama Image algorithm but also discovered some problems. The most serious problem is the occurrence of hotspots due the high radiation levels in LEO (mainly high energy protons). Hotspots are pixels that suffered radiation damage resulting in a significant higher dark current and are permanent ("hard error"). Bright hotspots can shadow all pixel of a line (and therefore stars) when looking for the brightest pixel of a line. Over time this can result in a reduced data set (Panorama Image) that contains more and more hotspots and the algorithm fails.

1.2 MAROC-TUBSAT

For following missions a solution for the hotspot problem was implemented. An example is MAROC-TUBSAT (Fig. 3) launched in Dec. 10, 2001 [2]. The primary mission apart from technology transfer and a hands-on training program for Moroccan engineers was earth observation (Fig. 4).

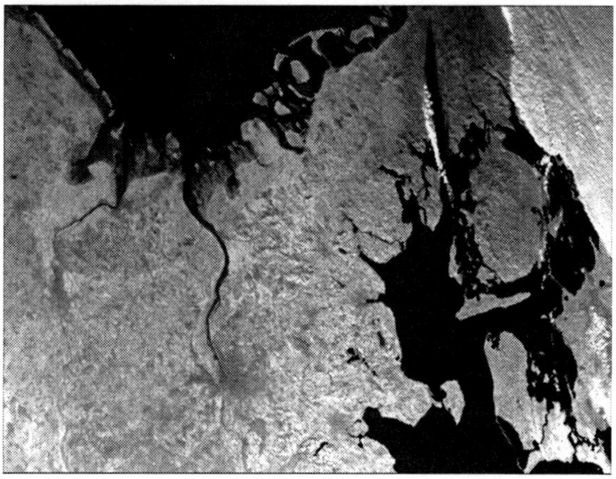

Fig. 3 Northern Germany

Fig. 4 MAROC-TUBSAT

The autonomous star sensor (Fig. 5) on board was an improved design of the TUBSAT-A device. Crucial to the hotspot problem is the observation that hotspots are static objects. They obviously have a fixed position and a stable brightness for a given exposure time and stable sensor temperature. The simple solution taken was the use of a reference image. The reference image is just a normal star sensor image taken in orbit (close to operation). If this image is subtracted from the actual star

Fig. 5 MAROC-TUBSAT star sensor

sensor images all hotspots will cancel since they are static objects. Stars on the other hand will in general not cancel since they differ in image position all the time due to rotations of the spacecraft. This procedure was tested extensively with the MAROC-TUBSAT. A weekly update of the reference image was quite sufficient to operate the star sensor for several years.

2 TUBSAT Star Sensor Design Principle

2.1 Hotspots and the Difference Image

There are several crucial problems or requirements to solve when designing a star sensor. The first one is the creation of hotspots on the image sensor due to the high radiation levels in orbit. An algorithm to exclude hotspots from the image data has to be implemented. TUBSAT missions proved that working with a difference image is sufficient to operate a star sensor with a commercial CCD sensor for more than 5 years. A standard sensor image is saved as reference image and subtracted from all the following images. Star extraction is done on the resulting difference images. The reference image has to be updated regularly (the attitude of the spacecraft has to be changed slightly afterwards so that stars not cancel after the update). Disadvantage is that the noise is increased by a factor of $\sqrt{2}$. The advantage compared to other solutions that filter hotspot pixel from the original image (ex. by means of lists) is that the pixel can still be used as long as it is not saturated by the increased dark current. Therefore the sensor is only blinded by strong hotspots that saturate the pixel and these are very seldom compared to weak radiation damages.

2.2 Star Extraction, Accuracy and the Panorama Image

To increase the accuracy of star sensors many designers are often defocusing the lens (Fig. 6). The star is spread over a large pixel area. By calculating the brightness centre by determining all pixels belonging to the star, the star position can be determined with an accuracy of less than half a pixel size depending on the degree

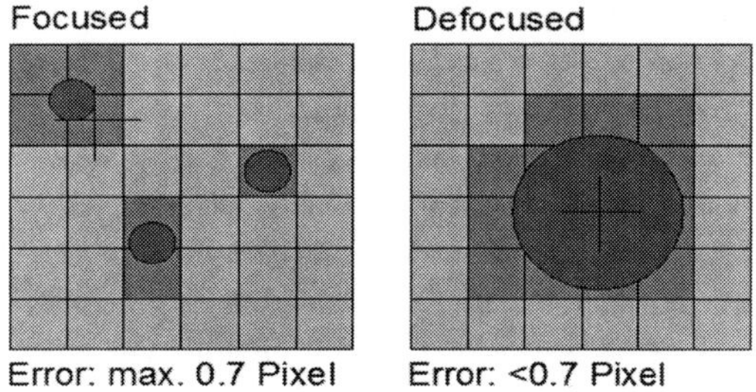

Fig. 6 Focus strategy

of defocusing. The disadvantage in computation is that the complete image must be processed to collect all pixels of the star. Further disadvantage is that the star light energy is spread over many pixels decreasing the signal to noise ratio. A very fast lens (small f-number) is necessary that comes with its own problems.

Because of these disadvantages TUBSAT derived star sensors are using a focused lens. To increase the accuracy small pixel are used (ex. 6–7 μm). The maximum error is 0.7 pixel size. The Panorama Image algorithm is used to significantly reduce the image data with a simple method. The extract of a Panorama Image of the Berlin night sky in Table 1 shows a bright star sitting in row 722 and 723 that can easily be identified against the noise background of about 100 digital units. The compression factor depends on the sensor size and can be enormous, ex. a factor of 1,000 for the latest model. The simplest algorithm to extract stars from a Panorama Image is just to take the 30 brightest pixel but more elaborate procedures can be used (ex. to remove pixel belonging to the same star).

Table 1 Panorama image

Row	Col	BV
718	757	99
719	323	89
720	960	114
721	472	106
722	1,286	1,110
723	1,285	2,275
724	185	211
725	605	108

2.3 Computation Speed, Software and the Star Catalogue

A very important item is the software design. As with the hotspot problem and star extraction procedure simple, small solutions and algorithms are developed and implemented. The initial star pattern recognition (or star identification) is based on standard techniques. The angular distances of stars on the image are compared with angular distances of the catalogue stars to identify all objects on the image. The algorithm is very robust allowing other objects than stars (ex. planets) to be present without failing. After the attitude is found the star sensor switches into a special self designed tracking mode allowing attitude update rates of up to 10 Hz.

In the tracking mode (Fig. 7) developed a star from the new image is taken to find a corresponding identified star in the old image (here ID1) within a given search radius. Since the star is now identified all neighbours can be taken from the catalogue. The theoretical star positions of these neighbours are calculated (on the image) with help of the old attitude. The next step is to find pairs of these neighbours and stars of the new image to identify them. This algorithm ensures identification all possible stars in the new image (just finding pairs of stars from the new and old

Fig. 7 Tracking mode

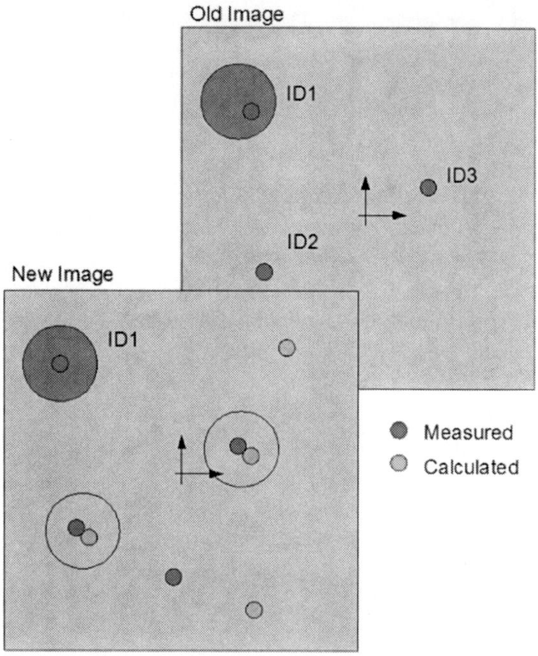

image directly would not be enough). After identification the new attitude is not recalculated but gained by a correction of the old attitude with a least square fit (much faster).

Special attention is given to the star catalogue. A big catalogue takes a lot of time to be searched for matching angular distances so a minimum catalogue is required to reduce hardware requirements. The star catalogue is based on the brightest stars of the sky. The final star selection depends on the lens used (in particular the resulting FOV and sensitivity). The general construction of the catalogue is always the same and involves covering the complete celestial sphere with identical cells. From each cell only the brightest star is taken into the catalogue. The size of the cells depends again on the lens and sensor used. The final catalogue (Fig. 8) contains only the

1st Star: Unit Vector	1st Neighbour: ID + Distance	...	Last Neighbour: ID + Distance
...	...	...	...
Last Star: Unit Vector	1st Neighbour: ID + Distance	...	Last Neighbour: ID + Distance

Star Catalogue File with min. Informations (No brightness)

Fig. 8 Star catalogue file

unit vector of the star (12 bytes) and its neighbours within a given angular distance whereas only the catalogue number (2 bytes) and the angular distance (2 bytes) of the neighbours is saved.

3 Latest Generation: STS-500

3.1 Design Requirements

The development of the latest generation of star sensors, the STS-500 system, was driven by the following criteria:

- Suitable for micro satellite missions
- Short development time and cost efficient
- Include TUBSAT experience
- Reliability by simplicity
- Update rate of 10 Hz
- Pointing accuracy: better than 30″
- Lifetime of 2 years

The designrequirements are partly driven by the TUBSAT mission requirements (earth observation with 6 m pixel ground resolution) as well as by current star sensor costumer needs.

3.2 Hardware Design

The most important step in actually designing the star sensor was to base the optical head on an existing industrial camera. A camera from the German company Kappa opto-electronics GmbH specialized in camera design and image processing was chosen (Fig. 9). The TUBSAT program and Kappa are cooperating for a long time and CCD cameras have also been successfully flown in the DLR-TUBSAT

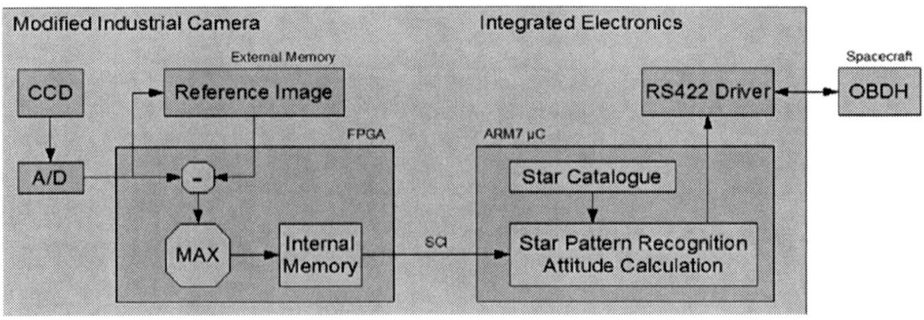

Fig. 9 STS-500 hardware

and LAPAN-TUBSAT [3] missions as primary payloads. Without even changing the hardware important algorithms could be integrated into the device. A detailed block diagram is shown in Fig. 10.

Fig. 10 Kappa camera

The difference image, Panorama Image and star extraction algorithms could be implemented on the existing FPGA (that before only controlled the CCD and A/D operation). The resulting Panorama Image or a suitable star list is then transmitted to an ARM7 micro controller. The micro controller (48 MHz, 512 kB Flash, 32 kB RAM) alone is sufficient to provide the star catalogue and attitude determination with 10 Hz using new developed optimized software and a small catalogue. The modified camera, the micro controller as well as the communication drivers for the spacecraft are integrated into one device.

The system is highly adaptable by using different lenses for different mission accuracy requirements. The STS-500 will be space qualified in two versions using either a 16 mm lens or a 50 mm lens. Each version will fly on the Indonesian LAPAN-A2 and AMSAT-ORARI missions. Every sensor comes with a diffuse reflection baffle.

3.3 STS-500 Performances

The STS-500-16 (see Fig. 11) uses a lens with a focal length of 16 mm and an f-number of 1.6. Together with the pixel size of 6.45 μm the pixel resolution is about $80''$. The large field of view (>30°) allows a robust star pattern recognition since many bright stars will be visible in average. The disadvantage is the low pixel resolution resulting into a lower pointing accuracy of about $30''$ (increased by averaging over a large number of stars).

The STS-500-50 uses a lens with a focal length of 50 mm. The advantage is the higher pixel resolution resulting into an accuracy of about $10''$. The second advantage of the smaller field of view is the possibility of small exclusion angles for earth/moon by use of a suitable baffle.

Fig. 11 Star sensor STS-500-16

Both lenses need good focusing properties but nothing special. In particular no brightness calibration is necessary since the pattern determination does not depend on the brightness of the stars. To see 15–30 stars for averaging and increased accuracy stars up to magnitude 6 are visible (result of high quality lens and low noise CCD sensor).

The roll angle accuracy does not depend on the focal length but on the sensor size since the distance between two stars is used to calculate the roll angle. It is therefore the same for both versions (see Table 2).

Table 2 STS-500 performance

	STS-500-16	STS-500-50
Power	9–18 VDC (3 W)	9–18 VDC (3 W)
Mass	700 g	750 g
Dimensions (inc. baffle)	$160 \times 60 \times 60$ mm^3	$180 \times 60 \times 60$ mm^3
Pitch/Yaw accuracy	30″	10″
Roll accuracy	200″	200″

The STS-500 star sensor will be part of the Indonesian LAPAN-A2 and AMSAT-ORARI mission successor to the LAPAN-TUBSAT satellite [3]. Each satellite will fly a STS-500-16 and a STS-500-50 device. The launch is scheduled for 2010. The missions will also be used to test several applications. The first application is a high precision mission using two STS-500-50 sensors mounted 90° to each other (see Fig. 12). This configuration can be used to replace the roll angle inaccuracy by the other star sensor pitch and yaw angle measurement.

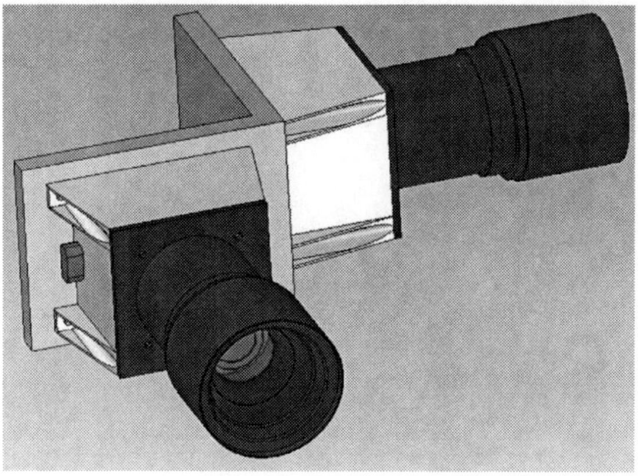

Fig. 12 High precision application

The second application is a momentum bias mission (3-axes stabilized, good precision) with a minimal attitude control system consisting of one momentum wheel, one star sensor and coils. The optical gyro mode that is also implemented on the STS-500 can be used for nutation movement determination and damping as well as for external torque response measurements of the spacecraft.

4 Summary

The TUBSAT program triggered the successful development of star sensors. Basic concepts like the difference image as solution to the hotspot problem and the Panorama Image have been developed and successfully tested in many TUBSAT missions. The latest generation of star sensors, the STS-500, is based on a modified industrial camera and will be space qualified in 2010. Restricting to a small optimized software code and a small catalogue proved to be valuable for cutting costs and development time by simultaneously increasing the reliability.

References

1. U. Renner. Flight results of TUBSAT-A, 42nd Congress of the International Astronautical Federation, Montreal, Canada, Oct. 05–11, 1991.

2. S. Roemer, U. Renner. Flight Experience with the Micro Satellite MAROC-TUBSAT, 54th International Astronautical Congress of the International Astronautical Federation, Bremen, Germany, Sep. 29–3 Oct., 2003.
3. U. Renner, M.Buhl, High Precision Interactive Earth Observation With LAPAN-TUBSAT, Proceedings of the 4S Symposium Small Satellites, Systems and Services, Rhodes, Greece, May 26–30, 2008.

Small Sensors Big Choices

J. Leijtens and C. W. de Boom

Abstract Accurate attitude control is as important for a small satellite as it is for a big one. TNO has been engaged in micro sunsensor developments for a couple of years now and has reached a situation where some critical decisions will have to be taken. Several options are still open but as research resources are limited some possible solutions will have to be discarded or further investigations postponed shortly. This chapter describes the current status as well as possible developments for the future which would lead to sensors that are small and reliable enough to provide a grown up attitude sensing solution for even the smallest satellites.

1 Introduction

TNO Science and Industry (formerly known as TNO TPD) has a long and successful history in the development and production of optical sensors, particularly sun sensors, for attitude measurement in AOCS subsystems in a variety of space missions. Contracts acquired for constellations like Galileosat and Globalstar-2 and requiring the production of larger series of sun sensors have prompted the cooperation between TNO and the Dutch SME company Bradford Engineering. While Bradford engineering will focus on the production of "standard" flight sensor units, TNO will mainly focus on special developments like miniaturization and special applications.

TNO has been engaged in micro sunsensor developments since 2004 through a nationally sponsored research program called MicroNed. This program has lead to significant insights related to appropriate manufacturing technologies for sensor miniaturization. Key to these miniaturization efforts is retaining the reliability present in the current generation sun sensors and trying to achieve at least the same level of accuracies. At this moment, several options are still under consideration that could lead to significantly smaller and even more reliable sunsensor products.

J. Leijtens (✉)
TNO Science and Industry, Stieltjesweg 1, 2628 CK Delft, The Netherlands
e-mail: info@tno.nl

R. Sandau et al. (eds.), *Small Satellite Missions for Earth Observation*,
DOI 10.1007/978-3-642-03501-2_37, © Springer-Verlag Berlin Heidelberg 2010

- Micro digital sunsensor on chip (this development should lead to albedo insensitive sunsensors that are both small, low power and accurate).
- Autonomous wireless sunsensors (the concept of full wireless operation can save extra weight and increase the flexibility in application as the sensors need not be verified extensively for each mission separately).
- Immersed sunsensor technology (Allows for wafer scale manufacturing and therefore has the characteristics needed to achieve a significant cost reduction).
- Albedo insensitive analogue (Uses a standard analog sunsensor which measures in a specific wavelength region and therefore allows the construction of far less complicated sensor which could lead to a solution that has a (in the order of 3 orders of magnitude) lower power consumption).

Each of these issues would require further research and qualification efforts before a fully qualified solution is available, but choices will need to be made as the available financial budget is limited.

2 Miniaturisation: A First Step

TNO already offers a mini sunsensor which basically is a fully qualified sunsensor with its own mechanical and electrical (micro-connector) interfaces (Fig. 1). This sensor is an example of a stand-alone analogue fine sun sensor module. This unit is currently offered as a commercial off the shelf mini-Fine Sun Sensor and combines a very high reliability with an attractive price.

Fig. 1 Mini-FSS (commercial off the shelf sunsensor)

Further miniaturization of this sensor is sought in applying immersed sunsensor technology. This technology implies that aperture masking and detector chip are directly adhered to a sheet of glass in between. To verify the capabilities of this technology a technology demonstrator has been build which uses a PSD sensor (Fig. 2). This sensor has exhibited some solderability problems but did show good linearity in the output versus incident sun angle (Fig. 3).

The graphs presented show the output signals for an –X to +X scan, a scan along the 45 degree axis and a –Y to +Y scan.

As can be appreciated from Fig. 4, the size of the prototype is mainly determined by the size of the mechanical fixture and the area used for wiring the sensor. In actual

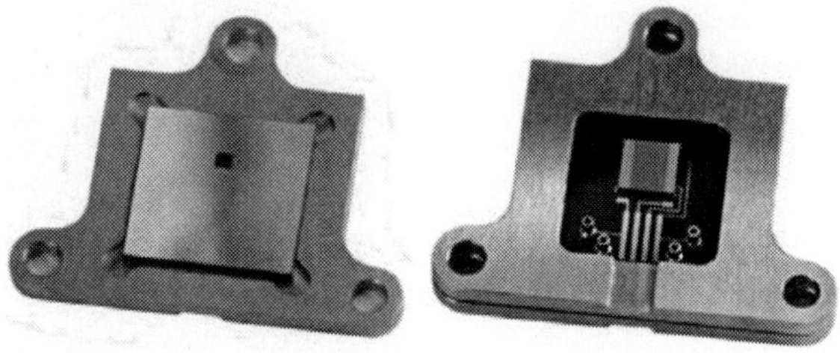

Fig. 2 Immersed sunsensor (ISS): first proto

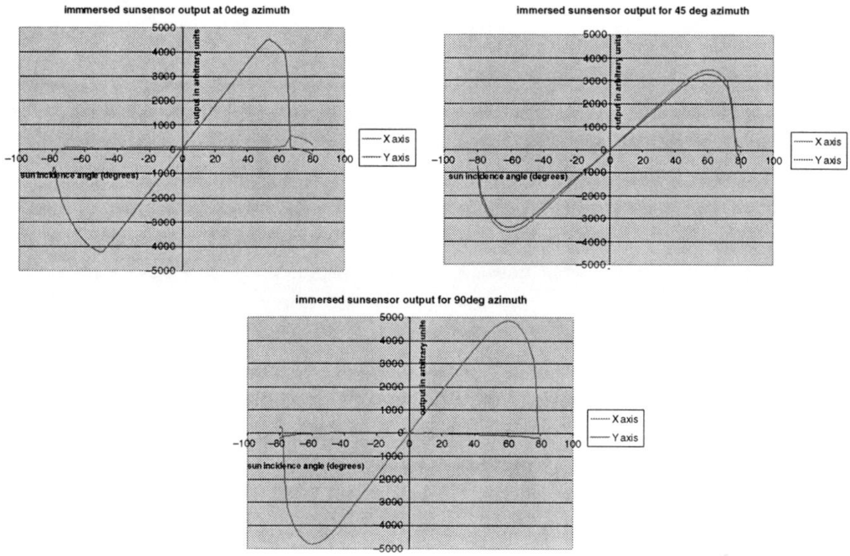

Fig. 3 Immersed sunsensor linearity measurements

fact the sensor core is significantly smaller. Future versions can probably have only about a quarter of the size of the depicted sensor, which will lead to a sunsensor core which is in the order of a couple of cubic millimeters. This type of sensor will be a small and very robust sensing element that can be used in very demanding environments.

Since a miniature sensor is a pre-requisite for, but doesn't necessarily lead to, a small sensor system, the miniaturization of the peripheral circuitry and the packaging, is an important step in the process of obtaining small and affordable sensor systems. Through a flight on board of the Delfi-C3 nano satellite, TNO has proven that systems can be build that don't need any wiring and use a dedicated (system integrated) solarcell for powering and a wireless link as the data interface (Fig. 4). As the AWSS uses a conventional sunsensor core, the current system height could potentially be reduced by an order of magnitude (from 3 cm to 3 mm) in case the wireless autonomous operation is combined with the immersed sensor technology.

Fig. 4 Autonomous wireless sunsensor (AWSS)

In order to mitigate the albedo influence, TNO has developed a digital sunsensor on chip for which the final hardware is expected April 2009. This sensor core can be used in combination with the immersed technology to build a sensor which is in de order of 1 cm^2 in surface area and about 2 mm high but still provides a ±60 degree FOV and an accuracy in the order of 0.03°. At a total consumed power of less then 25 mW this would be the lowest power high accuracy sunsensor available world wide. Although the complete chip will be produced in April 2009 only, tests performed on precursor versions (pixel arrays and complete chips without ADC) have proven a significant radiation resistance of >90 krad and full functionality of the design (with an external ADC). For the final chip the ADC which has been tested on a separate chip will be integrated with the APS imaging array and de sequencing, signal processing and I/O electronics. It is expected that this 5∗5 mm chip will be available for further integration into a complete micro digital sunsensor in Q4 2009.

A second approach towards these goals would be the use of a (patented) albedo insensitive analogue sunsensor. This sensor is basically an analogue sunsensor which operates in a carefully selected wavelength region. If the wavelength of operation coincides with a wavelength which is heavily absorbed by the Earth atmosphere (Fig. 5), the intensity of the operational wavelength in the reflected light will be largely reduced and consequently the albedo's effect on the sun sensor accuracy will also be reduced.

There seem to be two wavelength regions which are particularly interesting in that respect.

1) the wavelengths below 300 nm where ozone absorption largely reduces the intensities
2) the wavelengths around 1,300 nm where water absorption in the atmosphere largely reduces the intensity.

Both wavelength regions exhibit a high absorption in the atmosphere and have material and detector combinations that would allow to construct a high reliability sensor without reverting to exotic technologies. For the first range, UV enhanced silicon and quartz or sapphire can be used for instance and for the 1,300 nm range

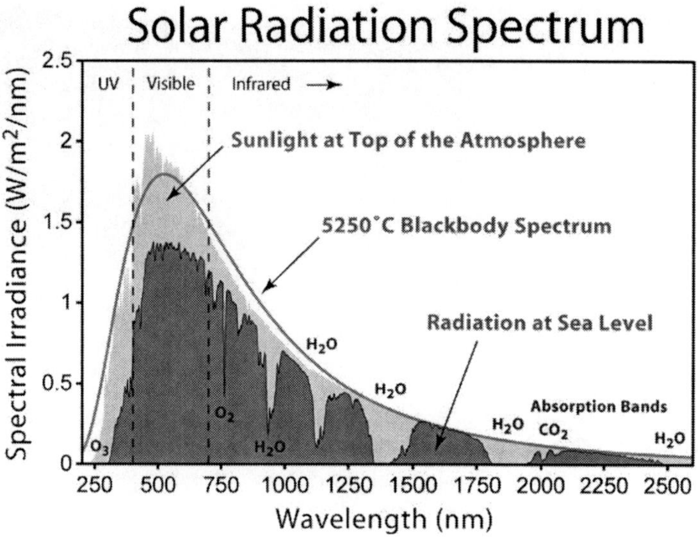

Fig. 5 Spectral irradiance at *top* and *bottom* of atmosphere

InGaAs and silicon. Due to the use of only four sensing elements the processing electronics required is much reduced as compared to a conventional digital sunsensor. Therefore this application is expected to lead to sensor systems which will only need in the order of 50 micro watts before a digital output is generated. This power level would allow for unprecedented energy harvesting schemes like tapping in on the stray radiation emitted by the down link antennas. The main issue in this case would be the power required for the datalink.

3 Miniaturisation: Implementation Options

Based on the above options TNO is currently investigating which route would lead to an optimized sensor. Recognizing that most missions could potentially do with a sensor that has an accuracy in the order of 1 degree if such errors would also include albedo effects, both the micro digital and the albedo insensitive analogue sunsensor seem very viable options. Although the albedo insensitive analog sensor has the advantage of an estimated lower power consumption and lower complexity, the micro digital sunsensor has a more progressed and enhanced development status. Furthermore the albedo insensitivity of a micro-DSS is inherent to the design, whereas the performance of the albedo insensitive analog variant may depend on atmospheric conditions like cloud cover etc. As these effects can only be demonstrated confidently in a real satellite application the possibilities to device a flight experiment which flies all three types of sensors are currently being investigated (Fig. 6).

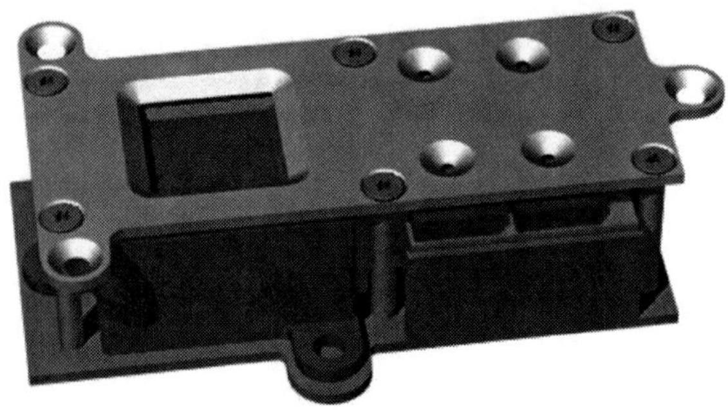

Fig. 6 Multi sunsensor experiment proposed for Delfi-n3Xt

If the funding can be arranged, TNO would like to fly the proposed sensor package as depicted in Fig. 6 on board of the next Technical University of Delft nano satellite called Delfi-n3Xt. The experiment consists of a standard mini-FSS which is a conventional but very high reliability sunsensor. A micro digital sunsensor, and one or two albedo insensitive analogue sunsensors. This experiment should verify the performance of the various sunsensors as a function of earth albedo and provide directions as to which sunsensor principle would be the most interesting one to pursue.

As for the final system implementation we are considering a number of options:

- Hybrid packaging (the conventional approach)
- Autonomous wireless operation where the solarcell for power generation and a wireless datalink are included in the sensor
- Integrated attitude and orbit control subsystems (IOPACS) where several sensors and sensor types share a common data interface and intelligent plug and play units provide a high level of flexibility and functionality while encouraging re-use of existing hardware.

The hybrid package (Fig. 7) will provide the most risk-free solution and could be very viable in both the short and the long term.

The hybrid packaging technology is very mature, and has proven a high reliability. The actual sensor would be in a well known environment (typically dry nitrogen) and by modifying the package (thinner or thicker walls and glass) the design can be optimized for the particular application. The problem associated with hybrid packaging is the fact that the sensor size will eventually be determined by the size of the connectors needed. Therefore this type of approach would require considering novel power supply schemes like powering the sensor over the datalink (much like a USB stick) given the low power consumption of for instance the APS+ chip this is seen as a very viable option. For any current implementation, the support circuits required to build a full functional sensor (like power supply and power on reset sub-circuits, clock circuits and buffers) by far exceed the actual sensor in both number and size. Therefore for a final device every attempt should be made to include as much of the

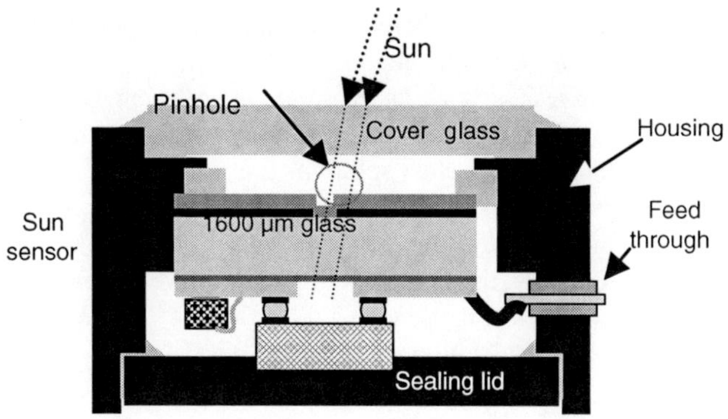

Fig. 7 Hybrid package

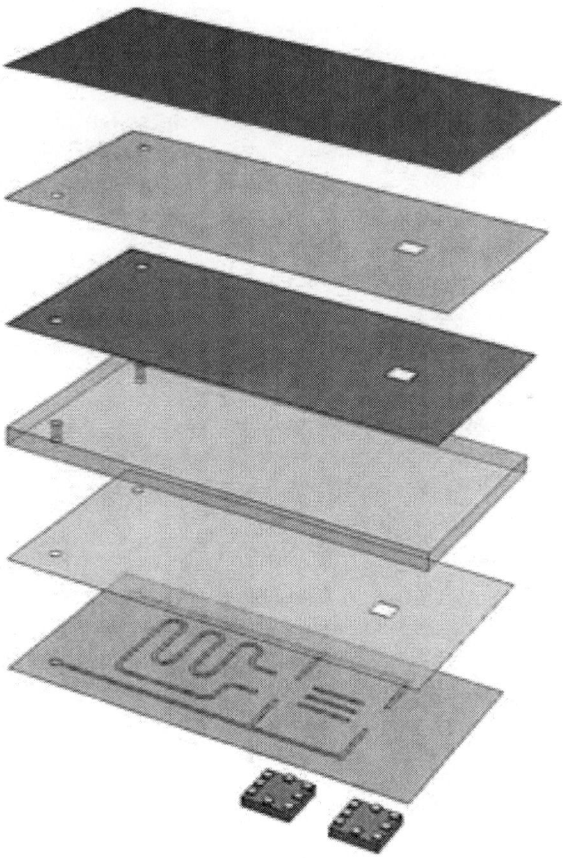

Fig. 8 Micro-autonomous wireless sunsensor with the various layers of coatings and metallisations

sub-circuits as possible on chip. Furthermore an autonomous power supply could be very advantageous even for a wired version of the sensor.

The autonomous wireless sensor (Fig. 8) would require some additional technology developments but could provide very small autonomous sensors.

For the micro-autonomous wireless solutions the size will be determined by the power requirements (or the size of the antenna required depending on the power consumption). For the current AWSS the power consumption of the signal processing part cannot be determined as the application uses a single RF enabled microcontroller. The power consumption however is in the order of 30 mW. The APS+ chip needed to build a micro digital sunsensor consumes in the order of 25 mW, as will the wireless link. Therefore the solarcell needs to be able to supply at least 50 mW under a sun incidence angle of something like 70° (to make sure enough power is available end of life as soon as the sun comes in the field of view of the sensor. If an albedo insensitive analogue sunsensor core is used, the power consumption will virtually be reduced to the power required for the wireless link and system size can be halved (presuming the required solarcell size is still larger then the antenna size for the wireless link). The power required for the wireless data link is this high because the sensors need to work in a real time control loop. Low power wireless solutions obtain these low powers by using a very short duty cycle, but the peak power consumption is mostly above 25 mW. As both the antenna size and the solar cell size required will be driven by the requirements for the data link, it can be stated that this would be a very interesting subject to investigate further.

When calculating the size of the solarcell required we could use the following data:

- Required output power 50 mW
- Efficiency of the power supply circuit 70%
- Switch on input angle 70°
- Solar cell efficiency end of life 25% (triple junction GaAs cell)
- Solar power top of atmosphere 1,350 W/m^2

Using this data the required size would be less then 4 cm^2, which is comparable to the antenna size used for the AWSS. (As additional space would be required to mount the electronics, lower efficiency solarcells which are directly deposited on the substrate can be used without driving the system size.)

Last but not least highly integrated attitude control subsystems could be an alternative for the packaging of single units, as the miniaturized modules allow to create rigid plug and play units where signal processing hardware and interfaces are shared and hardware is re-used to the largest extent possible (Fig. 9). These highly integrated units can play an important role in the design of modular spacecraft, where a quick turn around time for design and MAIT is an important issue.

As this modularity would require quite complex interface implementations it doesn't seem to make much sense to include the interfaces in each and every sensor

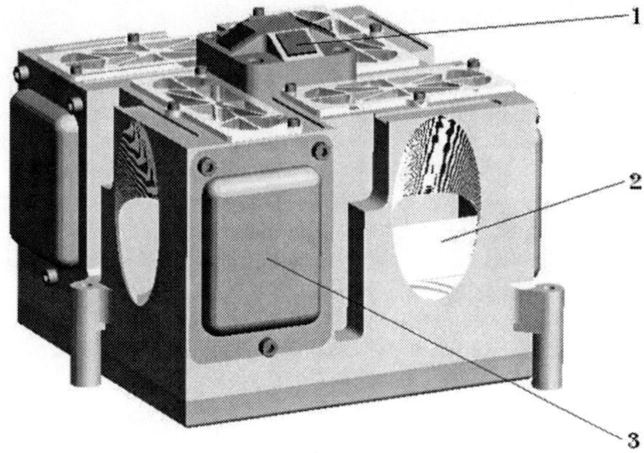

Fig. 9 Integrated optical attitude control subsystem (IOPACS)

used, but rather design fairly complex units dedicated to a certain function (attitude determination, propulsion, power supply etc.) with a certain scalability. When these units are qualified to the levels required to cover the majority (if not all) missions under consideration a large potential for re-use will be present. The depicted IOPACS systems consist of a hemispherical sunsensor unit (1), four startrackers (2) and (at least) two GPS receivers.

As the sun sensor developments are well on their way, but the startracker is a new component, a development program has been started (supported by the Dutch space agency NIVR) to demonstrate the feasibility of the startracker concept.

Which of these solutions is preferred is not known at this moment in time but will be determined on the opinion of the users (and the development money available).

4 Conclusion

TNO has produced reliable sun sensors for many years and is now striving to bring the same level of performance or more in a drastically reduced form factor. To this extent several developments have been initiated that are expected to bring mature sunsensor technology into the realm of micro and nano satellites. The actual implementation which will be used for future sensors is still under discussion and could vary from a hybrid packaged conventional analog or digital unit through autonomous wireless varieties till highly integrated attitude control subsystems. Due to the limited availability of funding some of the initiated developments will most probably have to be postponed or stopped all together but much of this depends on the interest which will be expressed by our customers for the solutions offered. In any event it can be confidently stated that attitude sensing solutions in a significantly smaller form factor will become available in the (near) future.

Further Reading

1. J. Leijtens, K. de Boom. Small Satellites for Earth Observation, ISBN: 3-89685-571-9, pp. 133–138, Small sensors for small (and other) satellites, TNO, The Netherlands.
2. R. Hamann, C. Verhoeven, A. Bonnema. Proceedings International Conference ICMENS, 24–27 July, 2005, Banff, Canada: Nano Satellites, a Fast way to Pre-qualify New Micro-Technology, Delft University of Technology, The Netherlands.
3. W. Jongkind. Proceedings International Conference ICMENS, 24–27 July, 2005, Banff, Canada: The Dutch MST program MicroNed and its Cluster MISAT, Delft University of Technology, The Netherlands.
4. N. Xie et al. Proceedings IEEE Sensors, 24–27 October 2008, Lecce, Italy: A CMOS Image Sensor with Row and Column Profiling Means, Delft University of Technology, The Netherlands.

Robust and Fault Tolerant AOCS of the TET Satellite

Zizung Yoon, Thomas Terzibaschian, Christian Raschke, and Olaf Maibaum

Abstract The chapter presents the design of the attitude and orbit control system (AOCS) of the small satellite TET-1 (Technology verification career) as an example of a cost effective but still robust and reliable AOCS for small satellites. The AOCS of TET-1 is fully three-axis stabilized. Particular attention is paid to the implementation of robust and fault tolerant design of the AOCS. The redundancy management concept and robust control algorithms are presented. Furthermore the fault detection, isolation and recovery (FDIR) mechanism, which are implemented in various functional levels, are shown. Finally the chapter presents the test strategy for verifying the proposed fault tolerant design.

1 Introduction

Pursuing design-to-cost strategy in small satellite missions often leads to a trade off in the satellite design thus resulting in a cut-off of dependability, failure tolerance and performance for the benefit of cost effectiveness. On the contrary cost reduction of the ground segment enforces higher level of autonomy on the space segment. This trade-off leads, especially for small satellites, to a significantly higher failure rate than the heavier and larger counterparts. Figure 1 categorized the mission success for NASA spacecraft missions from 1990 to 2001 subject to spacecraft launch mass.

At this period, NASA conducted a series of cost-effective frequent missions within the so called Discovery program with the goal to start many smaller missions using fewer resources and shorter development time. All missions with spacecraft mass greater than 800 kg were successful, while impaired/failed mission occurred for spacecrafts with less mass.

An extensive analysis in [2] categorized reported spacecraft anomalies into the responsible subsystems. Thirty five anomalies were reported for the AOCS (GNC

Z. Yoon (✉)
Department of Aeronautics and Astronautics, Berlin Institute of Technology, D-10587 Berlin, Germany
e-mail: zizung.yoon@tu-berlin.de

R. Sandau et al. (eds.), *Small Satellite Missions for Earth Observation*,
DOI 10.1007/978-3-642-03501-2_38, © Springer-Verlag Berlin Heidelberg 2010

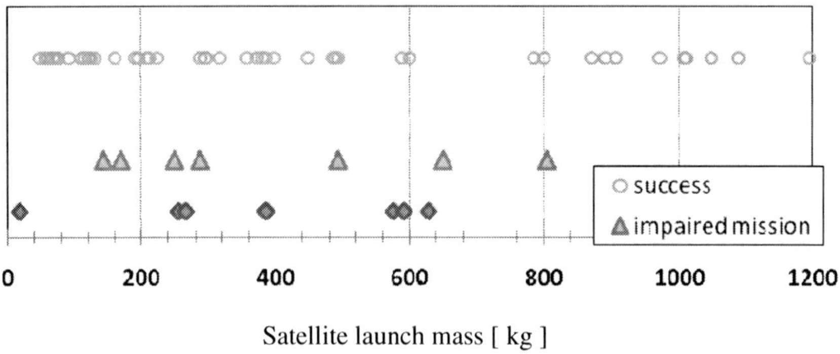

Fig. 1 Launch mass vs. mission success, NASA and US satellites, 1990–2001 [1]

and propulsion), which represents 29% of all recorded anomalies. Thirteen anomalies out of the 35 AOCS anomalies resulted in total loss, which are 37% of all catastrophic failures.

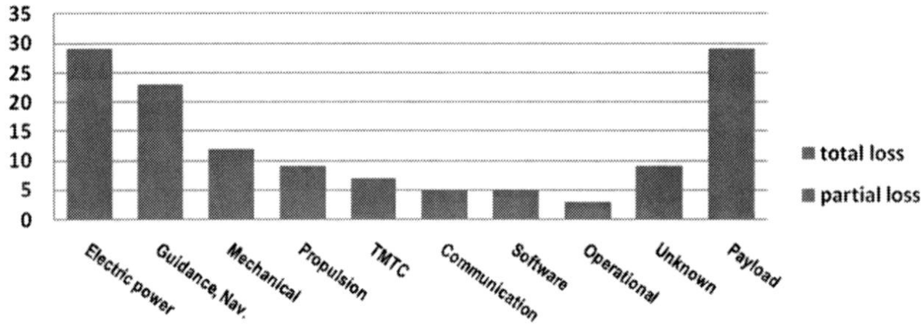

Fig. 2 Satellite anomalies and severity by subsystem (1990–2001) [2]

Therefore providing cost-effective space access further on with increased autonomy, robustness and fault tolerance within the AOCS will be a key for technical and economical break-through of small satellites. This chapter presents, as an approach for the above stated problem, the robust and fault tolerant design and implementation of a small satellite via the AOCS of the small satellite TET-1 as an example. The bus platform design of the TET-1 (start mass 120 kg) is based on the BIRD (Bi spectral Infra-Red Detection) satellite platform developed by DLR (German Aerospace Center). The AOCS is the most complex subsystem of TET-1 with most H/W components and a complex control S/W with a high level of autonomy.

2 Overview of TET-1 AOCS

The main design driving requirements for the AOCS are:

- Attitude determination < 0.5 arcmin
- Attitude control < 5 arcmin

- Pointing stability < 2 arcmin/s
- Position determination < 150 m
- Single failure tolerance at component level
- 21 h RAAN sun synchronous orbit

The robustness of the AOCS is realized by redundancy of H/W components and a "configuration management system" to handle the redundancy. The autonomy requires onboard failure detection and diagnosis. Another design aspect for robustness is a robust control loop to react against perturbation and component anomaly with sufficient stability margin. Also essential for a seamless filtering and processing of AOCS data is the autonomous monitoring of the "timings" of all AOCS thread and processes.

2.1 AOCS Hardware Overview

The H/W overview is illustrated in Fig. 3. The main torque actuators of TET-1 are four precise RWs in hot redundant tetrahedron configuration. This redundant configuration allows the RWs to work in lower rpm level which leads to less mechanical abrasion respectively. The three internally redundant MCSs are drivable independently from each other.

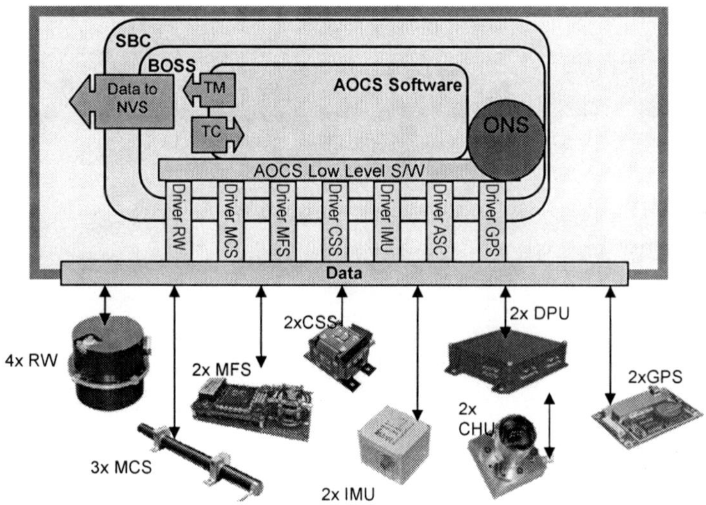

Fig. 3 Overview of the TET-1 AOCS

All sensors for attitude and orbit determination are available as redundant systems as well. The ASC is the primary sensor for precise attitude determination (<24 arcsec) and comprises a cold redundant DPU and two switchable CHUs. The CHUs boresight axes are tilted 70° away from each other to avoid simultaneous sun, moon or earth blinding.

The CSS consists of four internally redundant analog sensor heads which are mounted at the outmost area of the satellite's outer surface to ensure maximum signal availability. The analog signal processing board is arranged redundantly as well.

A further reference sensor is the fluxgate MFS (cold redundant) which is needed for magnetic field determination and MCS control input. The angular rate is determined besides the ASC mainly by the cold redundant IMU.

The orbit determination is performed by a cold redundant GPS system. The position and time data obtained by the GPS is introduced into the orbit estimator in the ONS module.

2.2 AOCS Software Overview

The AOCS S/W is build up as an object oriented and modular application with a static architecture. The break down into several layers (Fig. 4) allows an autonomous and robust handling of each feature and was extensively proven during the BIRD mission. The S/W runs as an application for the real time operating system BOSS (timing accuracy 1 ms) on the SBC.

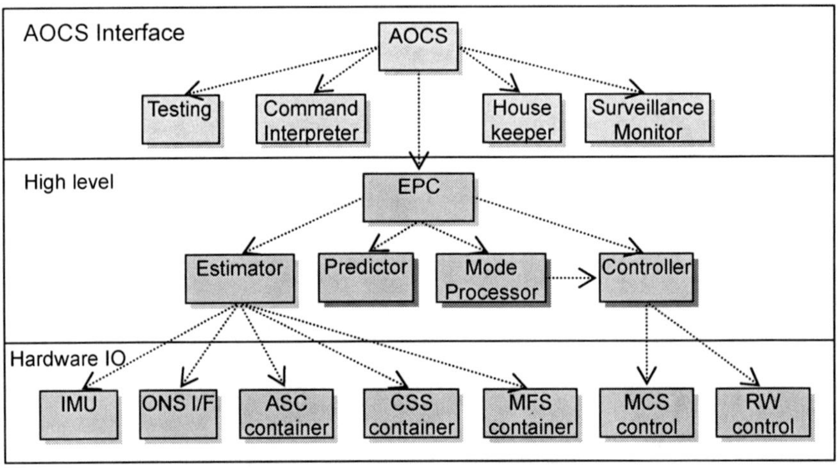

Fig. 4 AOCS software architecture [3]

The lowest layer consists of the interface modules to the hardware components and the ONS module. The low level data processing and formatting of actuator and sensor data takes place in this layer.

A layer above resides the so called EPC layer, in which the actual control task takes place. The encapsulating EPC module is a combination of state estimation, prediction (Kalman filter) and state control. The *e*stimator, *p*redictor and *c*ontroller modules form together with the mode processor the core of the control S/W (refer to Part III).

The topmost layer contains the TMTC (telemetry and telecommand) interface of the AOCS and its components. A very elementary role in the FDIR is taken by the surveillance module, which is responsible for thread timing monitoring among others.

3 Control Core – The EPC Module

The high level control core -the EPC module- resides in the middle layer within the AOCS S/W hierarchy. The following Fig. 5 illustrates the time discrete AOCS control loop.

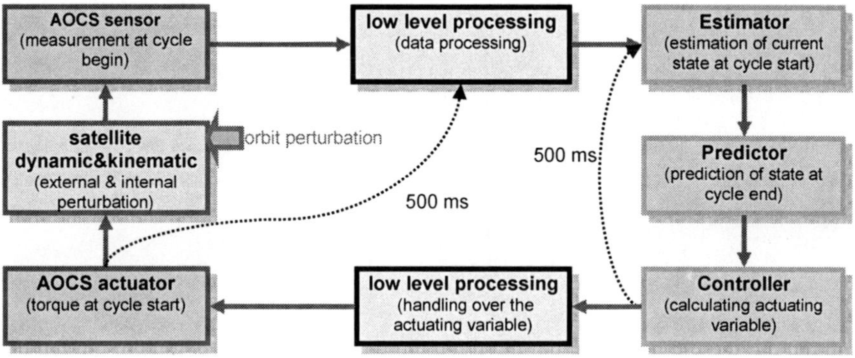

Fig. 5 EPC module within a control cycle

3.1 Estimator Module

The Estimator estimates the satellite state vectors (attitude, angular rate, angular momentum, perturbation torque) with respect to the start of the cycle. The Estimator is built up modularized and hierarchical, i.e. no collective estimator exists but each state vector is mapped to a corresponding sub-Estimator module. This breakdown of functionality ensures continuous availability at high level even when failure in sub modules occurs.

Actually the estimator modules are various Kalman filters using the available measurement values and status information from low level as well as predictor and controller values from directly prior cycle. Also a high level diagnosis is performed in the estimator module by comparing measured and prior predicted values. As result, the AOCS configuration flags are updated (refer also to Sect. 4.1).

3.2 Predictor Module

The Predictor predicts the state vector with respect to the cycle end using the results from the Estimator and the actual actuating variables (torque and dipole) given by the Controller. The prediction results are used in the current cycle by the controller to derive the actuating variables and in the next cycle by the estimator. The architecture of the Predictor is modularized in the same manner as the Estimator.

Even though the ONS is not contained in the middle layer, it provides prediction values for the upcoming control cycle. Besides providing the orbit position, it delivers the earth magnetic field vector and sun direction in inertial frame. This additional information gives even more redundancy to the AOCS: sun albedo effects in the sun sensor signal can be eliminated when inertial attitude is known, high level check of sun vector and magnetic field vector can be supported and furthermore an inertial attitude can be determined in combination with the sun sensors independently from the ASC.

3.3 Controller Module

The calculation of actuating variables (torque) is modularized as well. The actual attitude control is separated from the angular momentum dumping and even from compensating perturbation torque. This so called "virtual wheel" concept was already successfully applied at BIRD [4].

3.3.1 Attitude Control

The attitude and angular rate setpoints are obtained from the current attitude mode. The task to bring the estimated state to the setpoint counts for all attitude modes. Therefore works the algorithm within the controller irrespective of commanded attitude mode and used actuators. The Controller eventually computes the required torque, which leads the estimated attitude on a time optimized trajectory to the target attitude.

3.3.2 Momentum Dumping

The usual wheel desaturation concept is modified for TET into a general angular momentum dumping. An external "dumping torque" is computed, which reduces the total momentum. This dumping algorithm can be enabled/disabled independently from the attitude algorithm and do not affect the actual attitude control. When the satellite itself is motionless, the momentum in the RWs is reduced by the magnetic coil's external torque. Backwards when the satellite is rotating but the RWs are turned off or have failed, the satellite momentum is dumped till the satellite comes to a standstill.

3.3.3 Compensation of Perturbation

The Estimator estimates external (due to e.g. eigendipole) and internal (due to e.g. RW anomaly) perturbation torque. For the latter case even a so called "hidden" torque/momentum is estimated. The controller has to compensate the effect of these torques. This algorithm can be enabled/disabled as well.

3.3.4 Summarization of Control Torque

The sum of the above mentioned torque is the target torque for the actuators. The controller computes the respective RW torque and magnetic dipole moment depending on the available actuator configuration. The distribution of torque and dipole takes place on low level and thus the controller module is robust against reconfiguration of actuators.

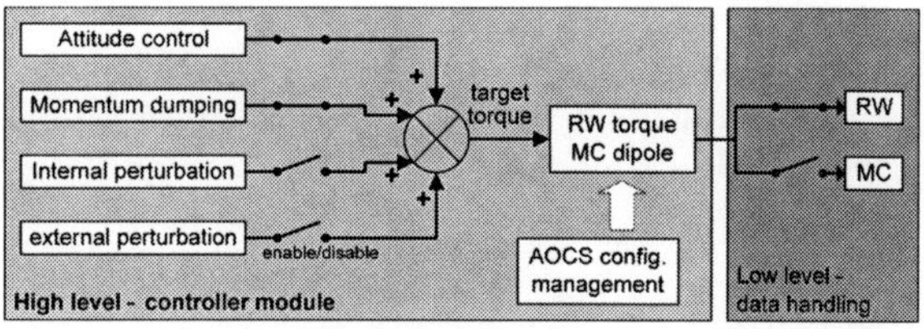

Fig. 6 Controller module

4 FDIR Architecture

The FDIR mechanism is distributed in numerous locations of the AOCS. The failure detection starts from low level checks on component level and moves up to high level checks in the EPC layer. On the topmost level the monitoring of received telecommands and surveillance of S/W thread timing takes place. Every check is straightforward without feedback from prior measurement. For isolation and recovery actions the onboard configuration management is used.

4.1 Onboard Configuration Management

The configuration management system consists of categorized status flags and handles the configuration of AOCS H/W components, EPC modules and satellite status information. In detail they are:

- satellite configuration
- sensor use and health configuration
- actuator use and health configuration
- Predictor and Estimator and Controller configuration.

The satellite configuration contains the power status of each AOCS component. Basically the low level communication is performed with all powered components. Additional satellite state information such as eclipse phase, and solar generator status is stored in this configuration.

The sensor and actuator “use” configuration manages whether a component is in the control loop or not. If this configuration is set to “not in use”, the respective components may provide data but is just linked to the telemetry. The “health” configuration signalizes any anomaly detected in the components or modules.

The EPC configuration enables/ disables each high level module. The following Table 1 illustrates the EPC configuration and the quality of output data.

Table 1 Estimator/Predictor filter configuration

Sensor	Estimator	Predictor	Output data	Data quality
In use	on	on	estimated	highest accuracy
Not in use	on	on	predicted	bridge data gap with accuracy drift over time
Not in use	off	on/off	no data	–
In use	on	off	raw data	each component quality

All flags are set new in every control cycle. The AOCS S/W is able to deal with any possible configuration autonomously and thus makes the AOCS insensitive against failures.

4.2 Fault Detection

Fault detection begins on the low level by a formal check of input data such as verifying checksums, correct timing and IDs. The low level status information is forwarded to the estimator modules. The high level failure detection is performed on the EPC layer applying following detection methods:

- Matching of measured state values with predicted state values.
- Comparing deviations with tolerance limits.
- Cross check of state values (e.g. IMU rate, sun rate, Magnetic field rate, quaternion rate).

All failure detection algorithms are accomplished impartially every single control cycle (500 ms, without any feedback from former cycles). Any anomaly is marked through the health flags in the configuration management system.

The implemented functional redundancy allows keeping the most important state variables always on a valid status even in cases of data gaps or sensor failures. The used state estimation inherits already a kind of functional redundancy [6]. Some state variables can additionally be computed on board by using mathematical models. Even a complete loss of some sensors (GPS, CSS and MFS) can be compensated by functional redundancy.

4.3 Surveillance Module

The AOCS monitoring is performed by the surveillance module (refer to Fig. 4). The progression of the EPC module is monitored as well as the defined configuration preconditions for each attitude mode. Severe anomalies may generate a reset of the SBC.

5 Verification of AOCS

The implementation of the robust AOCS design including control algorithm, configuration management and fault injection will be verified on different levels. The end to end tests will be performed on a three-axis high precision air bearing AOCS test [3, 5]. The low perturbation torque ($<10^{-5}$ Nm) occurring on the testbed permits even the testing of attitude manoeuvre and long term tests. In development phase simulation techniques [6–8] will be used as well.

6 Summary

The AOCS of the small satellite TET-1 was presented with focus on robustness and fault tolerance. The robust control core was described in Part III. The fundamental idea regarding the FDIR design was to decouple and modularize functionalities of H/W as well as S/W to prevent propagation of local anomalies to higher level.

The described AOCS design demonstrates how to implement a robust and fault tolerant AOCS with high reliability within a small satellite mission with limited budget.

References

1. D. Bearden. When is a satellite mission too fast and too cheap, 2001 MAPLD International Conference, Maryland, September 11–13, 2001.
2. B. Robertson, E. Stoneking. *Satellite GN&C anomaly trends*, AAS Guidance and Control Conference, 2003.
3. Z. Yoon, T. Terzibaschian, C. Raschke. Robuste und fehlertolerante des DLR-TET-1 Satelliten, DGLR Kongress, Darmstadt, 2008.
4. T. Terzibaschian. Experiences with the BIRD magnetic coil system for wheel desaturation, satellite rate damping and attitude control – the concept of virtual wheel systems, Small Satellites for Earth Observation – Selected Proceedings of the 5th International Symposium of the IAA, Berlin, S. 361–368, 2005.
5. S. Roemer, et al. Experiences with the BIRD ACS test facility and the resulting design of a state of the art mini- and micro-satellite ACS test facility, Small Satellites for Earth Observation – Selected Proceedings of the 6th International Symposium of the IAA, Berlin, 2007.
6. O. Maibaum, S. Montenegro, T. Terzibaschian. Robustness as Key to Success for Space Missions. In: Schuster, Alfons [Hrsg.]: Robust Intelligent Systems, Springer, pp. 233–250.

7. O. Maibaum, H. Schumann, A. Berres. Test of satellite control software with simulated faults, Small Satellites for Earth Observation – Digest of the 6th International Symposium of the IAA, Berlin, 2007.
8. O. Maibaum, , T. Terzibaschian. Lessons learned from the object-oriented design of the BIRD attitude control system, Automatic Control in Aerospace, Elsevier, pp. 156–161, 2004.

Implementation of the $T^3\mu PS$ in the Delfi-n3Xt Satellite

C. Müller, L. Perez Lebbink, B. Zandbergen, G. Brouwer, R. Amini, D. Kajon, and B. Sanders

Abstract This chapter presents the interfaces that exist between the TNO, TU-Delft and UTwente developed micro propulsion system experiment (T3μPS) and the Delfi-n3Xt satellite currently under development at TU-Delft. Delfi-n3Xt is successor to Delfi-C3, the first Dutch student nano-satellite. Delfi-n3Xt, like its predecessor is a 3-unit cubesat, but is more advanced in that it is equipped with an active three-axis attitude determination and control system, eclipse power and a high speed data link.

The $T^3\mu PS$ is especially developed to provide for a means of propulsion fit for use in nano- and microsatellites. Such a means is considered essential for e.g. future missions of clusters of satellites which require relative orbit control. It employs a cold gas thruster delivering a thrust in the order of a few up to several tens of millinewton. As propellant, the system uses cool nitrogen gas, stored in solid form in one or more cool gas generators.

The $T^3\mu PS$ hardware is assembled on a single printed circuit board easing integration of the system in Delfi-n3Xt.

When operating, the $T^3\mu PS$ may disturb the proper operation of Delfi-n3Xt.This is because of thrust misalignment. This disturbance needs to be compensated for to keep the solar panels of the satellite sun-pointed.

The chapter presents the Delfi-n3Xt satellite as well as the $T^3\mu PS$ experiment and describes the interfaces between the two. Furthermore, it investigates the implications on other satellite subsystems with particular emphasis on the ADCS. Finally, some Conclusions and recommendations are given.

1 Introduction

This chapter presents the interface control principles that are in place within the Delfi-n3Xt project. The practical implications of the interface control procedures

B. Zandbergen (✉)
Delft University of Technology, Delft, The Netherlands
e-mail: b.t.c.zandbergen@tudelft.nl

R. Sandau et al. (eds.), *Small Satellite Missions for Earth Observation*,
DOI 10.1007/978-3-642-03501-2_39, © Springer-Verlag Berlin Heidelberg 2010

are illustrated by describing the implementation of the T^3 micropropulsion system (μPS), one of Delfi-n3Xt's payloads. The $T^3\mu$PS is an appropriate illustrative example since it has far reaching implications on the satellite and its operations. Therefore it is given highest priority in terms of Interface Control.

The implications that the $T^3\mu$PS has on the satellite are described in detail. Through interface control procedures these implications are managed and solutions are described which are to ensure that the effects of the $T^3\mu$PS are accounted for.

Section 2 of this chapter describes the Delfi-n3Xt satellite and Sect. 3 the $T^3\mu$PS. The experiment which is to be carried out with the $T^3\mu$PS is described in Sect. 4. Section 5 describes the Interface Control procedures of the Delfi-n3Xt project while Sect. 6 describes the main interfacing issues between the $T^3\mu$PS and the satellite and how these are managed. Verification of the interfaces is described in Sect. 7. Section 8 provides Conclusions and recommendations.

2 The Delfi-n3Xt Project

The Delfi-n3Xt project is the second in line of a series of small satellite projects within the Delfi programme initiated by Delft University of Technology. It is the successor of the Delfi-C^3 nanosatellite which was launched successfully in April 2008. Like Delfi-C^3, it, is a 3-unit cubesat see Fig. 1. Its main objectives are to serve as a focal point for education and to demonstrate new technology and conduct scientific experiments [1]. Delfi-n3Xt carries on board five payloads that occupy roughly 50% of the spacecraft volume.

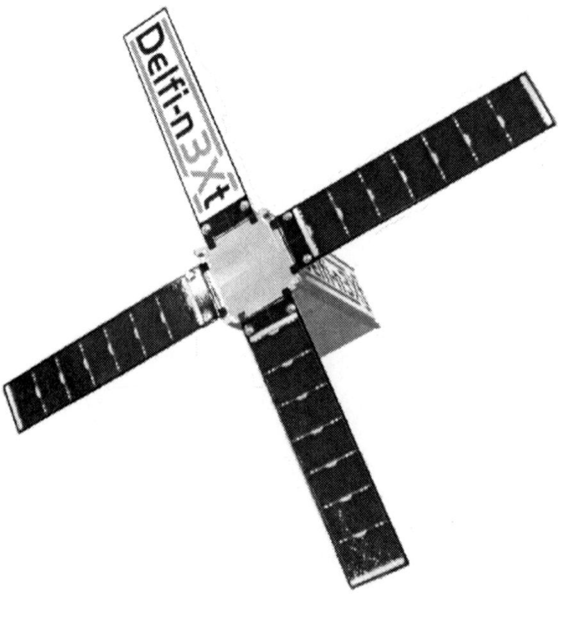

Fig. 1 Delfi-n3Xt artist model

The satellite facilitates:

- ♣ Pre-qualification of $T^3\mu$PS micro-propulsion system from TNO, TU Delft and UTwente
- ♣ Pre-qualification of a Multi-functional Particle Spectrometer from Cosine Research BV
- ♣ Scientific Radiation Experiment of Si-solar cells from DIMES
- ♣ Qualification of a high-efficiency communications platform from ISIS BV
- ♣ Proof-of-concept for a radiation risk-free implementation of commercial solid-state data storage devices, provided by NLR

Specifics of the Delfi-n3Xt satellite are given in the Table 1.

Table 1 Delfi-n3Xt characteristics

Delfi-n3Xt satellite	
Mass	3.5 kg
Dimensions	100 × 100 × 340.4 mm (span: 714 mm)
Orbit	Sun-synchronous
Attitude determination and control	3-axis, 3° determination accuracy, 10° control accuracy
Communication	1.2 kb/s downlink in VHF >9.6 kb/s (variable) downlink in S-band 0.6 kb/s uplink in UHF
Command and data handling	fail-safe I^2C
Electrical power	Photovoltaic array (4 deployable panels) and batteries providing 12 W average
Structural	Improved structure for better accessibility
Thermal	Passive

3 The T^3 Micropropulsion System

The $T^3\mu$PS experiment is developed within the scope of the Dutch MicroNed programme, which aims to advance micro-technology research in the Netherlands. A team of TNO, UTwente and TU Delft is developing a micropropulsion system to fit in a cubesat.

This system is referred to as $T^3\mu$PS and essentially consists of 4 different components including:

1. MEMS (Micro-Electro-Mechanical Systems) based valve and nozzle that provides for the thrust. This module also contains a MEMS-based pressure transducer and filter. The latter is to prevent flow blockage due to impurities. The pressure transducer is for control purposes;

2. Cool Gas Generator (CGG) which produces the nitrogen gas needed for thrust generation from a solid charge. A CGG can be ignited by a glow plug, avoiding the use of pyrotechnics;
3. Plenum that provides a buffer volume for the nitrogen gas produced by the CGGs and feeds this gas to the thruster. It also provides for the mounting of the MEMS valve and nozzle and one or more CGGs thereby reducing the amount of piping and brackets;
4. Printed Circuit Board (PCB) that holds the electronics needed to ignite the CGGs and to operate the valve and the pressure transducer. It also provides for mounting of the plenum.

An engineering model of the system is shown in the Fig. 2.

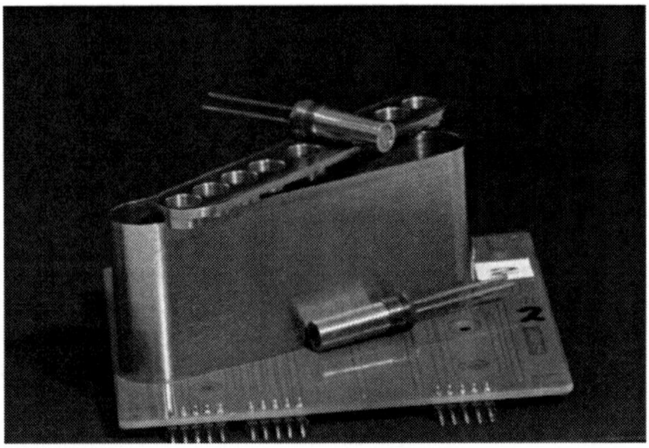

Fig. 2 Engineering model of the $T^3\mu PS$

The Fig. 2 shows two CGGs, plenum (middle) and the PCB (below). Not shown in the figure are the electronic circuits which are defined by Delfi-n3Xt and which regulate data traffic and power supply [6]. The model shown allows for eight CGGs to be mounted on to the plenum, which also acts as the main structure of the propulsion system, thus avoiding brackets and piping. The gas generators are ignited separately and each on its own will pressurize the plenum.

When the plenum is depleted another CGG can be ignited to re-pressurize it. It is noted that the system essentially is a blow down system with a decreasing pressure and mass flow rate during operation until the plenum is re-pressurized. Table 2 provides the main characteristics of the $T^3\mu PS$. A full description of the $T^3\mu PS$ can be found in [8]; initial test results are presented in [11].

Advantages of the system are:

1. Absence of pyrotechnics.
2. Absence of pressurized storage: System is pressurized in space by igniting the CGGs.
3. Limited propellant filling operations: Propellant is cartridge loaded.

Table 2 Main characteristics of $T^3\mu PS$

Characteristic	Value
Mass	<140 g
Dimensions	90 × 90 × 21 mm
Operation	Blow down mode
Thrust	Nominal thrust in range 6–150 mN, selectable by manufacturing
Propellant	Nitrogen from 8 miniaturised CGGs (extendable to 16 CGGs)
CGG dimensions and mass	Size: 8 mm diameter, 20 mm long Mass: 2 g (0.3 g propellant load), Gas volume: 0.1 normal litres of nitrogen per gas generator

4 Experiment Description

The $T^3\mu PS$ Project Team aims to evaluate the performance of the $T^3\mu PS$ in space. Main objective is to observe the behaviour of the propulsion system and more specific the ignition behaviour of the CGG, the functioning of the MEMS valve and nozzle and the leakage characteristics from the system. A secondary objective is to determine orbital changes due to $T^3\mu PS$ operations.

To deduce the performance of the $T^3\mu PS$ data is acquired on the pressure and the temperature of the gas within the plenum.

The experiment is set up to provide the team with as much useful information as possible in the least amount of time. Operating the $T^3\mu PS$ has some clear requirements on the satellite. For example, it is required that both igniting the CGGs as well as thrusting is done when in contact with the Delft ground station. This is to ensure proper processing of experiment data and to check the behaviour real time. To accommodate this, the experiment is implemented into the general satellite operations schedule together with the Delfi-n3Xt team. Table 3 provides an overview of essential mission events.

Table 3 Mission outline of Delfi-n3Xt satellite and $T^3\mu PS$ experiment

Event	Date (indicative)
Preliminary Design Review	September 2008
Interface verification tests	Summer 2009
Mid-term Design Review	Launch – 1 year
2nd detailed design phase (end)	Launch – 7 months
Launch	2010/2011
Launch and Early Operations Phase (LEOPS)	Launch + 20 days
Start $T^3\mu PS$ experiment	Launch +30 days
Basic experiment success (first combined thrust campaign + leak-tightness check)	Launch + 60 days
Intermediate experiment success (continuous and bursts thrust campaign completed)	Launch + 90 days
Full experiment success (ignition of CGG at end of mission)	Launch + 120 days

The experiment shall commence after in-orbit commissioning. It is planned to have one set of experiments per week which allows sufficient timeslots for all other on-board experiments. The T$^3\mu$PS Project Team foresees two different thrusting modes:

♣ Continuous thrust mode, wherein the system provides thrust over a prolonged time period.
♣ Burst thrust mode wherein the thruster provides short pulses. It is aimed for a minimum impulse bit of 30 μNs at a plenum pressure of 4.5 bar.

Figure 3 shows the current experiment timeline. It indicates a first thrust campaign (first CGG ignition) consisting of both continuous thrusting as well as thrusting in short bursts followed by a 10-day leak tightness evaluation. After the ignition of the second and the third CGG respectively a long and a bursts thrust programme are carried out.

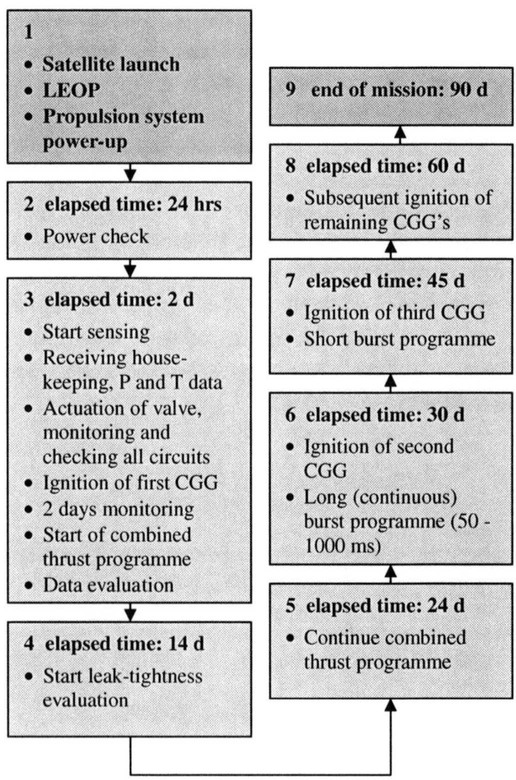

Fig. 3 Experiment outline

Three different levels of experiment success are defined. Basic success is reached after successfully completing the first combined thrust programme and after the leak-tightness check. Intermediate success is reached after successfully completing the short burst programme associated with the ignition of the third CGG. Full success is achieved when also a successful ignition of a CGG grain is performed at the end of the mission after 90 days.

5 Interface Control

To ensure proper interfaces between Delfi-n3Xt and its payloads [10], interface control is assigned to one Delfi-n3Xt team member. He/she is responsible for the generation of Interface Control Documents (ICDs), which contain information on interfaces in the categories *mechanical*, *electrical*, *command and data handling* and *environments,* and participates actively to ensure all information in the ICD is unambiguous, up to date, and consistent. In addition he/she is responsible for signalling interfacing problems, and solving these problems together with the involved system engineers.

Finally, to ensure compliance with defined interfaces [7], he/she is responsible for defining the requirements for the verification of the interfaces as well as the methods to be used.

6 Delfi-n3Xt – $T^3\mu PS$ Interfaces

Three issues are considered critical for the design of the interfaces with the $T^3\mu PS$:

Disturbance torque produced – when not aligned with the centre of mass (CoM) of the satellite the $T^3\mu PS$ produces a disturbance torque. The attitude control system must limit the effects of this disturbance so that the attitude of the satellite remains within the requirements.

Static envelope (incl. location) – due to the ambitious character of the Delfi-n3Xt project (five payloads in a 3-unit cubesat) the satellite offers only limited space to each payload. Secondly, to minimize the disturbance torque generated by the $T^3\mu PS$ experiment, the payload must be placed close to the satellite's CoM. Both restrictions are challenging and might influence other satellite systems or the satellite as a whole.

Data handling – the $T^3\mu PS$ produces a large amount of data during ignition or thrusting. Handling of this data is a challenge for the data bus especially when considering all other (house) keeping data that needs to be processed as well.

Below, the analysis performed in relation to the above and the solutions adopted are presented in more detail.

6.1 Disturbance Torque

Attitude control for Delfi-n3Xt is crucial since the solar panels require sun pointing to ensure sufficient power capacity. To ensure that the panels receive at least 90% of the available solar flux the angle between the pointing direction and the satellite-Sun line shall be kept within 25°.

Attitude control of Delfi-n3Xt is accomplished using miniaturized reactions wheels [2]. The next Table 4 gives the main characteristics of the reaction wheels.

Table 4 Reaction wheel characteristics

Parameter	Value
Torque produced	10^{-5} Nm (at least)
Max angular momentum storage	$7.8 \cdot 10^{-4}$ Nms
Max angular momentum storage available for thrust misalignment compensation	$1.7 \cdot 10^{-4}$ Nms or 13% of maximum storage capacity

To ensure that the T$^3\mu$PS experiment does not interfere with the correct pointing of the satellite, some restrictions are imposed by the attitude determination and control subsystem (ADCS) on the experiment. These restrictions are presented in Table 5.

Table 5 Restrictions on the T$^3\mu$PS experiment

Parameter	Value
Max impulse delivered over ½ orbit	0.034 Ns
Max thrust	6 mN
Max torque due to misalignment	$9 \cdot 10^{-6}$ Nm

The maximum allowable impulse bit of 34 mNs stems from the angular momentum storage capacity of the reaction wheels available for compensating for thruster disturbances. Over an ½ orbit this is maximum 13% of the maximum storage capability or an angular momentum L of $1.7 \cdot 10^{-4}$ Nms, see also Table 4.

The remainder is for storing the angular momentum resulting from other disturbances due to e.g. atmospheric forces and gravity gradient effect.

Based on the angular momentum storage capability of $1.7 \cdot 10^{-4}$ Nms and an estimated maximum total offset o of 5 mm of the thrust vector from the CoM of the satellite the maximum allowed impulse bit *Imp* as well as the limiting values for thrust F and thrust time t_{thrust} are determined:

$$\mathrm{Imp} = \frac{\mathrm{L}}{\mathrm{o}} = 0.034\ \mathrm{Ns} = \mathrm{F} \cdot \mathrm{t_{thrust}} \tag{1}$$

The offset of 5 mm stems from three factors:

- ♣ inaccuracy in determining satellite CoM: $^+$/–2.5 mm
- ♣ mounting inaccuracy of the thruster during integration: $^+$/–1 mm
- ♣ contribution due to thrust vector misalignment: $^+$/–1.5 mm.

A thrust level of maximum 6 mN is agreed upon with the T$^3\mu$PS team as this value allows for a reasonably long thrust period of 5.67 s, while limiting the disturbance torque produced by the T$^3\mu$PS experiment as of both lateral and angular misalignment to not more than $3 \cdot 10^{-5}$ Nm. It is noted that in principle a lower thrust value, limiting the disturbance torque to well below $1 \cdot 10^{-5}$ Nm, is preferred, but current design limitations of the T$^3\mu$PS experiment hardware dictate a minimum nominal thrust of 6 mN.

The effect of the disturbance torque being larger than the torque produced by the reaction wheels on the attitude of the satellite has been investigated in some detail. The angular velocity $\dot{\alpha}$ induced on the satellite by thrusting is calculated using Eq. (2) with N_{thruster} the torque produced by the experiment ($3 \cdot 10^{-5}$ Nm), I the mass moment of inertia of Delfi-n3Xt, $5.3 \cdot 10^{-2}$ kgm^2, and t_{thrust} the maximum allowed thrust time, i.e. 5.67 s:

$$\dot{\alpha} = \frac{N_{\text{thruster}}}{I} \cdot t_{\text{thrust}} = 3.2 \cdot 10^{-3} \frac{\text{rad}}{\text{s}} \tag{2}$$

Results show an angular velocity of 3.2 mrad/s or 0.184°/s and a total rotation angle of 0.52° at the end of the thrust period of 5.67 s. This is taking in account a further time lag of 2 s before actuation starts. It follows that the satellite rotates maximum over some 1.5°. Note that the angle over which the satellite is rotated is still within the attitude determination accuracy of 3° of Delfi-n3Xt, see Table 1. Corrective measures will only be taken when the satellite detects a change in attitude. In total it takes about 17.01 s to reduce the rotational rate down to zero. Including the accuracy in attitude determination of 3°, it is required that prior to thrusting off-pointing is less than 20.5°.

6.2 Static Envelope (Incl. Location)

The T$^3\mu$PS experiment flight hardware shall be mounted on a standard PCB, see Fig. 3, which in turn is stacked on to the other Delfi-n3Xt PCBs and fixed in position using rods and buses [4]. Preferred location of the PCB is close to the middle of Delfi-n3Xt to ensure that the nominal working line of the thrust vector points through the spacecraft's CoM with the thrust vector pointed parallel to the spacecraft x-axis (Fig. 4).

The T$^3\mu$PS experiment shall be designed and verified to withstand the static and dynamic stresses, strains, and vibrations associated with the launch vehicle. For initial design purposes, the T$^3\mu$PS flight hardware shall fit within a static envelope $90 \times 90 \times 26.6$ mm ($l \times w \times h$).

In Fig. 5 the working line of the thrust is drawn. It is considered to be in line with the geometric centre line of the nozzle which is situated 9.6 mm above the bottom plane of the T$^3\mu$PS PCB.

When mounting the T$^3\mu$PS experiment on board of Delfi-n3Xt, it must be ensured that the thrust vector points through the satellite CoM. This is to minimize disturbance torques when firing the thruster. The current position of the CoM with

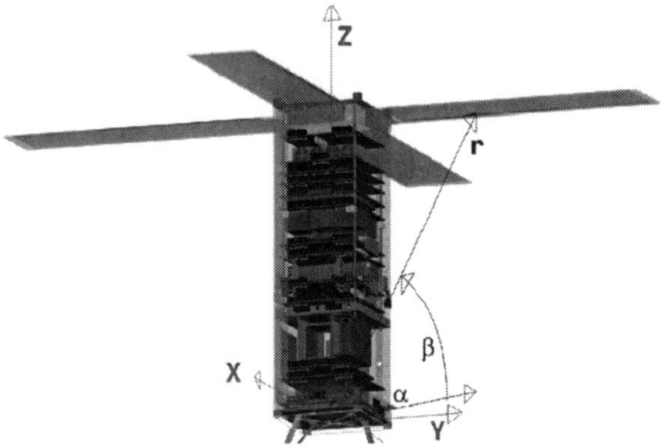

Fig. 4 Delfi-n3Xt body fixed reference frame

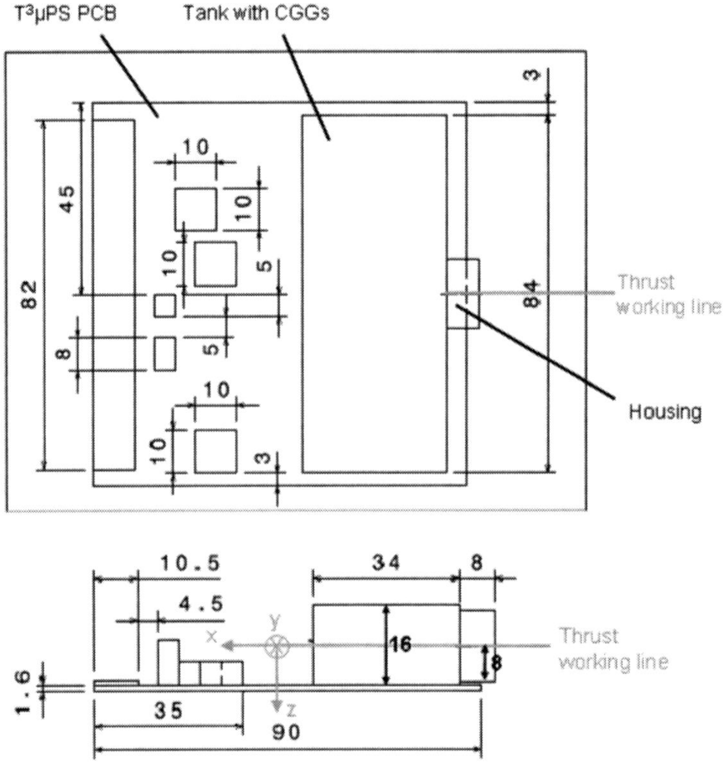

Fig. 5 Local reference frame of T$^3\mu$PS experiment

respect to the origin of the satellite reference frame is given in Table 6 (see also Fig. 4).

As the position of the CoM is not exactly known until late in the design, it is required that the position in z-direction of the T$^3\mu$PS experiment in the spacecraft

Table 6 Location of CoM of Delfi-n3Xt satellite

Centre of mass	Distance to origin (mm)
CoM_x	1.1
CoM_y	−1.1
CoM_z	200.6

can be adapted during integration. In case it is not possible to place the PCB at the desired location (z-direction) or if the thrust vector does not point through the CoM (y-direction), it is currently considered to use balance masses to influence the location of the spacecraft's CoM.

6.3 Command and Data Handling

The T$^3\mu$PS experiment shall communicate with the Delfi-n3Xt CDHS via an I^{2}C bus. This bus allows for a data rate of maximum 100 kbps. The T$^3\mu$PS experiment data rate is shown in the next Table 7.

Table 7 Data rate of T$^3\mu$PS when active (provisional)

Data type	Size (bits)	Minimum sampling rate (Hz)	Average bit rate (bits/s)
Pressure	10	100	1,000
Temperature	10	1	10
Current	8	1	8
Voltage	16	1	16
Total			1,034

When thrusting, the experiment produces 1,034 bits/s which are to be downlinked. To ensure that the experiment data is handled properly, several measures have been taken. One of these measures is to only ignite a CGG or perform thrusting when the satellite is in direct contact with the ground station in Delft. This way the data transfer does not need to rely on the radio amateur network. Additionally, it has been decided not to operate the experiment simultaneously with other payloads such that it can use the full available data band width. When operating the T$^3\mu$PS experiment without interferences from other payloads the data handling is more robust and the demand on the CDHS is minimized. Details on the CDHS can be found in [6].

7 Interface Verification

Interface control procedures only function if the payload partners adhere to the agreements within the ICD. Therefore, verification is an essential element within interface control [7].

Together with the T$^3\mu$PS team, Delfi-n3Xt has to ensure that every defined interface is verified for compliance. Different methods exist in verifying interface requirements and several factors play a role in this consideration. For example, an important factor in the verification process is costs. In [9] it is mentioned that especially within a thruster development project the test verification plan can have a significant influence on costs and schedule. In large projects, the difference between a small-scale verification plan and a full-out test campaign can be 2 years. Coste [3] adds to this discussion that costs can be minimized by setting up clear requirements early in the verification campaign. Changes in test parameters can be very costly, therefore, the verification campaign should be thoroughly considered.

In [5] four possible ways of verification are given:

- ♣ Inspection (I)
- ♣ Analysis (A)
- ♣ Demonstration (D)
- ♣ Test (T)

Also included in this Table 8 are the interfaces that have been defined in Sect. 6. The verification approach shall be executed in coordination of both the T$^3\mu$PS Project Team and Delfi-n3Xt.

Verification and test plans shall be set up by both parties to ensure that interface definitions are efficiently tested, leading to a minimum cost for testing purposes. These test plans are to be defined well before the actual testing is planned. In light of the Delfi-n3Xt project, the tests are to take place in the summer of 2009.

Considerations on which verification method to apply are based on amongst others costs, schedule and applicability. Some possibilities are better suitable to certain interfaces than to others. For example, testing requires quite some effort and would be rather pointless for verifying a payload's CoM. In this case, analysis would be a better suitable option. For all defined interfaces the most suitable verification method is evaluated. Table 8 provides an overview of selected interface definitions and which verification method is applied.

8 Conclusions and Recommendations

The incorporation of the T$^3\mu$PS experiment on the Delfi-n3Xt satellite poses several challenges to both the Delfi-n3Xt team as well as the T$^3\mu$PS team. To tackle these challenges, Interface Control procedures have been agreed upon, which centre up on the generation of ICDs and the active control of the interfaces.

Static envelope, disturbance torques, command and data handling have been identified as main interface issues. For these interfaces solutions have been defined jointly thereby ensuring that the integration of the T$^3\mu$PS into the satellite does not pose any complications and experiment operations do not threaten the satellite's operations. Considerations for verifying the interfaces have been presented as well.

Table 8 Verification of selected interfaces

Mechanical Requirements	Payload envelope	
	Static envelope	I
	Dynamic envelope	T
	Mass properties	
	Payload mass	I
	Payload mass variability	A
	Payload centre of mass	A
	Moments of inertia	A
	Disturbance	
	Torque profile	T
	Angular momentum	T
Electrical Interfaces	Signal characteristics	
	Frequency	T
	Voltage	T
	Load current	T
	Power requirements	
	Nominal operations	T
	Other operating modes	T
CDHS Interfaces	Data definition (all operating modes)	
	Data package size	T
	Data sampling rate	T
	Bit rate	T
	Software interface compliance	T
Environments	Electromagnetic environment	
	Payload EM fields	T
	Payload susceptibility to EM fields	T
	Material thermal characteristics	
	Surface finish	I
	Thermal conductivity	I
	Thermal expansion characteristics	I
	Produced/expelled heat	T
	Temperature range	T

Thorough verification plans need to be defined as the interface testing campaign is scheduled for summer 2009. Thoroughly defining these tests and the requirements are crucial to minimize testing costs.

References

1. J. Bouwmeester, E.K.A. Gill, C.J.M. Verhoeven. Advancing Nano-satellite Platforms: The Delfi Program, 59th International Astronautical Congress, 2008.
2. S. Brak, R. Amini, P. Massion. Student nanosatellite development: From passive stabilization to three-axis active attitude control, AAS-09-08-05, 32nd Annual AAS Guidance & Control Conference, 2009.
3. K. Coste. 2004. Qualification test methods for satellite ACS thrusters, JANNAF Conference, Las Vegas, Nevada, May 10, 2004.

4. S.Y. Go, J. Bouwmeester, G.F. Brouwer. Optimized three-unit cubesat structure for Delfi-n3Xt, IAC-08-B4.6.A1, 59th International Astronautical Congress, Glasgow Scotland, UK, 29 September–3 October 2008.
5. INCOSE. INCOSE Handbook of Systems Engineering v2a, International Council on Systems Engineering (INCOSE), 2004.
6. S. de Jong, G.T. Aalbers, J. Bouwmeester. Improved Command and Data Handling System for the Delfi-n3Xt Satellite. 59th International Astronautical Congress, Glasgow Scotland, UK, 29 September–3 October 2008.
7. V.R. Lalli, R.E. Kastner, H.N. Hartt. Training Manual for Elements of Interface Definition and Control, NASA Office of Management, 1997.
8. J.L.P.A. Moerel, H.M. Sanders, M.C. Louwerse, H.V. Jansen, J. Boscher, B.T.C. Zandbergen, F. La Torre, C.R. Kleijn. Development of Micro Propulsion System Technologies for Minisatellites in The Netherlands. 5th International Spacecraft Propulsion Conference, Heraklion, Greece, 5–9 May 2008.
9. P. Pempie, H. Vernin. 2001. Liquid Rocket Engine Test Plan Comparison, AIAA 01-3256.
10. R. Teuling, J. Bouwmeester, S. de Jong. Pragmatic system engineering on the Delfi-n3Xt nano-satellite project. 3rd International Workshop on System & Concurrent Engineering for Space Applications (SECESA 2008), Rome, Italy, 15–17 October 2008.
11. B.T.C. Zandbergen, M.C. Louwerse, M.N.W. Groen-endijk, F. Tardaguila, R.J. Koopmans, J. Boscher, J.L.P.A. Moerel. Initial Development of a Miniature Cold Gas Propulsion System for Nano/Micro-Satellites. 5th International Spacecraft Propulsion Conference, Heraklion, Greece, 5–9 May 2008.

A Novel AOCS Cold-Gas Micro-Propulsion System Design and Applications to Micro and Nano Satellites

E. Razzano and M. Pastena

Abstract Propulsion systems are required by several space applications, e.g. Formation Flying, when position and attitude control are foreseen and shall be very fine. The system design can present many challenges regarding system mass and performances: a great number of equipments and thrusters may be required and low and variable thrust values could be needed to manage disturbances on orbit. Throttleable thrusters often lead to an increase of complexity and system size, against small-spacecraft constraints. In this chapter a novel micro-propulsion system is proposed, based on MEMS micro-thrusters developed in house by Carlo Gavazzi Space S.p.A. The system design architecture refers to a baseline configuration where each thrust valve commands a torque around an assigned direction. Optimization for two Italian missions (micro and nanosatellite) is analysed. Both applications require high-accuracy attitude control, involving low thrust values (order of 10 mN) and low impulse bits (order of 10^{-5} Ns). With respect to the first application, a comparison with a reaction-wheel based system in terms of mass and power has been made, showing a power gain of about 5 W against a propellant mass of about 1.5 g for each attitude manoeuvre (200 s each) in LEO orbit.

1 Introduction

Propulsion systems are required by several space applications, e.g. Formation Flying, when position control is foreseen and attitude control shall be very fine. On the other hand, the System design can present many challenges regarding system mass and performances, when applied for Small-Spacecraft Position and Attitude control. A great number of equipments (valves, pressure regulators and transducers...) and thrusters are required to provide attitude control torques around each satellite axes and to apply thrusts along each direction. Conventional Off-the-Shelf thrusters are often not easily customizable for this kind of applications because of equipments high mass and size, and eventually also high power consumption.

E. Razzano (✉)
Satellites Project, Carlo Gavazzi Space SpA, Via Tiengo snc, 82100 Benevento, Italy
e-mail: erazzano@cgspace.it

R. Sandau et al. (eds.), *Small Satellite Missions for Earth Observation*,
DOI 10.1007/978-3-642-03501-2_40, © Springer-Verlag Berlin Heidelberg 2010

Moreover, low and variable thrust values are needed to rejection of disturbances on orbit. Throttleable thrusters often lead to an increase of complexity and system size, against small-spacecraft constraints.

Carlo Gavazzi Space S.p.A. (CGS) developed customized Cold Gas propulsion systems for Small Satellite applications based on MEMS micro-thrusters.

Applications to micro and nano satellite platforms have been investigated and are presented in this chapter: an Italian phase B project for a micro-platform for Earth Observation (MIOSAT) and a feasibility analysis for an Italian On Orbit Servicing nano-satellite (OLOS). Both applications require high-accuracy attitude control during operational phases, involving low thrust values (order of 10 mN) and low impulse bits (order of 10^{-5} Ns). A comparison with a reaction-wheel based system in terms of mass and power has been made and presented in this chapter.

2 Propulsion System Architecture

A typical blow-down cold gas propulsion system is mainly composed by the following elements: high pressurized tank; a gas loading valve; filtration; Pressure regulation; Pressure relief and a set of thrusters with valves.

Generally, thrusters and valves are univocally connected, and often they constitute unique equipment. System architecture are frequently different for orbit control and attitude control. System design is often redundant considering that a large number of failure occurs on the thrust valve. The architecture could be fully redundant (Fig. 1a), or split in two branches after pressure regulator (Fig. 1b).

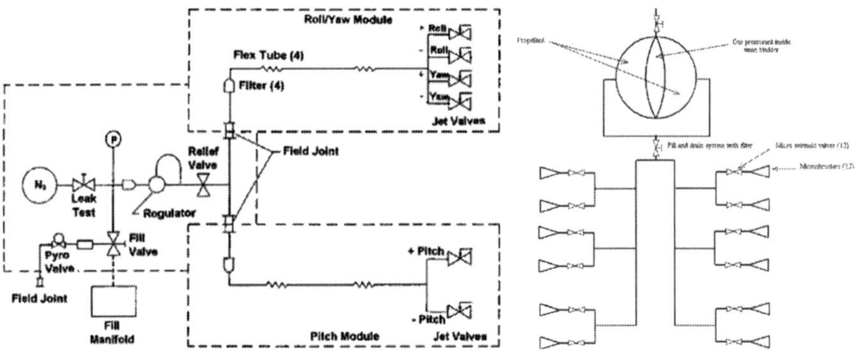

Fig. 1 Examples of architecture. (**a**) one of the two redundant Viking Orbiter reaction control systems ([1], p. 249); (**b**) two redundant branches propulsion system ([2], Fig. 4)

On the contrary, when nozzle are not fused with their valve, the architecture could be different and a better redundant philosophy that minimize system mass and power consumption could be found. The first possibility is to apply the same concept design as when nozzle and valve are a unique element (Fig. 1b). Another solution foresees an Introduction of the second branch only for orbit control, like ALMASAT system (Fig. 2).

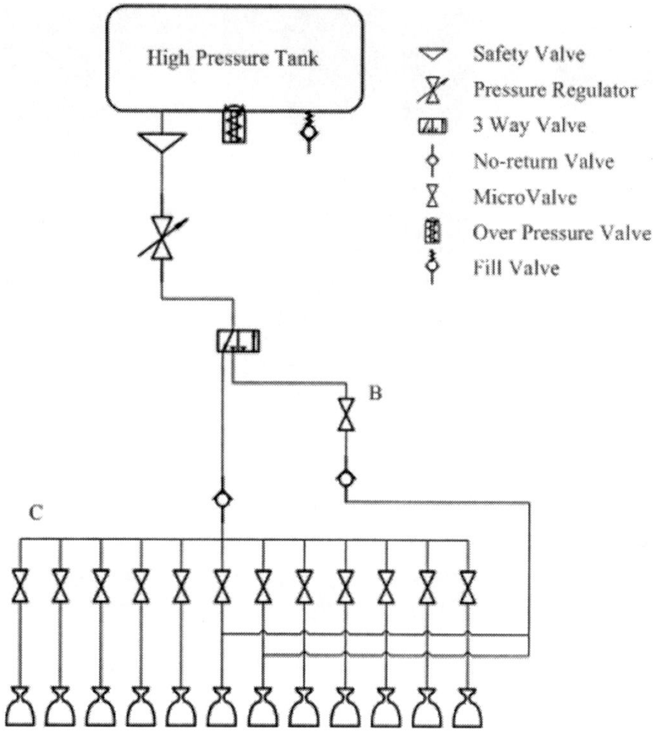

Fig. 2 ALMASAT propulsion system [3]

CGS Micro-thrusters are manufactured using MEMS technologies, reaching very small size throat nozzles (about 10 μm size), which reduced thrust level and achievable impulse bits [3, 4]. Thanks to their size of about μm, micro-thrusters can be accommodated in array to enlarge thrust modularity (Fig. 3): the nozzles can be fed in number related to the required thrust level, and the performances can be then customized to satisfy mission requirements; the thrusters can be operated in continuous or pulsed mode. The same system can then be used for disturbances torque rejection, attitude manoeuvres and orbit/position control, with a high reduction of the overall mass.

The proposed system architecture refers to a baseline configuration where each thrust valve commands a torque around an assigned direction. A lower number of thrusters can be involved. The use of the same valve for feeding two nozzles improves the synchronization of the thrust torque, while using two different valves lead to an increase of difficult of control law definition. When high accuracy is required, as in the following applications, a precise synchronism is demanded and the proposed first configuration achieves the condition better than the second one. On the opposite, more complex piping may occur to connect the redundant branches with the nozzles arrays.

Fig. 3 Example of CGS thruster array (during manufacturing process)

A ball latch valve protects each system branch against leakage of the thrust valves. The presence of micron filters integrated at three stages protects the contamination-sensitive seat elements of the "active" components.

A dedicated and customized board computer commands the valves, and it interfaces with the Satellite On Board Computer.

3 Applications

3.1 Microsatellite Attitude and Orbit Control System

MIOSAT program has been conceived as technological demonstrator ASI program with focus on Microsatellites and optical technology growth for satellite applications. The Space Segment includes a platform, class micro, which takes on board a set of optical instruments and technological experiments. Satellite lifetime will be 2 years from completion of the commissioning phase. The satellite will operate in a LEO sun-synchronous orbit at about 520 km, with a local time of descending node of 09.30 and with the capability to tilt both cross-track and along track from the nadir pointing. The mission is designed with a repeat cycle of 8 days. The overall mass is about 170 kg, the dimension $600 \times 900 \times 1{,}300$ mm and the average power consumption is about 120 W.

The propulsion system incorporates ten thruster arrays: six fully dedicated to attitude control and two fully dedicated to orbit control. Two arrays will be used both for attitude and orbit control. The accommodation of the arrays on the platform is illustrated in the schematic diagram of Fig. 4: the red thrusters are devoted to orbit control (the fourth thruster is symmetrically placed at the fourth corner of the satellite base panel). The system foresees two thruster arrays branches in a redundant configuration, with seven valves each (see Fig. 5). The dry mass is about 5 kg, tank excluded. The system architecture is single-point-failure tolerant. The propellant is N_2.

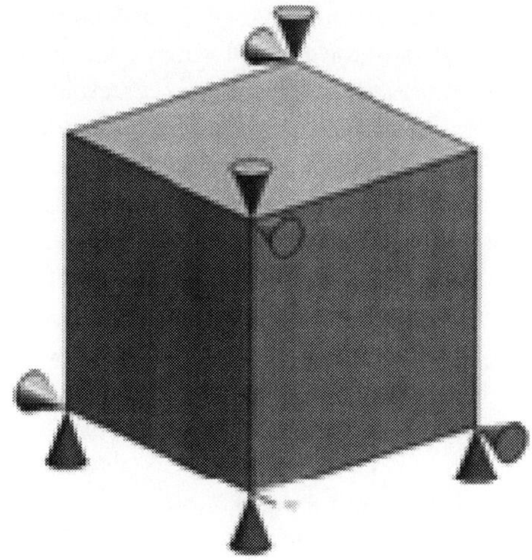

Fig. 4 Thrusters accommodation on MIOSAT

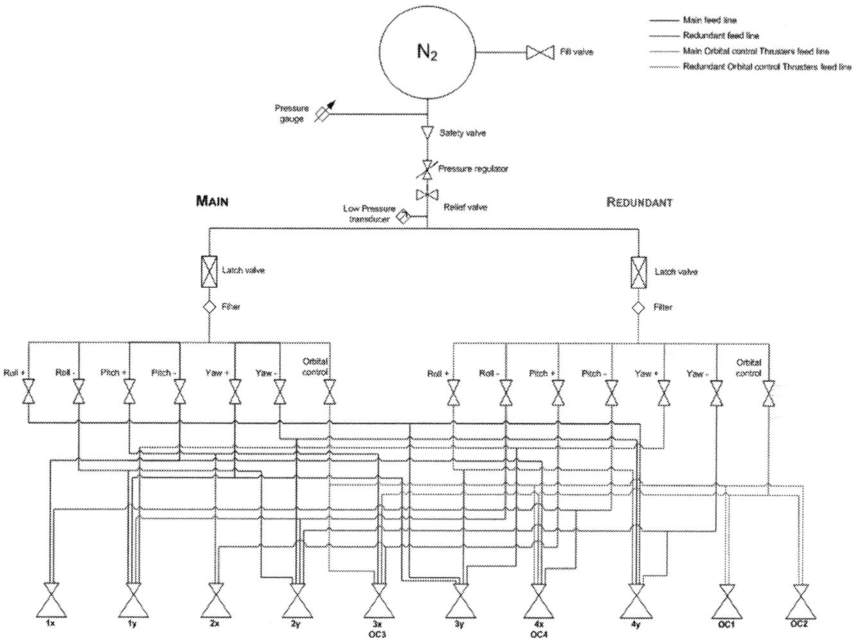

Fig. 5 Cold gas propulsion subsystem for microsatellite applications

The system design has been then optimized following a trade off between the configurations taking into account mass and power gain.

The sizing of the thrusters devoted to orbit control followed the requirement of performing the orbit manoeuvre in less than the 70% of an orbit period, i.e. Hohmann transfer with a $\Delta V = 5.54$ m/s in at least 3,970 s. Four thruster arrays of 100 mN each are foreseen to fulfil the requirement.

For the attitude control, a first consideration on the performances is worth of noticing: attitude control subsystem shall include the capability for both attitude manoeuvres and rejection of external disturbance torque affecting the attitude dynamics, due to disturbances. Nevertheless, the two requirements are very different in the MIOSAT application, and the sizing of a system which fulfill both is very challenging. System mass will increase more and more; different valves and nozzles must be used for the two applications, vanishing the original idea. Hence, the design criteria has been to distinguish two systems devoted one to attitude manoeuvres (and orbit control) and the other to disturbance rejection. In this chapter it is analysed only the Propulsion System dedicated to orbit control and attitude manoeuvres.

The performances of the Propulsion Subsystem are compared with those of a Reaction Wheels assembly, taking into account possible thrust-values of 10, 15 and 20 mN. The analysis considers also three values of Specific Impulse: 60, 65 and 70 s, following the achievable performances emerged in preliminary tests on CGS Micro-thrusters. The sizing requirement is to perform a pointing manoeuvre of about 65°. The manoeuvre time is a result of the subsystem design: the minimum duration ($t_{\min}$) is equal to 134 s (T = 10 mN) and it is determined from the relation (1), where the control torque is supposed to be applied by a bang-bang manoeuvre at the maximum acceleration.

$$t_{\min} = 2\sqrt{\frac{\vartheta \cdot I}{T \cdot b}} \tag{1}$$

θ is the manoeuvre angle, equal to 65° in this analysis; I is the Satellite's momentum of inertia (equal to 28 kg·m^2); T is the thrust and b is the Momentum arm (b = 0.7 m).

The comparison considers the variation of the power consumption of both thrusters (2) and wheels (3) with the manoeuvre length. The baseline Reaction Wheels are manufactured by ANTARES scarl (ARW-20.1-F839 Model, with the max torque equal to –/+ 20 mNm). The consumption includes mechanical power consumption, dissipating power and electronic board power. It also includes power consumption required for wheel desaturation. The wheel axis is considered aligned with the torque axis.

$$P_{TH} = \frac{n_{\text{burst/man}} \cdot n_{\text{thruster/burst}} \cdot (P_{\text{valve-open/close}} + P_{\text{transducers}} + P_{\text{coast}} \cdot t_{\text{coast}})}{t_{\text{manoeuvre}}} \tag{2}$$

$$P_{RW} = T_m(t)\, w(t) + RI(t)^2 + P_{\text{el}} + P_{\text{des}} \tag{3}$$

The following considerations have been made for thruster power consumption: each manoeuvre requires two bursts (to start and stop the manoeuvre), and each burst is performed by two thrusters in pairs: the relevant value, indeed, is the number of valves supplied per burst, and in the proposed configuration it is only one, with half power required with respect to conventional configuration. The power shall supply the valve opening and closing phase and the coasting time. Other consumptions (e.g. of pressure regulator and transducers) shall be taken into account. The higher the thrust, the lower the time to supply the valves to perform the manoeuvre in the same time (Fig. 6).

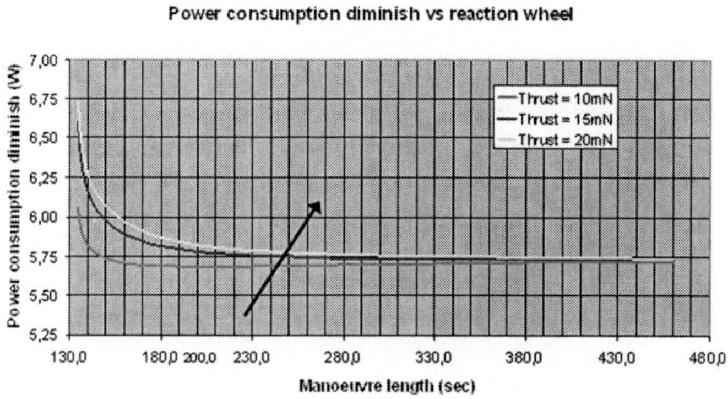

Fig. 6 Power consumption diminish of propulsion with respect to reaction wheels vs. maneuver length

Considering a single attitude manoeuvre of about 200 s, the gained power is about 5.7 W using the propulsion system instead of reaction wheels.

Following the mission scenario of MIOSAT, the gain in terms of power is evaluated also for the average consumption in a repeat cycle of 8 days. The value of 20 mN for the thrust has been fixed, due to the higher power gain, and a 60 s Specific Impulse (as worst case) is considered. The baseline configuration for Reaction Wheels is tethraedral configuration performing an attitude manoeuvre of 65° in 200 s around Body-X axis.

The analysis takes into account that, during the manoeuvre, the power is assumed provided to the platform only by the battery without any contribution from the solar panel, either in sunlight or in eclipse conditions. Even if a longer manoeuvre could require a lower power demand from the actuators, the reduction of the exposition time of the solar array to the sunlight requires an extra effort of the Power Subsystem. The pictures in Fig. 7 show the gain in terms of reduction of Power consumption of the Satellite AOCS using the proposed Propulsion subsystem instead of Reaction Wheels, when the manoeuvre is performed during the daylight part of the

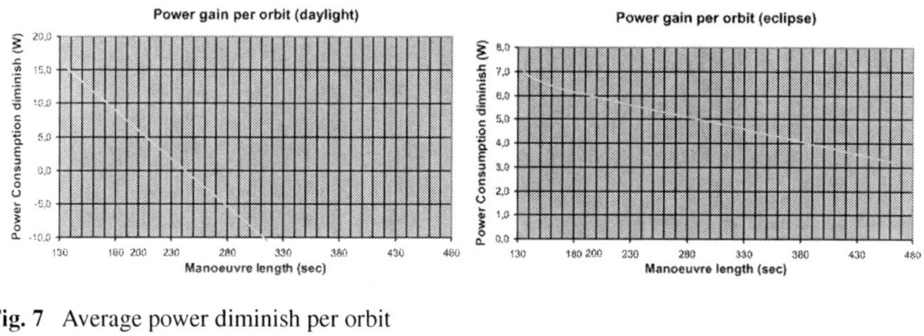

Fig. 7 Average power diminish per orbit

orbit or in the eclipse part. The consumptions are averaged out in the orbital period. It can be noticed that the use of Propulsion subsystem during eclipse periods lead to a reduction of Power requirement for any manoeuvre length. On the other hand, with a manoeuvre time over than 240 s, during daylight the Power required from thrusters becomes higher than the one required from Reaction Wheels.

Considering the repeat cycle of MIOSAT mission (with manoeuvres scheduled during daylight and eclipse periods), according to the baseline duration of 200 s per manoeuvre, the average power gain lowered to about 2.5 W per orbit (Fig. 8). This first consideration shows that for manoeuvre faster than 255 s the power consumption of propulsion system is higher than the one of Reaction Wheels assembly.

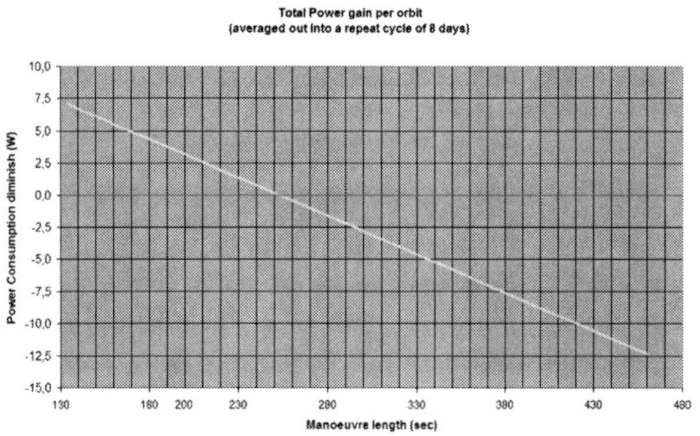

Fig. 8 Power gain per orbit (averaged out into a repeat cycle) using propulsion subsystem

Nevertheless, Satellite mass increases with the propellant: MIOSAT mission requires about 1.5 g of propellant/manoeuvre, with a global value of 6.5 kg of propellant to perform the scheduled manoeuvres for attitude control, as shown in Fig. 9 (i.e. about 4,200 manoeuvre in the lifetime). Table 1 compares the mass

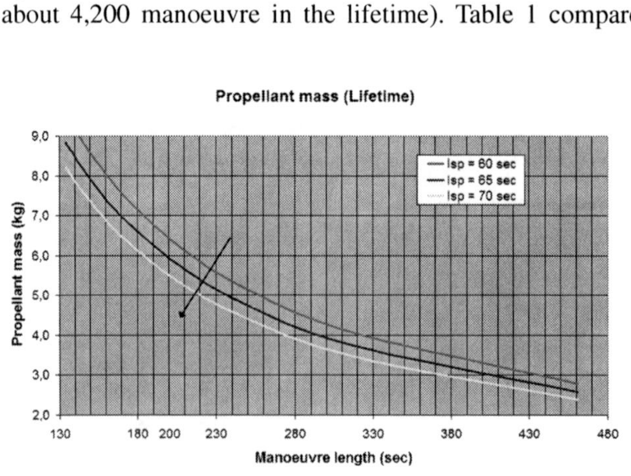

Fig. 9 Propellant mass for the MIOSAT lifetime

Table 1 Mass estimation for different subsystem configurations

Subsystem configuration	Dry mass (kg)	Propellant mass (kg)	Total mass (kg)
Micro-propulsion system (orbit and attitude control)	10.8	11.5	22.3
Micro-propulsion (orbit control) + Reaction wheel attitude control system	5.6 + 9.6	5	20.2

of the AOCS subsystem if Propulsion or Reaction Wheels are used for attitude control.

3.2 Nanosatellite Position and Attitude Control System

The objective of the mission conceived in OLOS phase A study is to provide a preliminary validation of some of the concepts and technologies necessary to perform the target operative monitoring missions of an OOS nanosatellite. The application can be considered as one kind of Formation Flying, where the nanosatellite shall fly around the target spacecraft with a specific flying profile. The presented system can be afterwards customized and adapted for other Formation flying applications.

The Space Segment includes a platform, class nano, which takes on board a stereocamera, for an overall mass of about 20 kg, dimension of 30 × 30 × 20 cm^3, and an average power consumption of about 20 W. Satellite lifetime will be 1 year from completion of commissioning phase. The satellite will operate in a LEO sun-synchronous orbit, at altitude of 800 km, with a local time of descending node 09.30 and with the capability to control and manoeuvre in/around every directions. A reaction-wheels attitude control system can not perform position control, so propulsion system is mandatory for this kind of application.

Twelve thrusters array are located within holes of the six OLOS panel, equally spaced (Figs. 10 and 11). The specific impulse is higher than 60 s. As a technological demonstrator, the system does not foresee any redundancy.

In this application, the system shall perform orbit control, position control, attitude manoeuvres and disturbance torque rejection. As a matter of fact, following the scheduled demonstration mission, the system is sized:

- to compensate disturbance torques. The sizing torque value refers to the Magnetic torque due to on-board residual dipole: maximum value of $5e^{-7}$ Nm.
- to perform the identified attitude control manoeuvres, with a maximum torque value of 2.1 mNm once in an orbit period.
- to perform the identified position control manoeuvre, with the worst case of a $\Delta v = 5.19$ mm/s in less than 30 s.
- for orbit maintenance, being able to perform a $\Delta v = 2.59$ mm/s in less than 15 s.

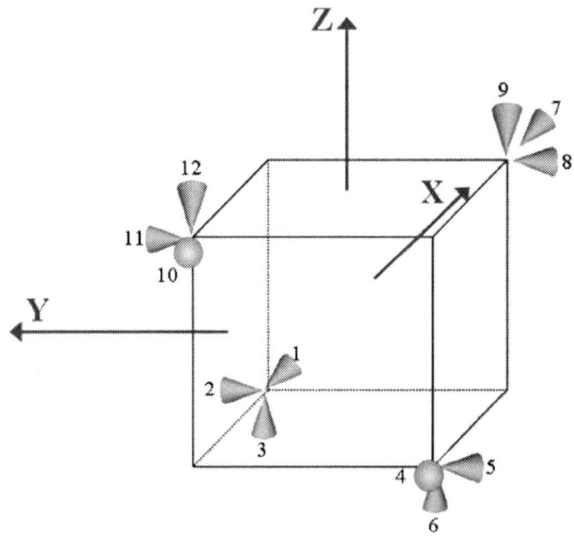

Fig. 10 Thrusters accommodation on OLOS

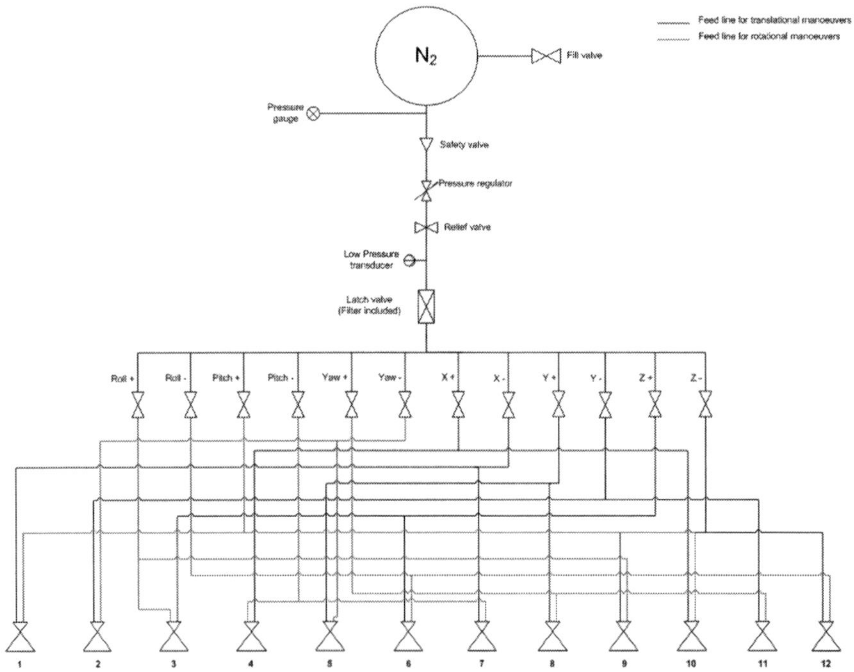

Fig. 11 Miniaturized cold gas propulsion subsystem for nanosatellite applications

The nominal thrust level has been determined after a trade-off. The disturbance torque rejection is strongly affected by the minimum impulse bit achievable, inversely proportional to thrust: the lower the thrust, the longer the impulse duration, while it can be difficult manufacturing valve with very short impulse duration.

Also regarding the power consumption, the lower the thrust, the longer the time to supply the thrust valves to provide the correct impulse. After those considerations, the baseline value of 7 mN is selected. The system shall hence provide:

- a $5e^{-7}$ Nm torque for an impulse duration of 0.06 s on every axis of OLOS body reference frame;
- a control torque of 2.1 mNm for an Impulse duration td = 0.1 s;
- an impulse of 22 s to perform the position control manoeuvre (worst case);
- an impulse of 11 s to perform each orbit control manoeuvre.

The system foresees an average power consumption lower than 2 W (margin included) per orbit period, and the mission lifetime needs a total propellant mass of about 1 kg of N_2 propellant (a 30% margin is included).

References

1. C.D. Brown. Elements of Spacecraft Design, AIAA Educ. Series, USA, pp. 153–254, 2002.
2. D. Platt. A Monopropellant Milli-Newton Thruster System for Attitude Control of Nanosatellite. 16th Annual USU Conference on Small Satellite, 2002.
3. P. Tortora, R. Cocomazzi, D. Modenini, A. Avanzi. Controllo d'assetto per microsatelliti mediante sistemi di micropropulsione a gas freddo, Università di Bologna, II Facoltà di Ingegneria, 2006.
4 J. Mueller. Thruster options for microspacecraft: a review and evaluation of state-of-the-art and emerging technologies. In M.M. Micci, A.D. Ketsdever (eds.), Micropropulsion for Small Spacecraft, Progress in Astronautics and Aeronautics, AIAA, Reston, VA, 187, pp. 45–137, 2000.

Part VII
Navigation

Navigation Needs for ESA's Earth Observation Missions

J. Roselló Guasch, P. Silvestrin, M. Aguirre, and L. Massotti

Abstract This chapter outlines GNSS receiver requirements driven by navigation needs identified from applications in future ESA Low Earth Observation (LEO) satellites. In the domain of operational EO, the GMES mission Sentinel-3 (S3) poses the most demanding positioning requirements in real-time in order to aid the altimeter tracking and also in on-ground post-processing for POD. The Sentinel-1 (S1) navigation requirements are also demanding in post-processing in view of differential interferometry applications with its SAR payload. Sentinel-2 has more relaxed navigation needs as compared to S1 and S3. A series of these Sentinels is planned to be manufactured with a common receiver and a common processing procurement approach. Sentinel-3, and possibly other Sentinels too, will require Galileo-compatible receivers in-orbit from 2017. The Sentinel-5 precursor (S5p) mission for air-quality and the candidate mission to the 7th Earth Explorer (i.e. Biomass, CoreH2O, Premier) require GNSS solutions, but the navigation requirements are less demanding than for S3 and S1. Within meteorological missions, the GNSS Radio Occultation instrument on Post-EPS will need to meet very demanding along-track velocity requirements and requires a high number of channels compatible with multiple GNSS. Regarding Earth science missions, the high-resolution gravity monitoring with Post-GOCE implies demanding post-processing relative position and velocity noise measurements. As a general trend, besides a multi-GNSS capability (GPS and Galileo, at least), the capability of on-board POD, already partly considered in S3, is likely to become more of a common feature for all missions. The robustness and reliability of the navigation solution will also become increasingly important. This chapter also describes the key technology "building blocks", like the AGGA-4 digital GNSS processor to fulfill the requirements in these future missions.

J.R. Guasch (✉)
ESA/ESTEC, European Space Agency, Keplerlaan 1, 2200 AG, Noordwijk, The Netherlands
e-mail: Josep.Rosello@esa.int

R. Sandau et al. (eds.), *Small Satellite Missions for Earth Observation*,
DOI 10.1007/978-3-642-03501-2_41, © Springer-Verlag Berlin Heidelberg 2010

1 Introduction

In the 1980s, the Global Navigation Satellite System (GNSS) techniques, devised for centimetre level global geodesy, were adapted to provide centimetre level orbit determination accuracy for Low Earth Orbit (LEO) satellites. Current applications include Precise Orbit Determination (POD) in support of altimetry, global geodesy and atmospheric sounding by Radio Occultation (RO). In the following sections we will review them, with examples from future ESA EO missions, emphasising that these applications may have very different data timeliness, robustness and accuracy requirements, which may impose different receiver architectures and operational approaches.

Possible future ESA EO missions and their navigation requirements are presented here below in the following sections of this document. Section 6 addresses the main drivers in GNSS receivers like the need for compatibility to many GNSS systems or single frequency versus dual-frequency receivers. Section 7 presents the basic building block technologies (e.g. AGGA-4 baseband processor) to fulfil these requirements.

GNSS Reflectometry (GNSS-R) has specific processing requirements for the elaboration of iso-Doppler maps that imply the need for a receiver-type as compared to most GNSS receivers and is not further described in this chapter. At present, there is no GNSS-R mission being considered under the ESA's Living Planet Programme, which includes the Sentinel missions in the domain of operational EO, the Earth Explorers in the domain of missions designed to address key scientific challenges and breakthrough technology in observing techniques, or in the domain of meteorological missions like Post-EPS.

2 Altimetry in Sentinel-3

The main objective of the mission is the observation with a radar altimeter of sea surface topography and sea ice measurements (see columns 3, 4, 5 in Table 1) in (enhanced) continuity of the Envisat and Cryosat missions. On-board Real-Time navigation (column 2 in Table 1) is required in order to aid the open-loop tracking of the altimeter when changing mode (land, coast, open ocean), hence improving mission data availability in particular in coastal zones.

Table 1 Error budget requirements in Sentinel-3 as a function of time with respect to measurement

Targets	Real-time	<3 h	<1~3 days	<1 month
Radial orbit error (RMS)	<3 m	<8 cm	<3 cm	<2 cm
Application	Support tracking mode changes	Atmospheric dynamics	Ocean	Global change

This level of accuracy requires dual-frequency receivers. Three models (A, B, C) will be manufactured in the first generation of Sentinel-3. For Sentinel-3-C (2019)

and Sentinel-3-B (2017), the receivers need to be compatible with modernized GPS and Galileo signals. This will also allow increasing the number of channels, minimize multipath and achieve the best possible performance and robustness.

Future additional altimetry missions under consideration (e.g. for mesoscale ocean sampling) will also require similar (or better) performance.

3 Other Sentinels and Candidate Earth Explorers

Sentinel-1 will fly a high-resolution (~5 m) Synthetic Aperture Radar (SAR) instrument in C-band. Orbit knowledge accuracy (<3 m RMS in each axis) in real-time for autonomous operations is not considered as demanding as the on-ground post-processing requirements (<5 cm 3D RMS) for the detection of (slow) land movements and deformations through the differential interferometry technique [1]. The latter is almost as demanding as for Sentinel-3 and requires dual-frequency receivers.

Sentinel-2 carries optical instruments and has more relaxed POD requirements than the other Sentinels. A single-frequency solution would suffice, however, due to common procurement approach in GMES, the same dual-frequency receiver will be used as for Sentinel-1 and Sentinel-3.

The Sentinel-5 precursor (S5p) mission for air-quality monitoring is likely to fly in formation with the NPOESS "afternoon" polar satellite, and requires POD for co-registration of the observations [2].

In the frame of the candidate mission to the 7th Earth Explorer (EE7), ESA is considering two medium-resolution SAR missions: Biomass and CoReH2O. The lower spatial resolution with respect to Sentinel-1 implies more relaxed real-time navigation requirements. Biomass (in P-band) is more demanding than CoreH2O (Ku- and X-band) because it will require interferometric post-processing to estimate biomass content and the height of forests, however this type of processing is less demanding in POD accuracy than for the POD needed in the "differential" interferometry of Sentinel-1. Premier is another EE7 candidate and would fly in tandem with MetOp, thus requiring lower resolution atmospheric data co-registration, however a single-frequency GNSS receiver is deemed sufficient.

4 Gravity Field Monitoring in Post-Goce

GOCE, the 1st ESA Earth Explorer, has been launched in 2009 and aims at modeling gravity field to microgravity level and uses GPS for POD.

The objective of a Post-GOCE mission, presently under study, is the investigation and monitoring of geophysical phenomena associated to mass distribution changes occurring in the geosphere, atmosphere, hydrosphere, cryosphere and biosphere. The gravimetric objective is the long-term monitoring of temporal variations of the gravity field at a spatial resolution as high as for GOCE, but extended also to longer wavelengths [3]. In the present concept, Post-GOCE would be composed of two satellites flying in formation. Accelerometers are utilized to measure non-gravitational forces, so to drive the on-board drag compensation system. The laser interferometer is designed to resolve an inter-satellite distance variation of ~1 nm over 10 km, but the formation control real-time requirements are not stringent (<30 m RMS, <0.03 m/s RMS) thanks to the utilization of a specific mechanism to control the orientation of the laser interferometer beam. That distance must be kept to within ±500 m along the inter-satellite line and ±50 m in the transversal directions. The post-processing position accuracy requirement is 3 cm, also within the state-of-the-art, however the required relative position (<2 mm/$\sqrt{Hz}$) and velocity (<0.005 mm/s/$\sqrt{Hz}$) measurement errors are more challenging. Use of GPS and Galileo signals is assumed.

5 Radio Occultation in Post-EPS

Within meteorological missions, Post-EPS (EUMETSAT Polar System) is based on the MetOp satellite series. The GNSS Radio Occultation (RO) instrument on the Post-EPS satellites, the first of which is to be launched in 2018, has a direct heritage from the GRAS on-board MetOp-A. Its main objective is to measure refraction angle profiles in the troposphere and the stratosphere, i.e. from ground up to 80 km, with high vertical resolution and accuracy [4].

Unlike most GNSS receivers, RO sensors observe signals from the Earth's limb and pose the most stringent POD requirement in the along-track direction, with maximum acceptable velocity error <0.1 mm/s. RO sensors also have to deal with very low SNR and highly dynamic signals when the signals traverse the lower troposphere. Directive antennas are required in the velocity and anti-velocity directions. An Ultra Stable Oscillators (USO) is also required to achieve good accuracy of retrievals in the stratosphere (i.e. at high altitudes). In addition, techniques of open-loop tracking, which use predicted Doppler profiles based on geometrical and geophysical information, allow one to recover the excess Doppler shift of the signal crossing the lowest part of the atmosphere when close-loop tracking would not be possible. OG post-processing of the measurements (namely code and carrier phase or I/Q receiver correlator outputs), in combination with POD, is required in order to reconstruct, in an iterative way, the geometry of the observation and the wave bending angles. Dual-frequency receivers are mandatory for adequate ionospheric compensation.

The RO sensor must be capable of observing as many rising and setting occultations as possible – at least 1,000 occultations/day are specified for Post-EPS when using both GPS and Galileo constellations. Assuming availability of specifications for Glonass and Beidou signals, an even higher number of occultations would be feasible. A secondary objective of the Post-EPS RO is also to provide vertical coverage of the ionosphere. For all these reasons, RO receivers need to be very flexible and powerful in their digital processing, e.g. including special capabilities to sample GNSS signals rather fast, and to allow the processing of as many GNSS signal types as possible.

6 Main GNSS Receiver Drivers

Taking into consideration the requirements for the future missions presented above, GNSS solutions need to address architectural aspects such as the need to use single frequency (SF) or dual frequency (DF) for ionospheric correction, compatibility with new GNSS constellation signals (e.g. from Galileo, with options available for highly accurate code phase measurements), and processing approaches, e.g. use of (reduced) dynamic POD [5].

Recent studies [6–8], describe the results achieved for using GPS POD of the GRACE satellites at 470 km altitude and of Metop-GRAS at 817 km altitude [8]. shows that the benefit of using a second frequency for ionospheric correction gets lower as the orbital altitude increases. It also shows that other orbital parameters, and in particular those related to solar activity and day-night changes, may have an even higher effect than altitude in the choice between SF and DF. Techniques like the use of carrier-code ionospheric divergence (GRAPHIC [9]) allow orbit accuracy up to at the very best 10 cm 3D RMS.

Consistency at the 5 cm 3D RMS level in Metop-A/GRAS for reduced-dynamics orbit solutions computed by four different software packages was demonstrated within ESA study Contract 20647 [6, 7], and is to be compared to the 10 cm obtained through only purely kinematic solutions. Reference [6] also shows the feasibility, in on-ground post-processing, of 1–2 cm RMS radial position for altimetry and 0.05 mm/s RMS along-track velocity for radio occultation. In [6], real-time (RT) ephemerides were combined with RT GNSS solutions, reaching 0.55 m 3D RMS (24 cm only in radial) with the publicly available IGS ephemerides [10], and 0.14 m 3D RMS (8 cm only in radial) with the commercial JPL GDGPS ephemerides [11], which makes GPS on-board RT navigation at least equivalent to other tracking systems, e.g. DORIS.

7 Technology Building Blocks to Achieve the Requirements

The typical GNSS receiver is composed of several "building blocks" as shown in Fig. 1.

In an application like Radio Occultation, it is very likely that three of these modules are used: one connected to the velocity antenna, one to the anti-velocity antenna, and one to the zenith antenna utilized for the POD. It is also very likely that only two frequencies (and not three as shown in Fig. 1) are needed for ionosphere correction.

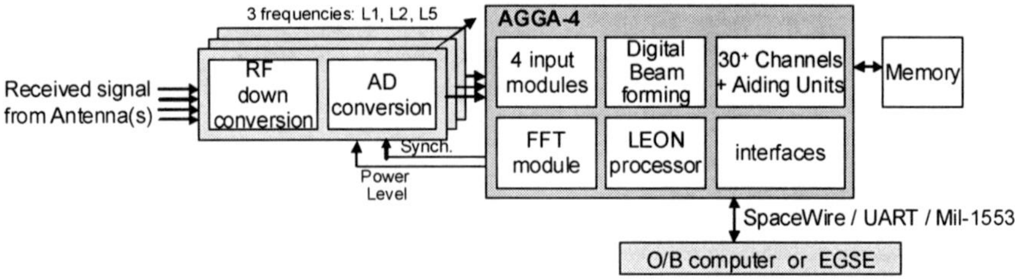

Fig. 1 AGGA-4 typical application module

7.1 RF Down Conversion and Analogue to Digital Conversion

In general, GNSS receivers just require narrowband and not more than 3 bits in the analogue to digital conversion (ADC). This is also true for Earth Observation applications, in particular if only public signals are processed, and it results in less complex technology. Requirements on signal quality and reproducibility impose the replacement of earlier approaches based on a large number of discrete components by dedicated and more integrated devices like the Saphyrion (former Nemerix) chipset (i.e. NJ1008 for RF down-conversion and NJ1017 for ADC) developed under ESA Contracts. These devices are programmable and capable of processeing all the relevant public GNSS bands (e.g. L1, L2, E5a and E5b)

7.2 AGGA-4: Core Device for the Next Decade

The development of the first AGGA (Advanced GPS/GLONASS ASIC) device was initiated in 1995 within ESA in order to support Earth Observation (EO) applications using navigation signals. After prototyping iterations, the flight component known as AGGA-2 was manufactured under Atmel product code T7905E [6] in 2000. The AGGA-2 is a space-qualified digital integrated circuit providing all the high-speed digital signal processing functionality for EO applications such as atmospheric sounding with Radio Occultation (RO) [7] and Precise Orbit Determination (POD). AGGA-2 devices are used by a very large number of ESA missions (e.g. Metop-Gras a/b/c, GOCE, Sentinels 1/2/3, Swarm) and non-ESA missions (e.g. ROSA, Radarsat-2, Cosmo-Skymed).

New GNSS signals are expected to become available by the middle of the next decade. ESA will rely on the flexible AGGA-4 [12] component to enable the digital processing of all the public signals in modernized GPS, Galileo (see Table 2) and possibly Beidou/Compass and Glonass too. AGGA-4 will provide much higher integration than its AGGA-2 predecessor, such as for Example:

- larger number of GNSS single-frequency/double-code channels,
- a LEON2-Fault Tolerance microprocessor embedded on-chip,
- a large number of interfaces like SpaceWire, Mil-Std-1553 bus, UART ports,

- other functionality like very versatile input modules compatible with I/Q and Intermediate Frequency sampling, FFT module for fast cold acquisition, and Beamforming capabilities for Radio Occultation.

Table 2 Modernized GPS and Galileo signals and number of channels needed in AGGA-4 to process them

GPS signals	Nb. Channels needed
GPS/SBAS L1 C/A	1
L2CM (Data) and L2CL (Pilot)	1
L5-I (Data) and L5-Q (Pilot)	1
Galileo signals	*Nb. Channels needed*
E1-B (Data) and E1-C (Pilot)	1
E5a-I (Data) and E5a-Q (Pilot)	1
E5b-I (Data) and E5b-Q (Pilot)	1

AGGA-4 aims at having not less than 30 (target is 36) highly configurable single-frequency double-code (e.g. one for the pilot and the other for the data component) GNSS channels, which include:

- Very flexible primary code generator units with:
  - ○ a Linear Feedback Shift Registers (LFSR) for very long spreading codes (e.g. for signals like GPS L2CL with 767,250 chips)
  - ○ memory-based code capabilities required in signals like E1b and E1c.
- Support of Binary Offset Carrier BOC(m,n) and secondary code sequences and secondary code stripping used in modernized GPS and Galileo signals.
- 5 complex (I/Q) code correlators, to allow the Early-Early, Early, Puctual, Late, Late-Late correlations required for the processing of BOC signals.
- Two configurable delay lines for tracking of double-code (pilot/data) signals *in one channel*
- hardware Aiding Unit (see Fig. 2), allowing autonomous code and carrier phase aiding to compensate for the Doppler rate (Hz/s) caused by high orbit dynamics.

Multiplex BOC (MBOC) signals in the form of TMBOC or CBOC are processed with the BOC codes generated by AGGA-4 at the expense of minor losses in code phase tracking sensitivity.

AGGA-4 does not support signals like AltBOC, codeless P-code or non-public signals like Galileo Public Regulated Service (PRS – i.e. E6 and E1a), Safety of Life (SoL) and Commercial Service (CS). Priority was given to optimising gate counting, and as a result maximising the number of channels on-chip rather than adding more complexity for signals that that will not bring substantial gains for Earth Observation applications with respect to the new E5a, E5b or GPS L2C signals.

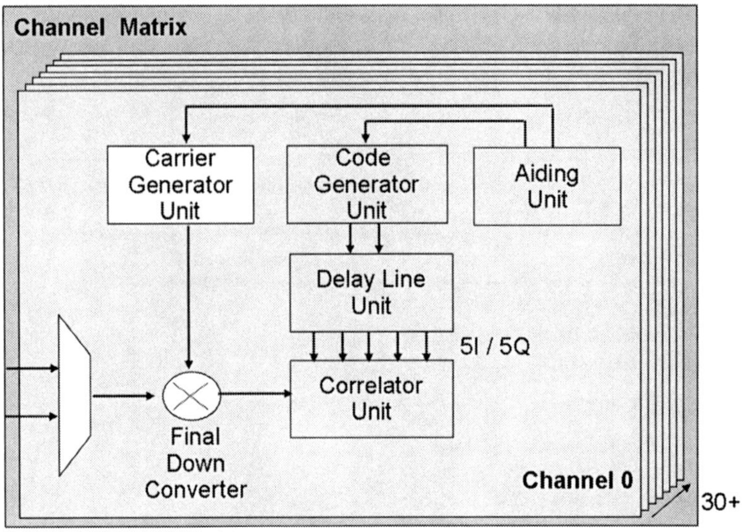

Fig. 2 GNSS channel matrix with 36 (target) GNSS channels

With respect to Table 2, and assuming 30 single-frequency dual-code channels, AGGA-4 allows to process 15 dual-frequency channels. Note also that it would also be possible to process only one component (e.g. Pilot) of these signals in one channel.

The AGGA-4 chip will be manufactured by ATMEL in the ATC18RHA 0.18 μm technology with a 352 MQFP package. The GNSS part of the AGGA-4 chip runs at max. 40 MHz and the LEON part runs at maximuM 80 MHz. The AGGA-4 is under development at Astrium GmbH and Austrian Aerospace under ESA guidance and contract. ASIC components are expected to become available for the whole European space industry by the end of 2010. Like for AGGA-2 and AGGA-3, the resulting Intellectual Property Rights (IPR) of AGGA-4 will continue to belong to ESA.

The ESA authors wish to thank the development teams at Astrium GmbH and Austrian Aerospace for the excellent work performed and outstanding cooperation.

8 Conclusions

Sentinel-3 (S3) poses the most demanding POD requirements, specially in the radial component, not only in post-processing on-ground, but also in real-time. Sentinel-1 navigation requirements are demanding only in post-processing for differential interferometry SAR. Other Sentinels and Earth Explorers have more relaxed requirements but would benefit from the common procurement receiver approach driven by Sentinel-3. S3-C (and possibly S3-B too) will already require Galileo-compatible receivers. A future Post-GOCE mission will require very stringent post-processing relative position and velocity noise measurements. Like GRAS on METOP, the Radio Occultation instrument of Post-EPS will pose very demanding along-track velocity requirements and will need to be compatible with as many GNSS systems as possible. With the availability of a larger number of signals and the new GNSS signal improvements, new challenges and requirements mainly on robustness, and to a less extent on accuracy too, will arise.

Overall, single-frequency receivers offer attractive POD solutions with around 10 cm RMS accuracy in post-processing, but the most demanding missions will keep a need for dual-frequency receivers. The capabilities already considered in Sentinel-3-C, which include accurate on-board POD and compatibility with new GNSS signals is likely to set the desired performance level for most missions, also for convenience at system level.

ESA relies in key technology "building blocks" like the AGGA-4 device and a miniaturized RF front-end to achieve these new challenges.

References

1. R. Hanssen. Radar Interferometry, Data Interpretation and Error Analysis, Remore Sensing and Digital Image Processing, Dordrecht, Kluwer Academic Publishers.
2. http://www.esa.int/esaLP/LPgmes.html > ESA's Sentinel satellites on left-bar.
3. System Support to Laser Interferometry Tracking Technology Development for Gravity Field Monitoring, ESA Contract No. 20846/07/NL/FF, Final report, September 2008.
4. P. Silvestrin, R. Bagge, M. Bonnedal, A. Carlstrom, J. Christensen, M. Hagg, T. Lindgren, F. Zangerl. Spaceborne GNSS Radio Occultation Instrumentation for Operational Applications, ION 2000 Conference.
5. O. Montenbruck, T. van Helleputte, R. Kroes, E. Gill. Reduced Dynamic Orbit Determination Using GPS Code and Carrier Measurements, Aerospace Science and Technology, 9 261–271, 2005. DOI: 10.1016/j.ast.2005.01.003.
6. ESA Contract 20647 /07/NL/GS, GRAS Tracking and Navigation Performance Analysis, with DLR, Technical University of Delft and University of Bern.
7. O. Montenbruck, Y. Andres, H. Bock, T. van Helleputte, J. van den Ijssel, M. Loiselet, C. Marquardt, P. Silvestrin, P. Visser, Y. Yoon. Tracking and Orbit Determination Performance of the GRAS Instrument on MetOp-A, GPS Solutions, 12, 289–299, 2008. DOI: 10.1007/s10291-008-0091–2.
8. H. Bock, A. Jäggi, R. Dach, S. Schaer, G. Beutler. GPS single-frequency orbit determination for low Earth orbiting satellites. Advance in Space Research, 2009, doi: 10.1016/j.asr.2008.12.003.
9. T.P. Yunck. Orbit determination. In B.W. Parkinson, J.J. Spilker (eds.), Global Positioning System. Theory and Applications, Washington DC, AIAA Publications, 1996.
10. International GNSS Service (IGS) – http://igscb.jpl.nasa.gov/
11. Global Differential GPS with JPL Real-time Ephemeris – http://www.gdgps.net/
12. J. Rosello, R. Weigand, G. Lopez Risueño, P. Silvestrin. AGGA-4 – Core device for GNSS space-receivers of the next decade, NAVITEC-2008.

Benefits of Galileo for Future Satellite Missions

Werner Enderle

Abstract This chapter outlines the benefits of Galileo for future satellite missions. In this context, it will address the Galileo services, signals and frequencies. Further, aspects of interoperability with other GNSS and their potential impact on space applications will be discussed.

1 Introduction

The number of GNSS-based space applications have significantly grown over the last decade. The areas of applications are wide spread and range from launchers, sounding rockets, automated transfer vehicles, and the space station, to single satellites, satellite formations and satellite constellations.

Future satellite missions will become more demanding and complex. However, at the same time it is expected that costs are reduced for infrastructure and operations. In this environment, on-board autonomy, as well as aspects of positioning and navigation (absolute and relative), will play a key role in the success of future missions. For this reason, new technology will have to be developed. In this context, GNSS-based technology will offer potential benefits of being; light weight, small-sized and consuming relatively low power. Furthermore, GNSS receivers can be used as scientific payload or as an integral part of the satellite attitude and orbit control system (AOCS).

In any case, the following types of orbits are considered of interest for GNSS space applications; Low Earth Orbits (LEO), Medium Earth Orbits (MEO), High Eccentric Orbits (HEO) and the Geostationary Orbit (GEO). Each of the four named orbits, or a combination of them, are interesting from commercial, scientific and military perspectives.

W. Enderle (✉)
European Commission, Galileo Unit G3, Rue De Mot 28, Brussels, Belgium
e-mail: werner.enderle@ec.europa.eu

R. Sandau et al. (eds.), *Small Satellite Missions for Earth Observation*,
DOI 10.1007/978-3-642-03501-2_42, © Springer-Verlag Berlin Heidelberg 2010

2 Galileo

2.1 Galileo Mission

Galileo's mission is to provide users with a state-of-the-art high accuracy global positioning and timing capability. This global capability will be designed to meet the needs of a wide range of applications [1].

Galileo is a civil system, operated under public control. Its mission includes the implementation of an independent Global Navigation Satellite System (GNSS) that will fit in with and operate alongside existing and planned navigation and communication systems to provide both enhancements and redundancy. Galileo is intended to be operational by 2013.

Concerning the use of Galileo for space applications, it has to be stated that Galileo has so far no service volume specifications for space users. However, it also has to be stated that the Galileo signals can be used for space applications. In any case, the future use of Galileo for space applications can be considered as a potential candidate in the context of Galileo Mission Evolution.

2.2 Galileo System Architecture

The overall Galileo System Architecture (see [1]) consists of three components, a space component, a ground control component and the user component. The global space component comprising the satellite constellation of 27 active satellites +3 spare satellites in Medium Earth Orbit. When fully operational, GALILEO will feature satellites spread evenly around three orbital planes inclined at an angle of 56° to the equator. The ground control component will consist of 30–40 globally distributed Galileo Sensor Stations, 10 Mission Up-link stations, 5 Telemetry and Telecomm and Up-link stations and 3 Control Centres (Fig. 1).

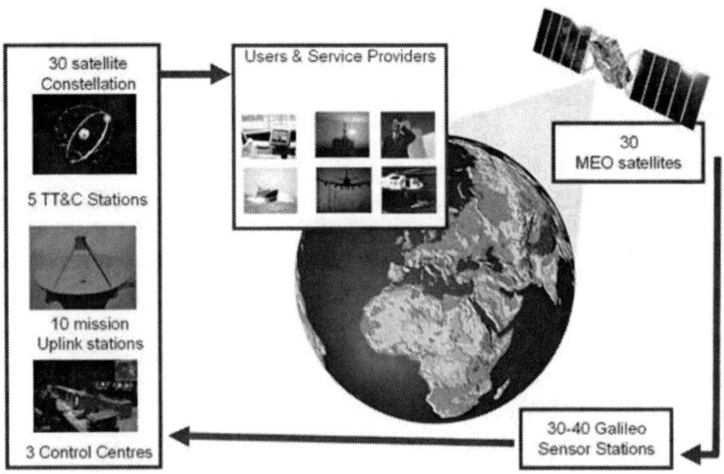

Fig. 1 Galileo architecture

2.3 Galileo Signals and Frequencies

The mapping between Galileo navigation services and the Galileo frequency bands is given in Table 1. The Galileo signal and frequencies plan is outlined in Fig. 2. As it can be seen, Galileo will provide 4 navigation services based on 4 frequencies.

Table 1 Mapping between Galileo services and its frequency bands

Band	E5a	E5b	E6	E1
Service	OS/SoL/CS	OS/SoL	PRS CS	PRS OS/SoL/CS

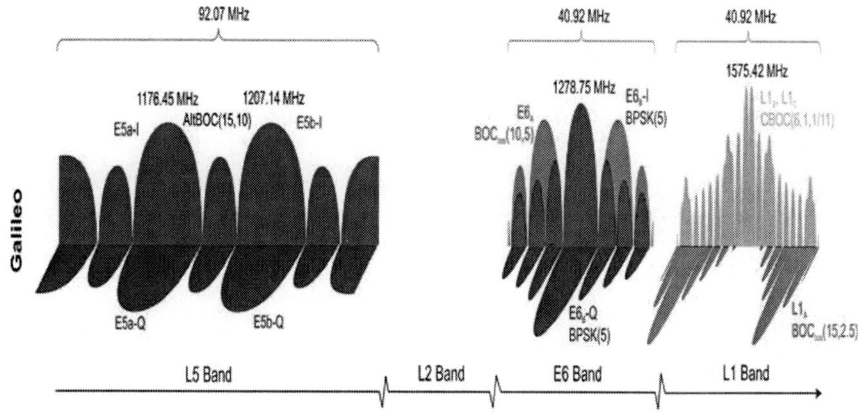

Fig. 2 Galileo signal and frequencies plan

2.4 Galileo Services

Galileo will provide 4 satellite-only navigation services. These services will be provided worldwide and independently from other systems:

Galileo *Open Service (OS)* – This service results from a combination of open signals, free of user charges, and provides superior position and timing performance compared to other GNSS systems.

Galileo *Safety of Life (SoL) Service* – Will be offered and guaranteed to the critical transport community, e.g. aviation, maritime, etc., delivering enhanced performance that includes the provision of the integrity function, i.e. a warning of system malfunction that will reach the user in a given alarm time.

Galileo *Commercial Service (CS)* – Provides access to two additional signals, to allow for a higher data throughput rate and to enable users to improve accuracy. The signals are encrypted. A service guarantee is envisaged for this service.

Galileo *Public Regulated Service (PRS)* – provides positioning and timing information to specific users requiring a high continuity of service, with controlled access. Two PRS navigation signals with encrypted ranging codes and data will be available.

In addition, Galileo's support to search and rescue services represents Europe's contribution of to the international COSPAS-SARSAT co-operative effort on humanitarian search and rescue activities. Galileo is to play an important part in the Medium Earth Orbit Search and Rescue system (MEOSAR).

2.5 Galileo Performance

The Galileo accuracy performance for the 4 navigation satellite-only services is outlined for the dual frequency case in Table 2.

Table 2 Galileo services – performance dual frequency

Galileo services	Horizontal accuracy (95%)	Vertical accuracy (95%)	Availability	Integrity
Open service	4 m	8 m	99.8%	No
Safety of life service	4 m	8 m	99.8%	Yes
Commercial service	Detailed performance requirement under elaboration			
Public regulated service	6.5 m	12 m	99.8%	Yes

2.6 Galileo Interoperability

Galileo interoperability with other GNSS is an important issue. The interoperability between Galileo and the civil GPS is implemented via the use of certain standards for time reference (TAI), the international terrestrial reference system (ITRS) and signals (codes and centre frequencies).

The first step of Galileo interoperability[1] with other GNSS was the agreement with the US (2004 and 2007-MBOC). Next steps will follow and will also cover GPS and other systems.

2.7 Galileo Implementation Plan

The Galileo In-Orbit Validation (4 satellites plus ground segment) is planned to start in 2010. Full Operational Capability (27+3 Galileo satellites plus ground segment) is expected around 2013, subject to ongoing FOC procurement process.

[1] ICG Definition

3 Galileo for Space Applications

3.1 GNSS Space Applications – General overview

The first tracking of GPS signals by a space borne receiver can be dated back to 1982 and has been performed onboard the Landsat-4 satellite. Tracking of GPS 1st side lobe signals and tracking of GPS signals in GEO altitudes was first done onboard the EQUATRO-S satellite in December 1997. In general, it can be stated that the use of GNSS technology for space missions has constantly increased, especially over the last decade.

GNSS is an enabling technology for many new space applications. This is not only reflected by the pure number of space missions using GNSS technology, but also in the diversity of the missions and applications.

The use of GNSS within space missions, ranging from launchers, sounding rockets, rendezvous and re-entry, Space station to the use onboard of satellites in Low Earth Orbits (LEO), Medium Earth Orbits (MEO), High Elliptical Orbits (HEO) and Geo Stationary Orbit (GEO), see Fig. 3.

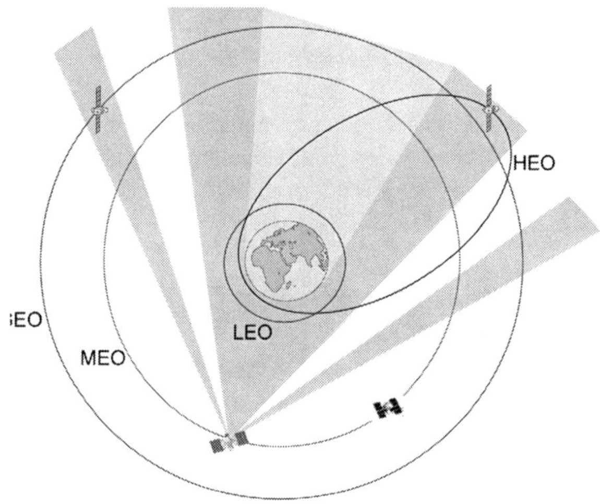

Fig. 3 Main lobe and 1st side lobe for GNSS and the four orbit types of main interest for GNSS space users

The main categories for applications are considered to be absolute positioning, relative positioning, Orbit -and Attitude determination, timing and atmospheric research.

The requirements for the above outlined categories of applications are wide spread. Requirements for the position accuracy in GEO are in the area of 100 m, whereas the position accuracy requirements for LEOs ranging from around 10 m level for absolute positioning to cm level for relative-navigation (e.g. formation flying). The requirements for scientific applications are even more stringent.

3.2 Galileo and Space Users

In the current Galileo design, no specific space user requirements have been taken into account. However, Galileo signals can be used for space applications in orbits below 1400 km altitude, based on the Galileo satellite antenna characteristic.

Despite the fact that in Europe, no detailed study for the identification of end user needs for space applications based on Galileo has been conducted so far, it can be stated that feedback from various GNSS R&D activities provided an indication that the momentum of this user community is growing. For this reason, the use of Galileo for space applications can be considered as a potential area for Galileo Mission Evolution.

In any case, it is considered that the main areas of discussions related to GNSS user needs for space applications will be related to Pseudorange accuracy, received power and signal availability for the subject orbits and interoperability issues between the various GNSS, which will exist.

In this context, it is also interesting to monitor the ongoing activities and discussions in the US, related to the space service volume definition for GPS III (see [2]).

3.3 Galileo Use in Space – Regulation

The use of Galileo for space applications falls under the Council Regulation (EC) No 1167/2008 amending and updating Regulation (EC) No 1334/2000 setting up a Community regime for the control of exports of dual-use items and technology.

3.4 Galileo – Potential Benefits for Space Users

As already mentioned in Chapter "EO Small Satellite Missions and Formation Flying", Galileo signals and services can be used for space applications. However, applicability of Galileo signals for space applications strongly depends on the type of orbit that is considered for the space applications. The most benefits will be obtained by users in LEOs, whereas users in other types of orbits will experience limitations in terms of availability of Galileo signals.

The benefits of Galileo for space applications outlined below strongly depend on the orbit type, but are certainly applicable for LEOs.

The benefits of Galileo for space users are:

- The number of available services, signals and frequencies associated to the services, provides flexibility in receiver design with respect to space mission requirements. This will allow mission tailored receiver concepts, ranging from single frequency receiver concepts to complex multi frequency concepts.
- Redundancy/Robustness of signals and services.

- Application of the Three Carrier Ambiguity Resolution (TCAR) technique for relative navigation, spacecraft attitude determination algorithms, precise orbit determination and scientific applications.
- High data rates for improved signal acquisition – Time To First Fix (TTFF) and reacquisition receiver behaviour.
- Quality of Pilot Tone ranging signals (no data on signal) for high ranging accuracy and also tracking of weak signals. This could support applications of GTO and GEO.
- Galileo wide band signals (AltBoc and MBOC) will provide improved accuracy and robustness against multipath effects.
- Integrity information will be available SoL and PRS.
- Authentication is available for SoL.
- High accuracy information available for CS.
- PRS provides a very robust service for authorized users.
- Combined use of Galileo and GPS (civil signal) plus other systems in future for improved visibility of GNSS satellites and signal availability for space users, especially for LEOs (see Fig. 4).

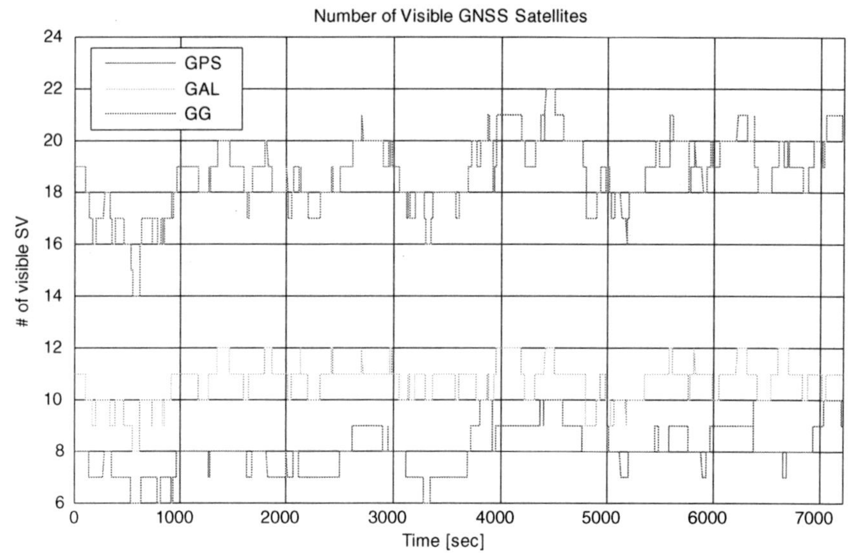

Fig. 4 Visibility of GNSS satellites for a space user in a typical LEO (550 km altitude) with a nadir pointing attitude and a GNSS antenna, which is directed away from the Earth (zenith direction)

4 Conclusions and Outlook

To date the Galileo Mission has not taken User Needs from Space Users specifically into account.

Galileo signals and services will be available and usable for space applications, particular in LEOs and Galileo will bring significant benefits to Space Users especially for LEOs, based on some distinctive features. In addition, space applications will greatly benefit from the interoperability (civil signals) between Galileo and GPS and in the future with other systems as well.

Galileo and the use of it in space will be subject to discussions in the context of Galileo Mission Evolution activities.

References

1. European Space Agency, Galileo Global Component System Requirement Document Issue 4 Rev 2 (SRD), 27 July 2004.
2. GPS Space Service Volume Increasing the Utility of GPS for Space Users, M. C. Moreau, Flight Dynamics Analysis Branch, NASA Goddard Space Flight Centre, USA, 16 October 2008.

Differential GPS: An Enabling Technology for Formation Flying Satellites

Simone D'Amico and Oliver Montenbruck

Abstract The differential processing of carrier phase measurements from Global Navigation Satellite Systems (GNSS) is a well know technique for relative positioning in terrestrial and aeronautical applications. Over the past decade intensive research has been conducted to demonstrate its suitability for high-precision relative navigation of spacecraft in low Earth orbit (LEO). This chapter describes the fundamental concepts of differential GNSS and its application to spacecraft formation flying. A review of past achievements is given and the practical aspects of differential GPS are discussed for real-time and offline navigation in the upcoming PRISMA and TerraSAR-X missions.

1 Introduction

Since many years, the Global Positioning System is a well known and indispensable tool for precision navigation [1, 2]. A wide range of terrestrial applications has been developed, which includes such diverse areas as precision farming, open pit mining, surveying, bridge deflection monitoring, race car tracking and Earth crustal motion measurements. In aeronautics, the Global Positioning System (GPS) nowadays supports the control of unmanned aerial vehicles, en-route, approach and landing navigation of airplanes and even enables the precision landing of fighter aircraft on carriers.

Common to all high precision applications is the use of so-called carrier phase measurements which measure the beat-phase of the received signal relative to a reference signal at the nominal frequency. Carrier phase measurements offer an extremely low measurement noise of typically 1 mm, which is two to three orders of magnitude lower than that of pseudorange measurements employed in more simple positioning techniques. The full strength of carrier phase based navigation is revealed in differential applications, which benefit from a high level of common error cancellation and enable a resolution of the unknown carrier phase ambiguity.

S. D'Amico (✉)
DLR, German Space Operations Center, D-82234, Munich, Germany
e-mail: simone.damico@dlr.de

R. Sandau et al. (eds.), *Small Satellite Missions for Earth Observation*,
DOI 10.1007/978-3-642-03501-2_43, © Springer-Verlag Berlin Heidelberg 2010

This enables mm to cm positioning accuracies in post-processing and even real-time applications.

Following the successful demonstration of spaceborne GPS receivers onboard the Landat-4 and Topex satellites, researchers soon recognized the potential of differential GPS for relative navigation of multiple satellites. Their work was strongly motivated by a growing interest in formation flying missions which are commonly considered to offer innovative, cost effective and flexible mission architectures as well as numerous new science opportunities [3]. Evidently, it is not possible to operate and tightly control a formation of multiple spacecraft without proper knowledge of their relative motion. The availability of adequate relative navigation systems and sensors is therefore a prerequisite for a successful implementation of formation flying mission. In this context, differential GNSS (DGNSS) offers high-accuracy relative (and absolute) navigation at adequate cost.

Spaceborne GPS receivers for single- and dual-frequency tracking are presently offered by a range of different suppliers [4]. This includes fully space qualified receivers as well as a variety of low cost receivers built around commercial-off-the-shelf (COTS) components. These have inspired numerous university level research projects in differential GPS navigation as well as various nano-satellite formation flying projects.Examples include the Orion/Emerald mission developed at Stanford University and MIT [5], the FASTRAC mission of the University of Texas [6] and the Dutch-Chinese FAST mission [7].

A successful in-flight demonstration of GPS based relative navigation with meter levels accuracies was first performed in 1999 as part of the Japanese ETS-VII technology demonstration mission [8] using differenced pseudorange measurements. Much better performances could later be demonstrated in hardware-in-the-loop simulations using GPS signal simulators due to the use of more sophisticated differential carrier phase techniques [9–11]. Subject to negligible mulTipath errors, these studies have demonstrated the potential of real-time relative navigation at the (sub-)centimeter level using single frequency GPS receivers.

Given the wide range of proposed formation flying projects it is obvious that a single navigation system will never be able to cover all performance and mission needs. In fact, the use of DGNSS is typically restricted to cooperative formation flying missions in low Earth orbit, while independent optical or radiometric sensors are required for the approach to non-cooperative targets, for deep-space formation flying and for interferometric space telescopes. Nevertheless, the use of GPS has already been considered for high-altitude missions [12] and even CIS lunar navigation [13] despite severe performance restrictions that have to be expected at large distances from the Earth.

In the sequel, focus will be given to near Earth applications of DGNSS navigation. Following an outline of the basic principles and the mathematical formulation of differential positioning with carrier phase measurements the aspects of real-time navigation and high-accuracy offline processing will be discussed. As an Example, the relative navigation of two upcoming formation flying missions (PRISMA and

TerraSAR-X) will be presented, which strongly rely on differential GPS for a successful mission performance.

2 Relative Navigation Using Differential GNSS

The measurements provided by a GNSS receiver comprise pseudorange (or code) measurements (P) and carrier phase measurements (Φ), which both provide information on the distance between the receiver and the GNSS satellite. The two measurements differ in their noise level (0.1–1 m vs. 1 mm) and the fact that carrier phase measurements are subject to an ambiguity $N\lambda$, which is an unknown (but constant) mulTiple of the carrier phase wavelength λ [1].

Depending on the available receiver and GNSS system, measurements may be collected on one, two, or (in the future) even three frequencies. In case of GPS, signals are nowadays transmitted at the L1 frequency (1575.42 MHz) and the L2 frequency (1227.60 MHz). Dual-frequency tracking is desirable to eliminate ionospheric path delays (I), which are inversely proportional to the square of the frequency and can thus be determined from the difference of L1 and L2 measurements. On the other hand, ionospheric path delays are highly correlated for receivers in close proximity. Thus, when considering only the difference $\Delta(\bullet) = (\bullet)_B - (\bullet)_A$ of measurements from two receivers A and B, the differential ionosphere (ΔI) is typically negligible for short distances. At spacecraft separations of less than a few kilometers this allows the use of less costly single-frequency receivers for differential navigation.

The four types of differential measurements depend in an equal manner on the difference $\Delta\rho$ of the geometric distances between the receivers and the GPS satellite as well as the differential receiver clock offsets $c\Delta\delta t$, but differ in the sign and magnitude of ionospheric delays and the ambiguity terms:

$$
\begin{aligned}
\Delta P_1 &= \Delta\rho - c\Delta\varepsilon t - \Delta I \\
\Delta P_2 &= \Delta\rho + c\Delta\delta t + (\lambda_2/\lambda_1)^2 \Delta I \\
\Delta \Phi_1 &= \Delta\rho + c\Delta\delta t + \lambda_1 \Delta N_1 - \Delta I \\
\Delta \Phi_2 &= \Delta\rho + c\Delta\delta t + \lambda_2 \Delta N_2 - (\lambda_2/\lambda_1)^2 \Delta I
\end{aligned}
\tag{1}
$$

For simplicity, various other terms such as carrier phase wind-up [14], phase center-variations, line-biases and inter-signal-biases have been ignored in the above equation but may have to be considered if utmost accuracy is desired.

For spacecraft navigation, the measurement model is complemented with a suitable dynamical model that describes the relative motion of the two satellites as a function of the initial conditions and selected dynamical model parameters. For real-time applications and close separations Earth gravity models with a representative degree and order of 10 are recommended while models of up to 70 $\times$ 70 may be required for large baselines and utmost accuracy. Differential drag may be ignored for spacecraft of common design and orientation but should be explicitly modeled in case of widely different ballistic coefficients.

Using appropriate batch or Kalman-filter estimation techniques, the relative state, the differential clock offsets and, where applicable, the differential ionosphere can be determined in the relative orbit determination process. By incorporating additional empirical accelerations in the estimation state vector it is furthermore possible to compensate deficiencies in the deterministic part of the force model, which may, for Example result from an incomplete knowledge of the atmospheric density, the ballistic properties of the spacecraft, or intentional force model simplifications to reduced the computational complexity.

To full exploit the high accuracy of carrier phase measurements, use can be made of the fact that the double differences $\nabla\Delta N_i$ of the ambiguities for two tracked GNSS satellites and two receivers is always an integer number. When this integer number is known, the ambiguous carrier phase measurements effectively become extremely low noise pseudoranges, which then enable a relative positioning at the carrier phase noise level. Various strategies for integer ambiguity resolution have been developed in geodesy, among which the LAMBDA (Least-Squares Ambiguity Decorrelation) technique [15] is probably most famous and widely used. Without integer ambiguity resolution, the ambiguity can still be adjusted as a float parameter inside the navigation filter. However, the achievable positioning accuracy is typically smaller by a factor of two or more and the estimation result is more sensitive to measurement and modeling errors.

Relative navigation using differential GNSS measurements can be performed either off-line or in real-time and various studies have addressed the specific challenges and prospects of these two areas. Off-line applications certainly benefit from a higher computing power and the availability of accurate auxiliary data such as GNSS orbit and clock values. In the context of the GRACE formation flying mission various authors [16–18] have in recent years demonstrated the feasibility of 1 mm level relative navigation over a 200 km distance using GPS measurements from a high-grade dual-frequency BlackJack receiver. To achieve this accuracy a careful mitigation or calibration of mulTipath errors caused by signal reflections is mandatory. In case of GRACE a choke ring antenna is employed and the antenna environment is essentially free of obstructing and reflecting elements.

In real-time applications, an inter-satellite link is evidently required to exchange GNSS measurements between different spacecraft. This may necessitate adequate communication architectures and systems when working with more than two spacecraft. As an Example, real-time relative navigation of up to four spacecraft has been demonstrated in hardware-in-the-loop simulations [11] using a time-mulTiplexing scheme. Even though a mm-level relative navigation accuracy might, in principle, also be achieved in a real-time application, the corresponding mission requirements are typically less stringent than for post-processing. In practice, computational restrictions, code complexity and robustness considerations or a restricted knowledge of the spacecraft attitude may restrict the achievable accuracy to the centimeter or even decimeter range.

3 GPS Based Navigation of the PRISMA Mission

PRISMA is a Swedish micro-satellite mission aiming at the demonstration and flight validation of key technologies for formation flying [19]. The mission comprises two spacecraft that are launched together in a stacked configuration and separated after an initial checkout of all systems. The main satellite, named MANGO, has a mass of 154 kg at a body size of $80 \times 83 \times 130\ \text{cm}^3$ and offers accurate three-axis attitude determination and control as well as orbital maneuvering capabilities. The smaller target spacecraft (named TANGO) has a mass of only 42 kg at a size of $80 \times 80 \times 31\ \text{cm}^3$ and is equipped with a simpler magnetometer, sun-sensor and magnetorquer based attitude control system. A bi-directional intersatellite link allows commanding of the target satellite from the main spacecraft and the transmission of telemetry data in the reverse direction. In accord with the mission objectives, PRISMA carries different sensors for relative navigation including a differential GPS system (contributed by DLR, Germany), a Formation Flying Radio Frequency Sensor (FFRF; contributed by CNES, France) and a Vision Based Sensor (VBS; contributed by DTU, Denmark). The mission is presently scheduled for launch by a DNEPR rocket in the second half of 2009 and will fly in a polar orbit at 700 km altitude.

The differential GPS system serves as the primary source of absolute and relative navigation in the PRISMA mission. It comprises a cold-redundant set of Phoenix-S single-frequency GPS receivers on each spacecraft as well as a real-time navigation software operated in the onboard computer of the main spacecraft. For offline-analyses, the GPS measurements will also be transmitted to the ground and processed in a precise orbit determination system. The Phoenix-S receiver provides C/A code and carrier phase measurements of up to twelve GPS satellites with a representative noise level of 0.5 m and 1 mm, respectively. Each spacecraft is equipped with a redundant pair of GPS antennas which are mounted at the end of the solar panels of MANGO and the side walls of TANGO (Fig. 1). The antenna with the best sky coverage (i.e. with its boresight closest to the zenith) can be autonomously selected by the on-board software and connected to the active GPS receiver via an R/F switch.

The design of the PRISMA real-time navigation process [20] is driven by tightly limited computational resources as well as a need to provide robust results over a wide range of operational scenarios. The first constraint is accounted by operating the navigation filter with a reduced update rate of once per 30 s rather than the native attitude and orbit control system cycle of 1 Hz. This is made possible by an accurate orbit model, a special numerical integration and interpolation technique as well as a consistent incorporation of thruster information into the predicted trajectory. For robustness reasons, the PRISMA navigation filter is designed to work both with an undifferenced, ionosphere-free code-/carrier combination as well as differenced carrier phase measurements. In this way, the filter can operate even if no common GPS satellites are observed by the two PRISMA spacecraft but can make best use of common observations whenever available.

Absolute positioning accuracies are expected to be around 2 m in real-time and 0.5–1 m in post-processing. The relative navigation accuracy depends critically on

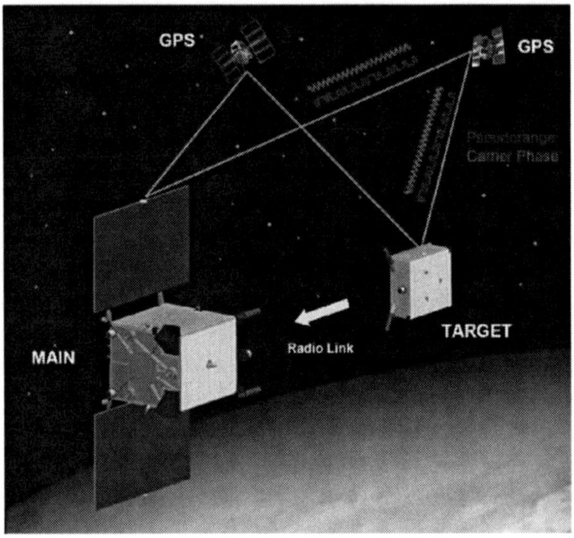

Fig. 1 GPS based relative navigation of the PRISMA satellites

the attitude profile during specific mission phases and may range from a worst case accuracy of 0.5 m if the GPS antennas of MANGO and TANGO point in different directions down to a few cm if a sufficient number of common GPS satellites is tracked by the two spacecraft. In the latter case the accuracy is limited by the restricted attitude knowledge of the target spacecraft and the lacking knowledge of GPS antenna phase pattern variations in the spacecraft environment.

4 TanDEM-X

The TerraSAR-X add-on for Digital Elevation Measurements (TanDEM-X) mission comprises a set of twin spacecraft that will fly in close formation to collect interferometric SAR measurements for the generation of a global, high-resolution digital elevation model (DEM) [21]. While TerraSAR-X is already in orbit since mid 2007 and provides SAR imagery on a routine basis, the second spacecraft will only be launched in October 2009.

TerraSAR-X and TanDEM-X are equipped with IGOR dual-frequency GPS receivers that represent a commercial follow-on of the BlackJack receiver flown on GRACE and have been contributed by the German GeoForschungszentrum (GFZ), Potsdam. Measurements from the IGOR on TerraSAR-X satellite are routinely processed in DLR's GPS High-precision Orbit determination Software Tools (GHOST [22]) to obtain rapid and final science orbit products with a representative 3D rms accuracy of 10 cm and <5 cm, respectively. Inter-comparisons with precise orbit determination (POD) solutions of other agencies solutions show a consistency of about 2 cm, while satellite laser ranging (SLR) residuals from high-grade stations are at the level of 1.5 cm rms [23, 24].

Based on experience form the GRACE mission, a simple differencing of absolute POD solutions can deliver a relative navigation accuracy of slightly better than 1 cm, which is not, however, enough for the interferometric SAR processing. In fact, the baseline vector between the two spacecraft has to be determined with an accuracy of 1 mm (1 D rms) in the TanDEM-X mission. In particular, an accurate determination of the line-of-sight (LOS) component is considered critical for the generation of the Digital Elevation Models (DEMs). Errors in the knowledge of the LOS component result in a scene dependent vertical and, more importantly, lateral shift of the resulting DEM. For a direct matching of overlapping DEMs the lateral errors should be maintained to less than one pixel, which in turn limits the tolerable LOS baseline error to typically 1 mm.

To meet this challenging requirement, a precise baseline determination software has been developed, which directly processes single-difference code and carrier phase to in a forward-backward filter-smoother [16]. For TanDEM-X the FRNS filter has been extended to handle either single- or dual-frequency measurements and to describe the differential ionospheric path delays as exponentially correlated random variables. In this way a priori information on the expected magnitude of differential ionospheric delays at a given spacecraft separation can be incorporated in the estimation process, which in turn improves the robustness of the dual-frequency ambiguity fixing process.

So far, the software has extensively been tested with GPS measurements from signal simulator tests (Fig. 2) as well as using real flight data from the GRACE

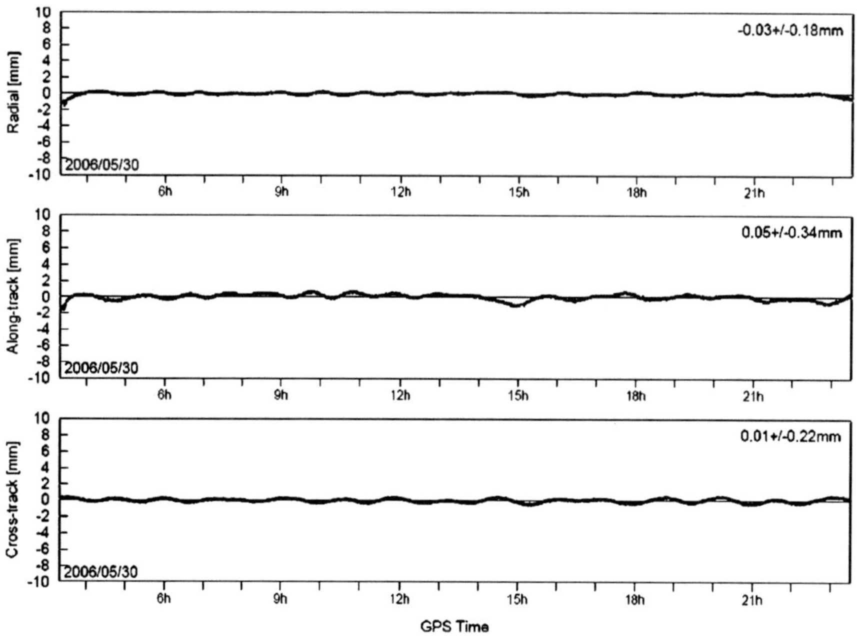

Fig. 2 TanDEM-X baseline determination accuracy in a signal simulator test bed

mission. The results confirm that a better than 1 mm 3D rms baseline accuracy can indeed by achieved, if the employed carrier phase measurements are free of systematic errors. On the other hand, great care must be taken to properly calibrate the phase center variation of the TerraSAR-X and TanDEM-X GPS antennas. Besides a robotic ground calibration of an engineering model of the antenna, supplementary phase patter variations caused by the spacecraft environment are calibrated from carrier phase residuals observed in the POD process. Preliminary values are already available for TerraSAR-X [24] and a similar calibration will be carried out for TanDEM-X as part of the commissioning phase. In order to safeguard against remaining systematic error sources, dedicated SAR calibration data takes will we conducted during the mission, which enable a verification of the system performance and a calibration of possible residual biases.

Acknowledgements The TanDEM-X project is partly funded by the German Federal Ministry for Economics and Technology (Förderkennzeichen 50 EE 0601).

References

1. P. Misra, P. Enge. Global Positioning System (GPS): Signals, Measurements and Performance, USA, Ganga-Jamuna Press, 2nd edition, 2006.
2. B. Parkinson, J. Spilker, P. Axelrad, P. Enge (eds.), Global Positioning System: Theory and Applications, Washington, DC, AIAA, 1996.
3. J. Leitner, F. Bauer, D. Folta, R. Carpenter, M. Moreau, J. How. Formation Flight in Space. GPS World, 13(2), 22–31, Feb. 2002.
4. O. Montenbruck, M. Markgraf, M. Garcia, A. Helm. GPS for Microsatellites – Status and Perspectives. 6th IAA Symposium on Small Satellites for Earth Observation, April 23–26, 2007, Berlin, Germany, 2007.
5. Ph. Ferguson, F. Busse, B. Engberg, J. How, M. Tillerson, N. Pohlman, A. Richards, R. Twiggs. Formation Flying Experiments on the Orion-Emerald Mission; AIAA-2001-4688, AIAA Space 2001, Albuquerque, NM, Aug. 28–30, 2001.
6. G. Holt, T. Campbell, E.G. Lightsey. GPS, Distributed Communications, and Thruster Experiments on the University of Texas FASTRAC Mission, AMSAT 22nd Annual Space Symposium, Washington, DC, Conference and Exposition, Oct. 2004.
7. D. Maessen, E. Gill, J. Guo, E. Laan, S. Moon, G.T. Zheng. Mission Design of the Dutch-Chinese FAST Micro-Satellite Mission. 7th IAA Symposium on Small Satellites for Earth Observation, May 4–8, 2009, Berlin, Germany, 2008.
8. I. Kawano, M. Mokuno, M. Kasai, T. Suzuki. First autonomous rendezvous using relative GPS navigation by ETS-VII. In Proceedings of the 12th International Technical Meeting of the Satellite Division of the Institute of Navigation, pp. 393–400. The Institute of Navigation, Nashville, TN, 1999.
9. T. Ebinuma. Precision Spacecraft Rendezvous Using Global Positioning System: An Integrated Hardware Approach. PhD Thesis, University of Texas, Austin, TX, 2001.
10. F.D. Busse. Precise Formation-State Estimation in Low Earth Orbit Using Carrier Differential GPS, PhD thesis, Department of Aeronautics and Astronautics, Stanford, CA, Stanford University, March 2003.
11. S. Leung, O. Montenbruck. Real-Time Navigation of Formation-Flying Spacecraft using Global Positioning System Measurements. Journal of Guidance, Control and Dynamics, 28/2, 226–235, 2005.
12. M.L. Psiaki, S. Mohiuddin. Relative Navigation of High-Altitude Spacecraft Using Dual-Frequency Civilian CDGPS. Proceedings of the ION-GPS-2005, Long Beach, CA, September 13–16 2005.

13. W.A. Bamford, G.W. Heckler, G.N. Holt, M.C. Moreau. A GPS Receiver for Lunar Missions, Proceedings of the 2008 National Technical Meeting of the ION, San Diego, CA, pp. 268–278, January 28–30, 2008.
14. J.T. Wu, S.C. Wu, G.A. Hajj, W.I. Bertiger, S.M. Lichten. Effects of antenna orientation on GPS carrier phase. Manuscripta Geodaetica, 18, 91–98, 1993.
15. P.J.G. Teunissen. The least squares ambiguity decorrelation adjustment: a method for fast GPS integer ambiguity rounding and bootstrapping. Journal of Geodesy, 72, 65–82, 1995.
16. R. Kroes, O. Montenbruck, W. Bertiger, P. Visser. Precise GRACE baseline determination using GPS. GPS Solutions, 9, 21–31, 2005. DOI: 10.1007/s10291-004-0123-5.
17. A. Jäggi, U. Hugentobler, H. Bock, G. Beutler. Precise Orbit Determination for GRACE Using Undifferenced or Doubly Differenced GPS Data. Advances in Space Research, 39(10), 1612–1619, 2006.
18. S.-Ch. Wu, Y.E. Bar-Sever. Real-time sub-cm differential orbit determination of two Low-Earth Orbiters with GPS bias fixing; ION GNSS 2006, September 26–29, Fort Worth, Texas, 2006.
19. S. Persson, P. Bodin, E. Gill, J. Harr, J. Jorgensen. PRISMA–an autonomous formation flying mission. ESA Small Satellite Systems and Services Symposium (4S), Sardinia, Italy, September 2006.
20. S. D'Amico, E. Gill, M. Garcia, O. Montenbruck. GPS-Based Real-Time Navigation for the PRISMA Formation Flying Mission, NAVITEC'2006, Noordwijk, 11–13 December 2006.
21. G. Krieger, A. Moreira, H. Fiedler, I. Hajnsek, M. Werner, M. Younis, M. Zink. TanDEM-X: a satellite formation for high resolution SAR interferometry. IEEE Transactions on Geoscience and Remote Sensing, 45(11), 3317–3341, 2007. DOI: 10.1109/TGRS.2007.900693.
22. O. Montenbruck, T. van Helleputte, R. Kroes, E. Gill. Reduced dynamic orbit determination using GPS code and carrier measurements. Aerospace Science and Technology, 9/3, 261–271, 2005. DOI: 10.1016/j.ast.2005.01.003.
23. Y. Yoon, M. Eineder, N. Yague-Martinez, O. Montenbruck. TerraSAR-X precise trajectory estimation and quality assessment. IEEE Transactions on Geoscience and Remote Sensing, 47/2, 2009. DOI: 10.1109/TGRS.2008.2006983.
24. O. Montenbruck, M. Garcia-Fernandez, Y. Yoon, S. Schön, A. Jäggi. Antenna phase center calibration for precise positioning of LEO satellites. GPS Solutions, 13/1, 23–34, 2009. DOI: 10.1007/s10291-008-0094-z.

GPS-Based Relative Navigation in Earth Observation Missions Relying on Cooperative Satellites

Alfredo Renga, Urbano Tancredi, and Michele Grassi

Abstract The problem of relative navigation of formation flying satellites for remote-sensing applications is addressed in this chapter. Specifically, a filtering scheme is proposed for the determination of relative position and velocity in real time, relying on the cascade combination of an Extended Kalman Filter and a kinematic filter. Since high measurement accuracy is desired both filters process dual-frequency double-difference carrier-phase GPS observables. In addition, the dynamic filter state is augmented with the Vertical Total Electron Content. Filtering scheme performance in terms of accuracy and robustness are evaluated by Monte Carlo simulations on orbital scenarios relevant to considerable Earth remote sensing applications.results shows that centimetre-level average accuracy can be achieved in real-time and it can be maintained in presence of poor GPS satellite geometry.

1 Introduction

In the recent years several space mission concepts have been developed that rely on more satellites flying in formation. Indeed, a distributed system of platforms offers a number of scientific and technological advantages with respect to a single monolithic platform, as the possibility of synthesizing large antennas and optics in space or performing scientific experiments by integrating the measurements from payloads dislocated on more platforms. Nevertheless, a number of technology advancements are connected to the use of formation flying, many of which are aimed at increasing the level of autonomy in the relative navigation and control of the satellites to maintain or change the formation geometry.

Earth science missions based on distributed architectures, like spaceborne gravimetry, bistatic/multistatic synthetic aperture radars, or atmosphere and ionosphere profiling demand for precise and robust relative navigation technologies. In

M. Grassi (✉)
DIAS – Department of Aerospace Engineering, University of Naples "Federico II", P.le Tecchio 80, 80125, Naples, Italy
e-mail: michele.grassi@unina.it

R. Sandau et al. (eds.), *Small Satellite Missions for Earth Observation*, 467
DOI 10.1007/978-3-642-03501-2_44, © Springer-Verlag Berlin Heidelberg 2010

this ambit, using Carrier-phase Differential GPS (CDGPS) is of considerable interest, although problems as finding the unknown integer number of carrier cycles have to be faced.

Many studies have addressed the problem of identifying and testing novel CDGPS filtering techniques for determining, in real time or post-processing, the relative position of formation flying satellites in different applications, with accuracy requirements ranging from the centimetre to the millimetre level [1, 2]. However, these techniques are usually developed and optimized to a specific application, where the inter-satellite distance does not experience large variations during the mission. Instead, in many of the mentioned Earth science applications the inter-satellite distance largely varies due to the exploited relative orbits [3, 4].

Based on this premise, this chapter aims at developing a filtering approach capable of providing in real-time relative positioning accuracy at the centimetre level in applications where the inter-satellite distance largely varies. The approach aims at being accurate and robust. Accuracy is achieved by exploiting the integer nature of Double-Difference (DD) CDGPS measurements taken on L1 and L2 carriers, whereas robustness is achieved by the use of an Extended Kalman Filter (EKF) that provides accurate floating point estimates of the DD integer ambiguities. To allow real-time application, the EKF relies on a keplerian model of the satellites' relative dynamics instead of propagating independently the absolute orbits of the two satellites by using quite complex models. In this way it is possible to get good performance by modelling the small differential perturbations (e.g. differential J2 and drag) as process noise while keeping low the computational effort. Indeed, formation flying satellites usually lie on orbits whose parameters slightly differ and have similar ballistic coefficients so to minimize control efforts to maintain the formation.

To get accurate estimates of the floating ambiguities the EKF state is augmented with the Vertical Total Electron Content (VTEC). This allows estimating the DD ionosphere delay terms with enough accuracy. Filtering approach performance is verified by Monte Carlo simulations of orbital scenarios relevant to considerable Earth remote sensing applications.

2 Filtering Approach

As illustrated in Fig. 1, the proposed approach combines in cascade a dynamic filter with a kinematic filter both processing DD observables. This scheme meets requirements for high accuracy and robustness in relative navigation. The former is achieved by exploiting the integer nature of double-difference carrier-phase observables on L1 and L2 carriers within the kinematic filter, the latter is met by the Introduction of an Extended Kalman Filter. This last one outputs a robust estimate of the floating point values of the DD carrier-phase ambiguities, other than of the relative position and velocity of the two satellites. These estimates are processed within a least square integer search algorithm to compute the corresponding integer values [5], that are then used to correct the DD observables for the baseline determination with the kinematic algorithm. This is a Weighted Least Square (WLSQ) algorithm that process ionosphere-free combinations of the DD observables in case large baseline values are attained during the mission [5].

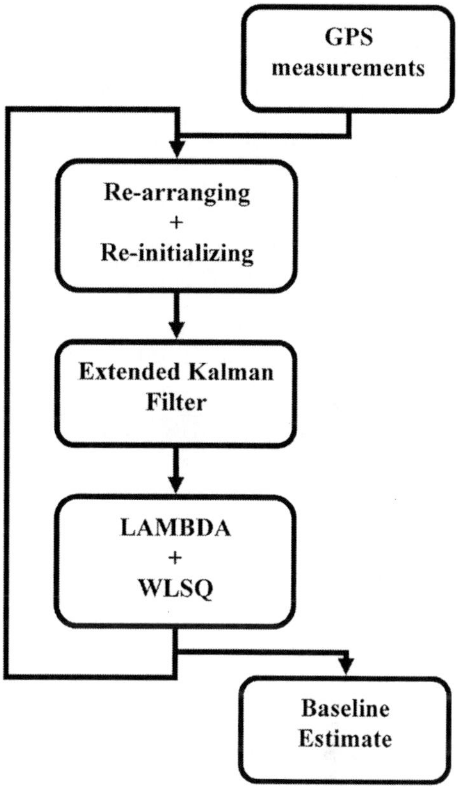

Fig. 1 Filtering scheme

Two major drawbacks are introduced when implementing a dynamic filter processing DD observables: the number of available measurements reduces and the pivot satellite can change several times during one orbit. This of course may affect filter accuracy and convergence. In the proposed scheme an original procedure is implemented for fast re-arrangement/re-initialization of the filter in case the pivot satellite and/or the number of GPS satellites commonly viewed change. Details about the adopted procedure can be found in [5, 6] and are not reported herein for the sake of conciseness. The effectiveness of this procedure is tested by numerical simulations.

The filter model is developed in the Earth-Centred-Earth-Fixed (ECEF) reference frame (see Fig. 2), with reference to a formation of two satellites (master, #1, and slave, #2). As well known, the relative navigation problem can be described by a nonlinear discrete stochastic model with additive process and measurement noise,

Fig. 2 Geometry for filter model

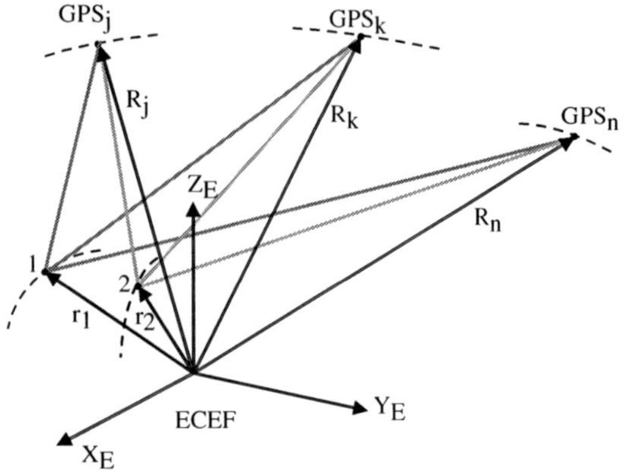

that are assumed to be Gaussian and mutually independent [5]. The state and the measurement vectors are given as [5]:

$$x = \begin{bmatrix} \underline{B} \\ \underline{\dot{B}} \\ VTEC_1 \\ VTEC_2 \\ {}_{L1}AM \\ {}_{L2}AM \end{bmatrix}_{8+2(N_{\text{sat}}-1)} ; y = \begin{bmatrix} {}_{L1}PR \\ {}_{L2}PR \\ {}_{L1}CP \\ {}_{L2}CP \end{bmatrix}_{4(N_{\text{sat}}-1)} \tag{1}$$

where N_{sat} is the number of commonly viewed GPS satellites, $VTEC_1$ and $VTEC_2$ are the vertical total electron content from, respectively, the master and slave receiver to the upper bound of the ionosphere, ${}_{L1}AM$ and ${}_{L2}AM$ are the DD integer ambiguity vectors, respectively, on L1 and L2 carriers, ${}_{L1}PR$ and ${}_{L2}PR$ are the DD pseudorange vectors, and ${}_{L1}CP$ and ${}_{L1}CP$ are the DD carrier phase vectors. As for the introduction of VTEC in the state vector it is worth outlining that achieving high accuracy in baseline determination requires correctly fixing the DD integer ambiguities. This, in turn, requires an accurate estimate of their floating values in the EKF, that can be achieved only if the DD ionosphere delay terms are estimated with enough accuracy. These terms are usually inserted in the filter state and modeled as random walk [1, 2]. Instead, in the proposed approach, the filter estimates *VTEC,* that is related to the ionosphere delay by the well-known mapping formula [7]. In this way it is possible to describe more accurately the variation of the DD-ionosphere delay terms as a function of the continuously changing relative position and GPS satellite viewing geometry of the two receivers [5]. To this end, the computation of the elevation angles of the GPS satellites above the single receiver horizon is needed and can be easily performed in real-time [5].

To trade-off real-time implementation and accuracy requirements, the EKF performs time update by using a Keplerian model of the satellites' relative dynamics [5, 6], with the remaining state vector components treated as random walk plus random constant processes [7]. The update stage of the filter is performed by processing dual-frequency, double difference (DD) observables formed by subtracting two single difference equations of the same type and frequency, taken by the same two GPS receivers at the same time from two different GPS satellites, one of which, named pivot and denoted as j, is taken as a reference (see Fig. 2):

$$\begin{aligned}
{}_{L1}PR_{12}^{jk} &= \rho_{12}^{jk} + \left({}_{L1}J_2^{jk}VTEC_2 - {}_{L1}J_2^{jk}VTEC_1\right) + {}_{L1}^{PR}v_{12}^{jk} \\
{}_{L2}PR_{12}^{jk} &= \rho_{12}^{jk} + \frac{f_{L1}^2}{f_{L2}^2}\left({}_{L2}J_2^{jk}VTEC_2 - {}_{L2}J_2^{jk}VTEC_1\right) + {}_{L2}^{PR}v_{12}^{jk} \\
{}_{L1}CP_{12}^{jk} &= \rho_{12}^{jk} - \left({}_{L1}J_2^{jk}VTEC_2 - {}_{L1}J_2^{jk}VTEC_1\right) + \lambda_{L1L1}AM_{12}^{jk} + {}_{L1}^{CP}v_{12}^{jk} \\
{}_{L2}CP_{12}^{jk} &= \rho_{12}^{jk} - \frac{f_{L1}^2}{f_{L2}^2}\left({}_{L1}J_2^{jk}VTEC_2 - {}_{L1}J_2^{jk}VTEC_1\right) + \lambda_{L2L2}AM_{12}^{jk} + {}_{L2}^{PR}v_{12}^{jk}
\end{aligned} \tag{2}$$

where f is the GPS signal frequency, λ is the GPS signal wavelength, k indicates the generic GPS satellite, and

$$\rho_{12}^{jk} = \left\|\underline{R}_k - (\underline{r}_1 + \underline{B})\right\| - \left\|\underline{R}_k - \underline{r}_1\right\| - \left\|\underline{R}_j - (\underline{r}_1 + \underline{B})\right\| + \left\|\underline{R}_j - \underline{r}_1\right\| \tag{3}$$

is the DD signal geometric path.

3 Numerical Simulations

Filter performance is evaluated by using a numerical code running in Matlab® that is based on the Satellite Navigation Toolbox 3.0 for Matlab® developed by GPSoft Inc. [8] to simulate the GPS satellite constellation, the propagation environment and the receiver measurements. Since this Toolbox is developed mainly for Earth-based GPS receivers, to adapt it to LEO receivers it was necessary to introduce a suitable ionosphere delay model [5, 6]. In addition, the Total Electron Content (TEC) values are evaluated by using the International Reference Ionosphere [9].

Filter performance is evaluated by comparison with a simulated truth obtained by running an orbit propagator including a seven-order model of the gravity field and the aerodynamic drag [5, 6], and with reference to orbit scenarios relevant to considerable Earth remote sensing applications. Specifically, pendulum-like and helix-like [10] configurations are considered that are relevant to interferometric and bistatic SAR applications, such as the generation of digital elevation model (DEM) [11] by SAR interferometry and stereo-radargrammetry, and reflects next generation monostatic/bistatic spaceborne SAR mission needs in LEO [10]. As shown in Table 1, the considered orbits share inclination, semi-major axis and anomaly of perigee. For the pendulum configuration, the baseline attains a maximum value of 200 km at the equator, and a minimum value of a few kilometers at the highest latitudes.

In all the simulation the Italian COSMOSkyMed mission [12] is taken as a reference for the master satellite and the LAGRANGE receiver nominal performances

Table 1 Simulated orbits

	Master	Slave	
		Pendulum	Helix
Semi-major axis (km)	6997.94	6997.94	6997.94
Eccentricity	0.00118	0.00118	0.001308
Anomaly of perigee (°)	90	90	90
Inclination (°)	97.87	97.87	97.87
Ascending node right ascension (°)	0	1.680	0.0049
Initial mean anomaly (°)	0	0.3406	0

are used [5, 6]. This is an integrated GPS receiver for spaceborne applications, characterized by 12 dual-frequency channels and 0.5-m and 1.2-mm representative measurement uncertainty, for code and carrier phase, respectively. In addition, 1-Hz GPS data frequency update is considered.

The filtering scheme was successfully tested on satellite formations that are relevant to spaceborne bistatic SARs: interferometric missions characterized by a few-kilometer baseline and pendulum formations conceived for Large Baseline Bistatic (LBB) SARs.

Figure 3 shows the estimate error in the radial component of the baseline for the Helix configuration superimposed to the 3-σ bounds estimated by the kinematic

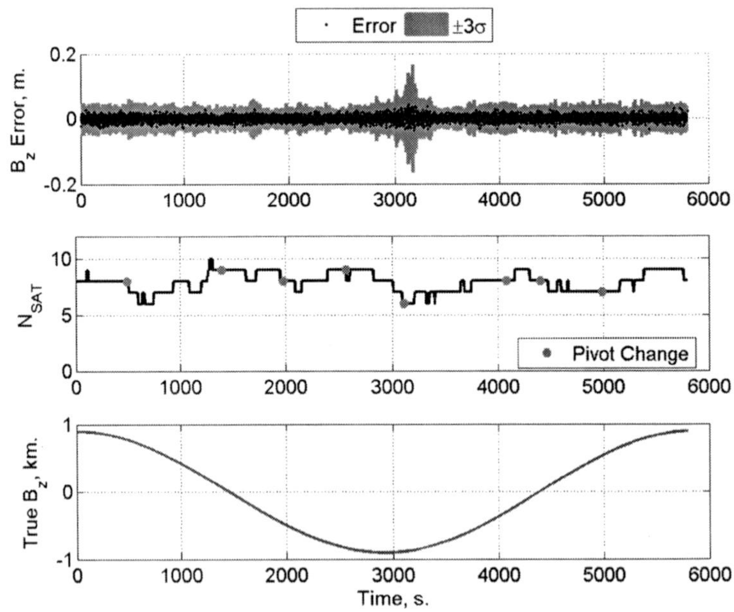

Fig. 3 Radial error, number of commonly viewed (pivot change marked by points) GPS satellites, and true baseline radial component for Helix

filter. As common in GPS navigation the vertical error is higher, remaining however at the centimeter level. Figure 3 allows verifying also filter robustness with respect to pivot satellite variations and integer ambiguity estimation.

Filter performance is assessed by Monte Carlo analysis. In this regard, Fig. 4 shows the estimate error in the baseline components (as evaluated in the orbiting reference frame of the master satellite) resulting from 1000 runs of a single orbit scenario, characterized by an orbital period of 97 min. Showed results are for the pendulum configuration in which the baseline varies from 2 to 200 km in one orbit and are achieved with a mask angle of 5°. This configuration is relevant to Large Baseline Bistatic SAR applications. It can be seen that the filter exhibits good average performance in spite of a significant variation of the baseline length along the orbit, as reported in Fig. 5. This figure shows also the variation in the number of commonly viewed GPS satellites and in the pivot satellite (marked in black), thus confirming the effectiveness of the procedure developed for filter re-initialization and re-arrangement. Indeed, spikes in the vertical component estimate error are not correlated to pivot satellite changes.

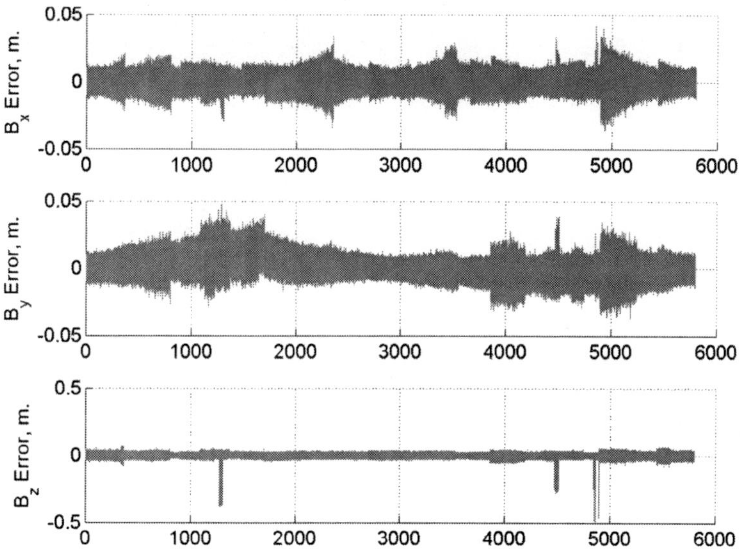

Fig. 4 Baseline components for pendulum

From Fig. 4 it can be seen that the behavior of the along-track and cross-track error is more uniform if compared with the radial error that, instead, shows some spikes with precision degrading up to tens of centimeters. These spikes appear in points of the orbit where poor GPS observation geometries and Vertical Dilution Of Precision values (not reported for the sake of brevity) are observed.

Table 2 summarizes the results of the Monte Carlo analysis in terms of mean and RMS errors (average values over 1,000 repetitions) and maximum errors. It can be seen that the average performance is within 1 cm, whereas the maximum error is

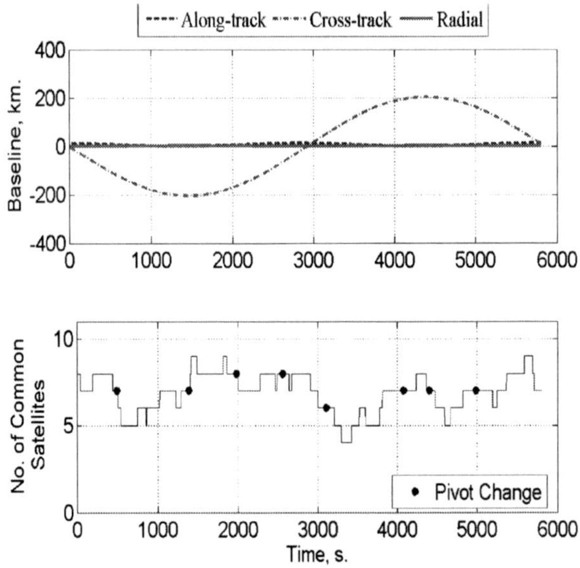

Fig. 5 Variation of baseline and number of GPS satellites (pivot change marked by points) for Pendulum

Table 2 Monte Carlo analysis results

	Mean Error (mm)	RMS Error (mm)	Max. Error (mm)
Bx	−0.08	4.09	41.89
By	0.86	6.45	47.67
Bz	−0.45	10.18	498.57

within 5 cm, for the along and cross-track components, and one order of magnitude higher for the radial component. However, it is worth outlining that spikes are present only in 10% of the simulated orbits and one spike per orbit is observed at most, representing less than 0.002% of all the simulated conditions, as reported in Fig. 6, that shows the radial error distribution as achieved with Monte Carlo analysis. It is expected that measured performance can be improved by improving the dynamic filter tuning and the model used to describe the VTEC variation.

4 Conclusions

This chapter showed that an effective filtering approach processing double-difference pseudorange and carrier-phase measurements on two frequencies can be developed to estimate in real time the baseline vector of formation flying satellites with high accuracy. The proposed approach combines an Extended Kalman Filter with a kinematic filter that processes double-difference ionosphere-free combinations.

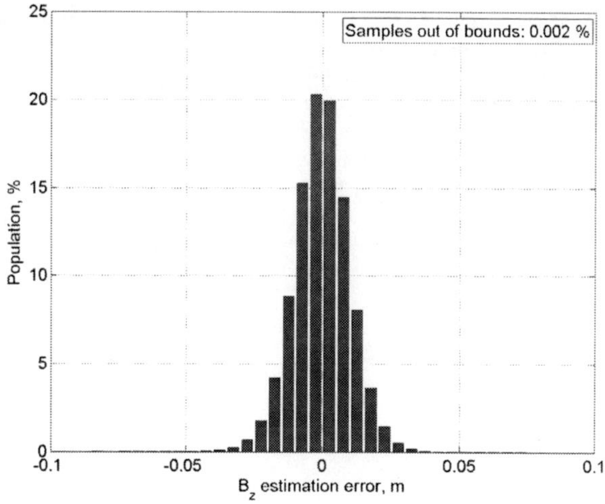

Fig. 6 Radial error distribution in Monte Carlo analysis

Monte Carlo analyses of formation flying configurations relevant to considerable Earth science missions showed that centimetre-level performance in the estimate of the baseline components can be achieved in the average. However, maximum errors of about 5 cm, in the long and cross-track components, and of about 50 cm, in the radial direction, are observed when the baseline length attains values of hundreds of kilometres. Finally, results confirm filtering approach robustness with respect to baseline variation, pivot satellite change and poor GPS satellite observation geometry.

Acknowledgements The present work has been carried out under the support of the Italian Space Agency (ASI) and the Italian Ministry for University, and Research (MIUR).

References

1. R. Kroes, O. Montenbruck, W. Bertiger, P.N.A.M. Visser. Precise GRACE baseline determination using GPS. GPS Solutions, 9(1), 21–31, 2005.
2. R. Kroes. Precise relative positioning of formation flying spacecraft using GPS, Ph.D Thesis, Publications on Geodesy 61, Delft, 2006.
3. G. Krieger, A. Moreira. Spaceborne bi- and multistatic SAR: Potential and challenges. Proceedings of the Institute Electronics and Engineering—Radar, Sonar Navigation, 153(3), 184–198, 2006.
4. F. Caltagirone, A. Capuzi, A. Coletta, C. Galeazzi, E. Lopinto, A. Moccia. SABRINA the Italian Mission for Endowing COSMO-SkyMed with Bistatic and Interferometric Capabilities. Proceedings of EUSAR 2006, Dresden, Germany, May 16–18 2006.
5. A. Renga, U. Tancredi, M. Grassi. Analysis of GPS-Based Relative Navigation Schemes for Earth Observation Missions Relying on Cooperating Satellites, ENC-GNSS 2009, Naples (Italy), May 3–6 2009.
6. A. Renga and M. Grassi. Precise Relative Navigation for Highly Variable Baselines using Carrier-Based Differential GPS, paper IAC-08 –C1.6. 50th International Astronautical Congress, Sept. 29–Oct. 3, Glasgow, Scotland, 2008.

7. O. Montenbruck, M. Garcia-Fernandez. Ionospheric Path Delay Models for Spaceborne GPS, DLR-GSOC TN 05-07, 2005.
8. http://www.gpsoftnav.com/satnav.html
9. nssdc.gsfc.nasa.gov/space/model/models/iri.html
10. G. Fasano, M. D'Errico. Relative motion model including J2 derivation and application to INSAR, 2006 IEEE Aerospace Conference, Big Sky, MT, USA, March 2006.
11. A. Renga, A. Moccia, M. D'Errico et al. From the expected Scientific Applications to the Functional Specifications, Products and Performance of the SABRINA missions, IEEE Radar Conference, May 26–30, Rome, Italy, pp. 1117–1122, 2008.
12. http://www.cosmo-skymed-ao.asi.it

Index

Note: The letters f and t following locators refer to figures and tables respectively.

R. Sandau et al. (eds.), *Small Satellite Missions for Earth Observation*,
DOI 10.1007/978-3-642-03501-2, © Springer-Verlag Berlin Heidelberg 2010

Breinigsville, PA USA
05 April 2010
235343BV00006B/30/P

9 783642 035005